Earthquake Research: Issues and Developments

Earthquake Research:
Issues and Developments

Edited by **Rosalina Peters**

R CALLISTO REFERENCE

New York

Published by Callisto Reference,
106 Park Avenue, Suite 200,
New York, NY 10016, USA
www.callistoreference.com

Earthquake Research: Issues and Developments
Edited by Rosalina Peters

International Standard Book Number: 978-1-63239-157-5 (Hardback)

Printed in the United States of America.

Contents

Preface

An earthquake is essentially a sudden release of energy in the Earth's crust that creates seismic waves. The cause for an earthquake can be both natural and man-made. Some of the most frequent causes are geological faults, landslides, incessant construction, nuclear tests among others. The occurrence of any earthquake is a seismic event, which is measured using observations from seismometers. Depending on the part of the world in which they occur, earthquakes have been classified into different categories.

Historical sources suggest that earthquakes have been associated with legends. For instance, in Greek mythology, earthquakes are associated with the wrath of a God. Detectable with current instrumentation, today, around 500,000 earthquakes occur each year out of which, approximately 100,000 can be felt.

An earthquake's point of initial rupture is called its focus or hypocentre. The onset of an earthquake releases a lot of energy. This energy released by the earthquake, is measured in terms of its magnitude, from a recording device called a seismograph. Earthquakes can trigger landslides, occasionally volcanic activity and tsunami.

I would like to thank all the contributors who have shared their knowledge, expertise and researches in this book. I would also like to thank the publishing house for giving me this opportunity.

Editor

An Improved Method for Seismic Site Characterization with Emphasis on Liquefaction Phenomena

Abbas Abbaszadeh Shahri[1], Roshanak Rajablou[2], Abdolvahed Ghaderi[3]

[1]Department of Geophysics, Hamedan Branch, Islamic Azad University, Hamedan, Iran
[2]Young Researcher Club, Department of Geophysics, Hamedan Branch, Islamic Azad University, Hamedan, Iran
[3]Department of Engineering, Science and Research Branch, Islamic Azad University, Tehran, Iran

ABSTRACT

Iran is an active seismic region and frequent earthquakes and because of the active faults, often leads to severe casualties caused by structural destruction. Earthquake damage is commonly controlled by three interacting factors, source and path characteristics, local geological and geotechnical conditions and type of the structures. Obviously, all of this would require analysis and presentation of a large amount of geological, seismological and geotechnical data. In this paper, nonlinear geotechnical seismic hazard analysis considering the local site effects was executed and the soil liquefaction potential analysis has been evaluated for the Nemat Abad earth dam in Hamedan province of Iran because of its important socioeconomic interest and its location. Liquefaction susceptibility mapping is carried out using a decisional flow chart for evaluation of earthquake-induced effects, based on available data such as geological, groundwater depth, seismotectonic, sedimentary features, *in situ*, field and laboratory geotechnical parameters. A series model tests were conducted and then on base of the achieved data the idealized soil profile constructed. A $C^{\#}$ GUI computer code "NLGSS_Shahri" was Generate, developed and then employed to evaluate the variation of shear modulus and damping ratio with shear strain amplitude to assess their effects on site response. To verify and validate the methodology, the obtained results of the generated code were compared to several known applicable procedures. It showed that computed output of this code has good and suitable agreement with other known applicable procedures.

Keywords: Nemat Abad Earth Dam; "NLGSS_Shahri"; GUI Computer Code; Earthquake Record

1. Introduction

G The interfacing software "NLGSS_Shahri" has developed on base of the geological information, seismic data, earthquake records and geotechnical database. The provided code is capable of reading geotechnical data from database, performing calculations of dynamic parameters for dynamic site response analyses. This version can also prepare input files corresponding to used several software packages in this study. Execute and performing the liquefaction analysis computation of the post liquefaction settlement is the main factor of generated code and in the other word it is the improved version of "Abbas Converter 3.01" that has proposed by Abbaszadeh shahri *et al.*, 2011 [1,2] with added properties to provide efficient logic relation with other applicable software packages for dynamic site response analyses. Such as "Abbas Converter" versions, this code also provides a graphical user interface (GUI) in order to link the constructed databases and lets to user to select type of input data. By click on library and samples, it is possible to access the prepared data. Borehole locations on digital map are the other ad-

vantages of this computer GUI code and by this point of view "NLGSS_Shahri" is stronger than can be regarded as moderate scale seismic geotechnical software. This code is able to perform liquefaction and post liquefaction settlement analyses and includes subroutine forms of C#. Connection between geotechnical properties and strong ground motion databases and dynamic analysis is provided in this program. **Figures 1** and **2** are showing the start screen of the code and its modular structure.

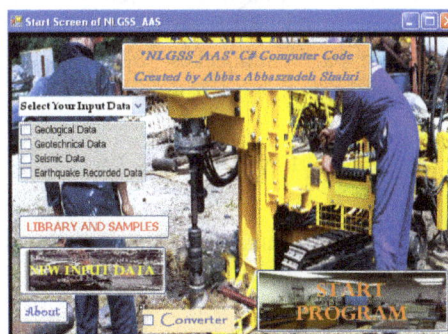

Figure 1. Start screen of the generated computer code.

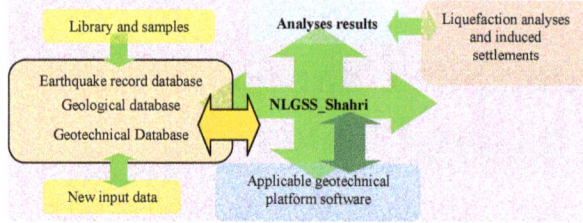

Figure 2. Modular Structure of generated code.

2. Liquefaction Phenomena

Serviceability of Earths dams is sometimes compromised during an earthquake because of development of large deformations as a result of inertial load and/or reduction of material shear strength because of popre water rise or collapse of structure. Earthquake induced liquefaction is a major concern for earth dam safety in seismically active regions of the world. Many liquefaction induced embankment failures or near-failures have been reported around the world during various earthquakes. Such embankment damages were particularly destructive when the underlying saturated granular soils liquefied.

A liquefaction susceptibility analysis determines whether a given soil deposit is in a contractive state, *i.e.*, susceptible to undrained strain-softening behavior and flow failure. Numerous investigators have proposed susceptibility boundary lines between penetration resistance and effective confining stress to separate contractive from dilative soil states [3-6]. Similarly, procedures are available to evaluate the liquefied shear strength for use in a post-triggering/flow failure stability analysis. Olson and Stark (2002) proposed a procedure to estimate the liquefied strength ratio using the corrected CPT or the SPT resistance [7]. The liquefied strength ratio can be used in a post-triggering stability analysis. In contrast, few procedures are available to evaluate the triggering of liquefaction in ground subjected to a static shear stress.

The procedure for assessing liquefaction potential typically uses the Cyclic Resistance Ratio (CRR) as a measure of the liquefaction resistance of soils and the Critical Stress Ratio (CSR) as a measure of earthquake load. For cohesionless soils, CRR has been related to normalized SPT blow count, (N), through correlations that depend on the fines content of the soil from field performance observations from past earthquakes [8]. The normalized SPT blow count is given by:

$$(N_1)_{60} = N \times \left(\frac{P_a}{\sigma'_{v0}}\right)^{0.5} \times ER$$

N: raw SPT blow count, P_a: Atmospheric pressure (≈100 Kpa), σ'_{v0}: Effective vertical stress, ER: Energy Ratio.

For a preliminary evaluation or small project, the mean stress could be calculated from the effective overburden stress and coefficient of lateral pressure at rest. The effective mean stress can be calculated by the following equation:

$$\sigma'_m = \frac{1}{3}\left(\sigma'_{v0} + 2K_o\sigma'_{v0}\right)$$

$$K_o = \begin{cases} 1 - Sin\varphi & For\ Sand \\ K_o = 0.4 + 0.007(PI) & 0 \le PI \le 40 \quad For\ Clay \\ K_o = 0.68 + 0.001(PI - 40) & 40 \le PI \le 80 \end{cases}$$

σ'_m: Effective mean stress, K_o: Coefficient of lateral pressure.

CSR is used to define seismic loading, in terms of the Design Peak Ground Acceleration (DPGA) and Design Earthquake Magnitude (DEM). CSR is defined as:

$$CSR = 0.65\left(\frac{DPGA}{g}\right)\left(\frac{\sigma_{v0}}{\sigma'_{v0}}\right)\left(\frac{r_d}{r_{MSF}}\right)$$

g: gravitational constant, σ_{v0}: Total vertical stress, r_d: Stress, reduction factor, r_{MSF}: The magnitude scaling factor.

The procedure for assessing liquefaction potential uses the CSR as the measure for earthquake load, thus:

$$CSR = 0.65 \times \frac{a_{max}}{g} \times \frac{\sigma_{v0}}{\sigma'_{v0}} \times r_d \times K_m^{-1} \times K_a^{-1} \times K_\sigma^{-1}$$

a_{max}: PHGA, $K_m^{-1}, K_a^{-1}, K_\sigma^{-1}$: Correction factors for the earthquake magnitude, the presence of initial static shear and depth of the layer.

The cyclic shear stress on the horizontal plane is used to calculate CSR. The relationship is as followed:

$$CSR = \frac{0.65 \times \tau_{peak,cyclic,horz}}{\sigma'_v}$$

Liquefaction potential can be evaluated by CSR and CRR (normalized resistance stress of the soil material is called Cyclic Resistance Ratio). This comparison can be considered in form of factor of safety against liquefaction as follow:

$$FS = \frac{CRR}{CSR}$$

The liquefaction resistance of the soil (CRR) can be estimated by various laboratory and field methods. The cyclic triaxial shear test and the cyclic simple shear test are common test used to characterize the CRR. The SPT, CPT and shear wave velocity test are the tests that are most frequently used for determining the liquefaction resistance (CRR) of the soil.

Liquefaction of fill in the dam may occur. Liquefaction is the large drop in stiffness and strength of soil due to seismic movements [9]. As a result, part of a dam may slump and slides off the structure. Liquefaction is the most important cause of instability of earth embankments during earthquakes and may cause large deformation, loss of capacities and even complete failures. Liquefaction is initiated when cyclic ground motions causes loose soil particles to attempt to rearrange into a denser configuration. The rapid nature of the loading of the saturated soil results in an undrained condition, and the soil particles cause an increase in excess pore pressures as they try to densify.

3. Testing Program and Methodology

The June 22, 2002 Avaj-Changureh earthquake with $m_b6.5$ occurred in a region of northwestern Iran which is crossed by several major fault lines. The focal depth of the event, according to the USGS report, was approximately 10 km. The epicenter coordinates of the earthquake was estimated at 48.93 longitude and 35.67 latitude. The maximum horizontal and vertical accelerations were recorded at approximately 0.5 g and 0.26 g, respectively at Avaj station. The fault plane solution indicates that the seismic event was occurred on a reverse fault having trend about N115° [10]. **Figure 3** shows the recording stations, earthquake epicenter and the studied area.

Nemat Abad dam is a homogenous earth fill dam with a maximum height of 50 m and crest length of 633 m on Shahab River with the aim of providing the required water for agricultural lands of Asadabad plains. This dam is situated at a distance of 45 km from west of Hamedan city and 12 km northwest of Asadabad in 34°43'45" north latitude and 48°02'41" east longitude.

A total of 16 boreholes were drilled but the data of 9 of them were available for analysis that presented in **Tables 1** and **2**. A high accuracy correlation between the drilled boreholes by taking into account of geological conditions, laboratory testing and *in situ* tests to propose the idealized soil profile of the target area was done by "NLGSS-Shahri" as shown in **Figures 4** and **5** and the contour map of the obtained characteristics is presented in **Figure 6**.

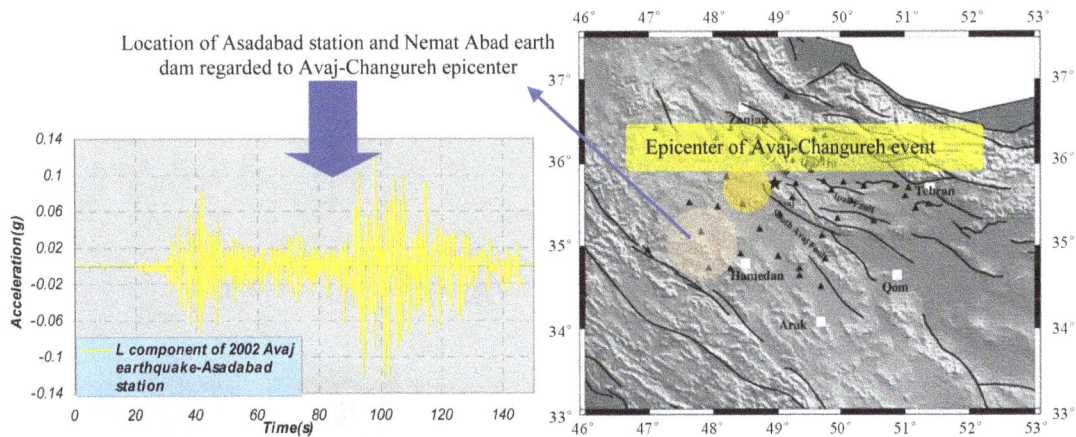

Figure3. Location of strong ground motion stations and epicenter of 2002 Avaj-Changureh event and studied area.

Table 1. Available average obtained parameters of drilled boreholes in site of Nemat Abad earth dam.

Borehole	Depth (m)	Soil parameters			GWT (m)
		LL	PI	W%	
Bh#201	50	33	5	-	14.7
Bh#202	50	35	10	10.6	13.6
Bh#203	50	30	10	10.6	5.95
Bh#204	60	31	10	17.6	1.2
Bh#206	60	33	7	11.8	0.6
Bh#207	50	34	5	18	2.3
Bh#211	30	29	10	18.3	2.0
Bh#212	30	29	5	9.3	1.0
Bh#214	30	31	10	22.11	2.50

Table 2. Grain size distribution of available picked up data from the boreholes.

Borehole Size	Bh#2 02	Bh# 203	Bh#2 04	Bh# 206	Bh# 207	Bh#2 12
75	100	100	100	100	100	100
19.05	97.8	100	94	94	94	100
4.75	95.1	88	88	86	80.5	89
2	89.6	80	81	80	70.5	78
0.85	85	71.5	73	73	63	66
0.425	82	66	67	67	57	54
0.15	78	60	62	62	53	53
0.075	74.3	52	56	58	49	48

In the selected area no attempts were made for developing the regression correlation based on the entire data and N for this study 40 pairs of N value and V_s were applied and a formula which explained V_s as a function of N values from locations where tests were conducted; thus value was determined for the selected area as shown in **Table 3**. The results of these trials were compared to existing field and laboratory relationships, and appropriate adjustments were made to the model parameters.

The methodology for dynamic site response analysis is based on the nonlinear standard hyperbolic model. The parameters G_{max} (maximum shear modulus) and ξ (damping ratio) are used to describe the dynamic behavior of soils in site response analysis. These pa-

rameters are calculated with "NLGSS_Shahri" utilizing geotechnical data collected at geotechnical properties database. G_{max} can be calculated from empirical relationships for clays [11] and for sands [12,13]. G_{max} can be also determined from corrected SPT-N values [14,15]. The variation of the modulus ratio (G/G_{max}) and ξ with shear strain (γ) is computed from various formulations such as [16]. Modulus ratio and damping ratio values for each layer of the soil profile are calculated for shear strains varying between 0.0001 and 10 percent using the generated computer program. The modulus reduction and damping curves can determined for each characterized material during this process.

By referred to **Figure 7**, which indicate the flowchart

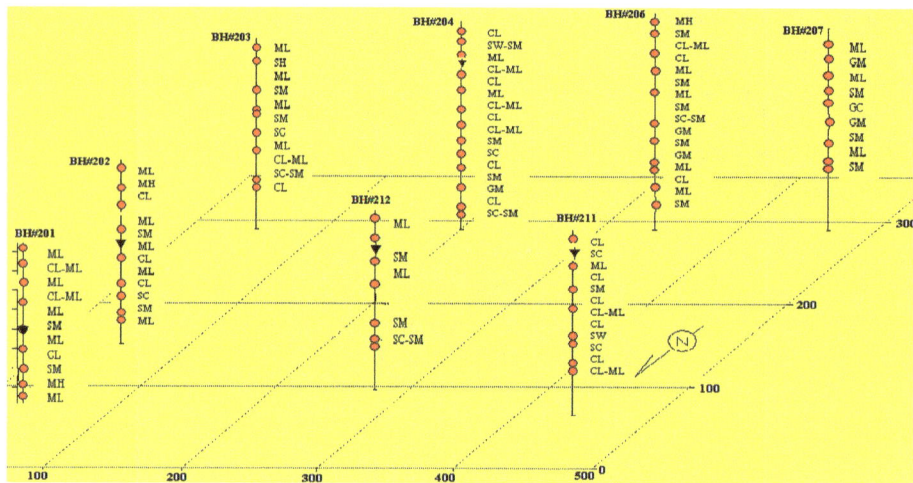

Figure 4. Boreholes location and layer classification of the area.

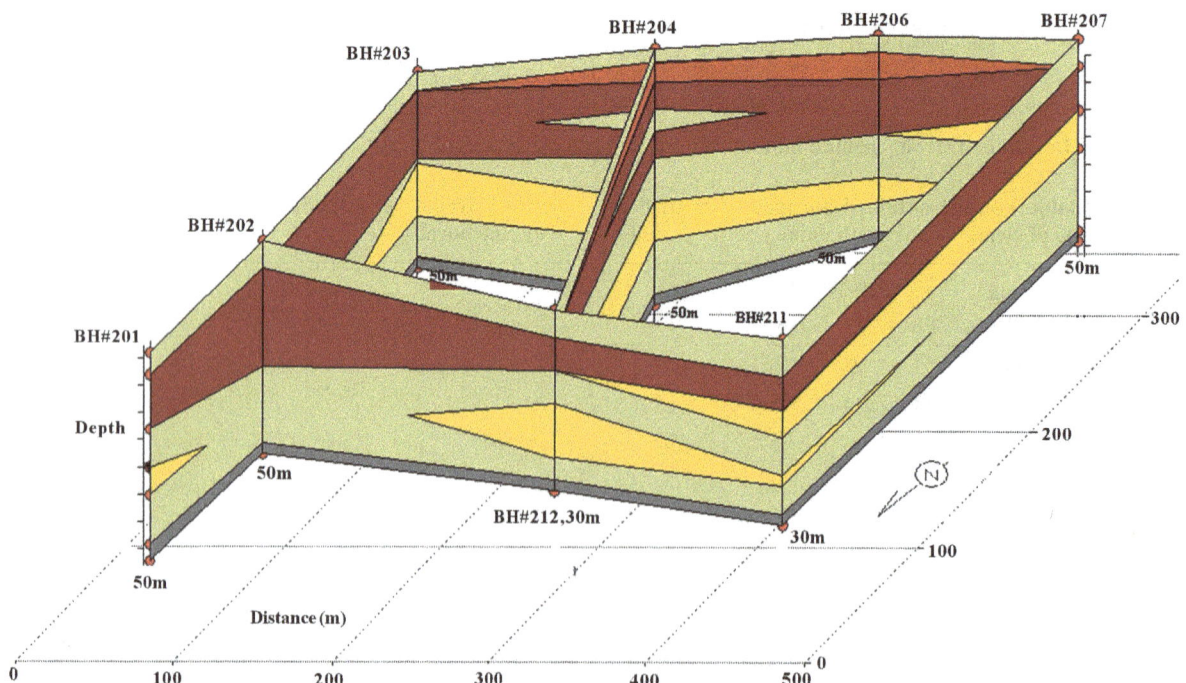

Figure 5. Layer correlation of the drilled boreholes in target area to propose the idealized soil profile.

Table 3. Correlation results of V_s - N for the selected region.

Model	a	b	c	R	S
$V_s = aN^b$	87.4926	0.3563		0.9907	6.6301(X)
$V_s = a + bN$	201.5882	2.9549		0.9769	10.4586
$V_s = a + N^b$	224.5296	1.2296	**a, b and c:** Constant parameters	0.9669	12.4898
$V_s = a + b^N$	201.5882	2.9549	**R:** Correlation coefficient	0.9769	10.4586
$V_s = aN^{b/N}$	516.4297	−4.8224	**S:** Standard error	0.9809	9.5181
$V_s = ae^{bN}$	224.1843	0.0086		0.9615	13.4561
$V_s = a + bLnN$	−74.988	109.4829		0.9873	7.7501
$V_s = aN^2 + bN + c$	−0.0242	5.0461	162.4514	0.988	7.5656

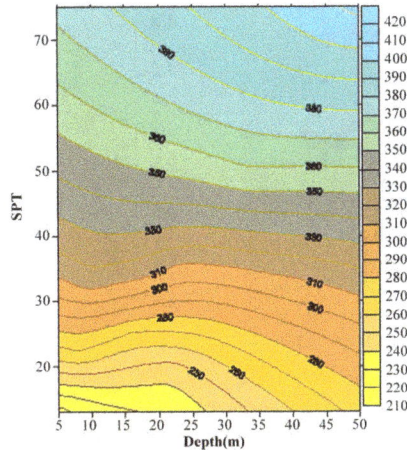

Figure 6. Contour map of Vs on base of SPT and depth of sampling in selected area.

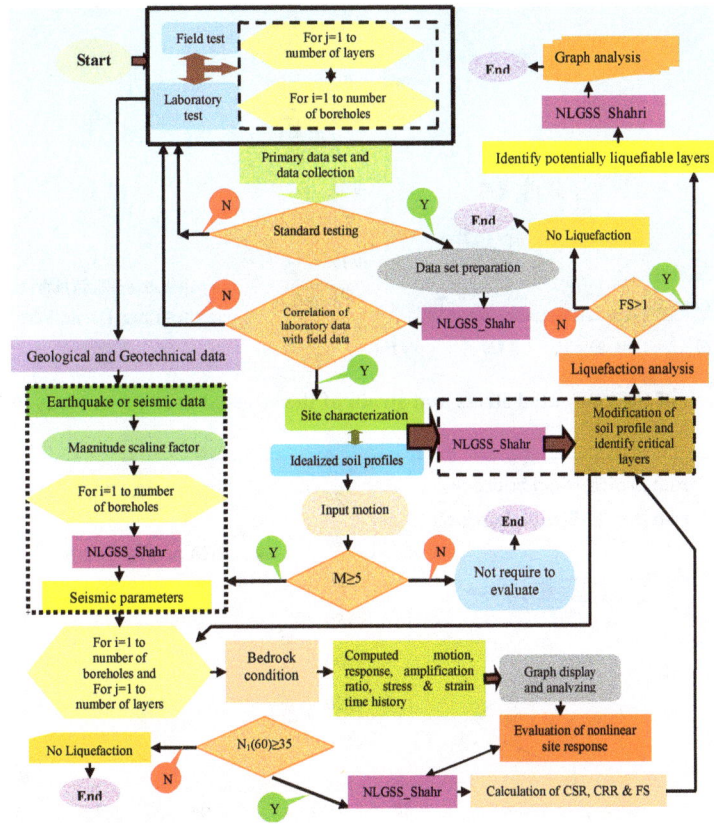

Figure 7. Summarized proposed flowchart for this study.

of the study, the experimental estimation of ground motion can be carried out using earthquake data, in areas having a sufficient seismicity and an adequate cover age of recording stations. In order to evaluate the effect of particular geological conditions on the change of the local seismic response of ground motion, geophysical surveys were performed for the lithotypes with a significant extent in the territory. The obtained results of this study by taking into account the fine correction factor (FC) were compared with the previous proposed proce-

dures [17-23]. To analyze the liquefaction potential of the region subjected to Avaj-Changureh earthquake, a comparison between computed motion, stress, strain and response spectra were executed and shown in **Figures 8-10**. To prove and verification of the applied method in this study, a comprehensive comparison between the liquefaction resistance factors, safety factor, shear modulus reduction curve and damping ratio curves were performed for the idealized soil profile, and the resulting liquefaction potential, for this area was determined and

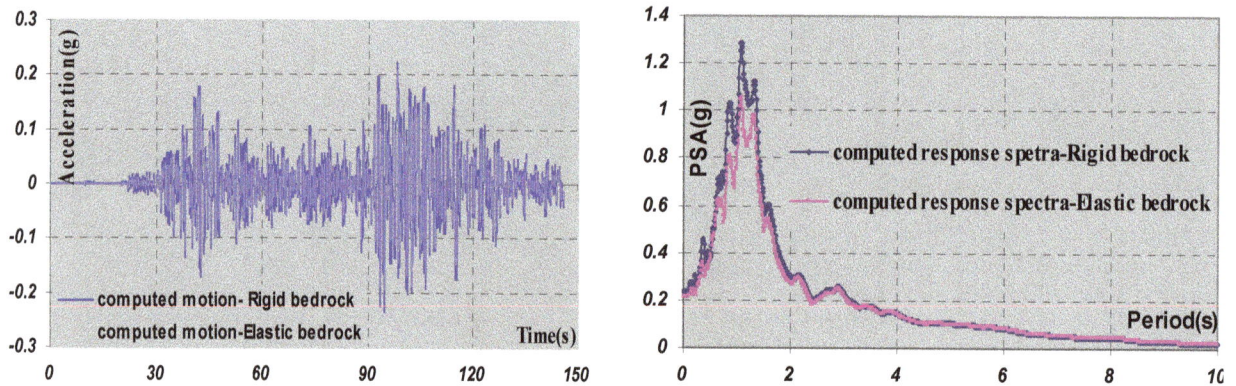

Figure 8. Overlay of the computed motion and surface response spectra for studied area for various bedrock conditions.

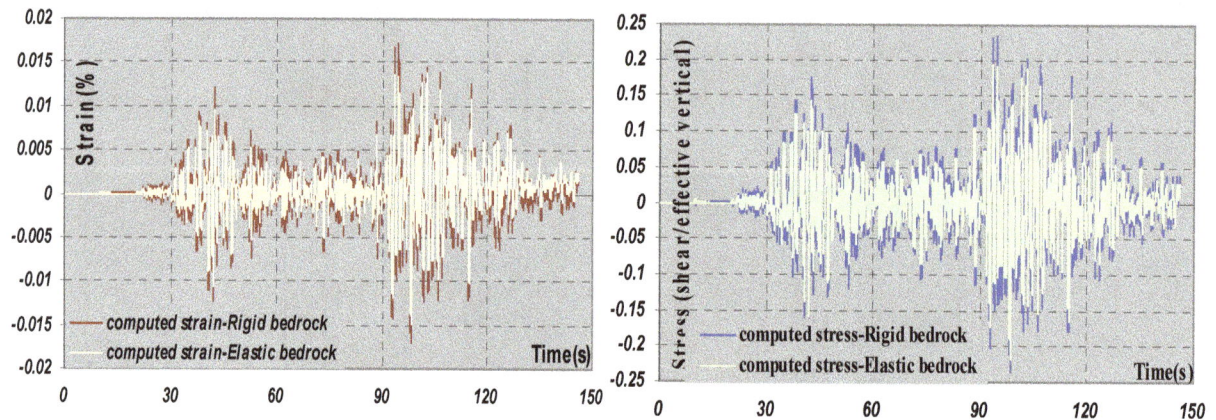

Figure 9. Comparison of the computed stress and strain of the selected area for various bedrock conditions.

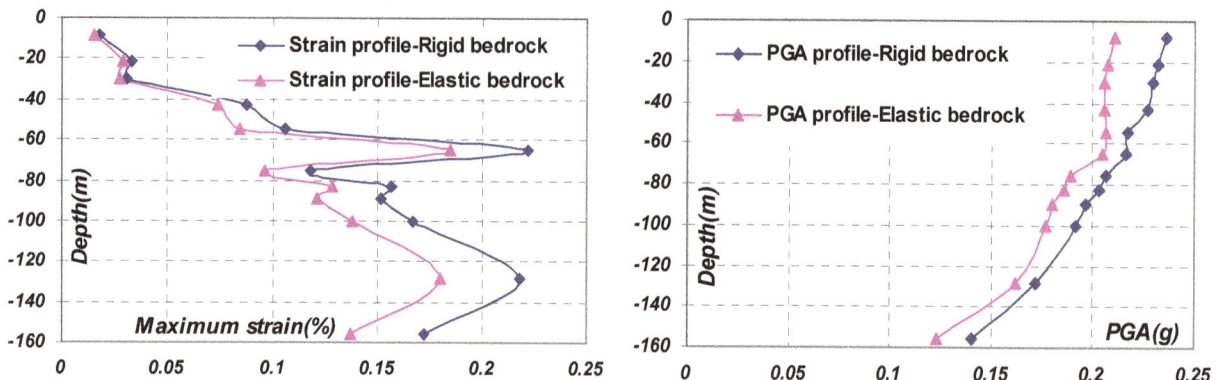

Figure 10. Variation of PGA and strain profile *Vs* depth in rigid and elastic half space bedrock.

compared with known procedures as indicated in **Figure 11** and contour map and 3D view of safety factor are given in **Figure 12** respectively. At last by this method the numerical analysis of this study for main parameters which is computed by the generated code was pointed in **Tables 4** and **5**.

4. Discussion and Conclusions

Most of the early constructed earth dams in Iran were-built with no consideration of earthquakes and were not designed for earthquake forces because designers did not consider earthquakes probable threats. As more informa-

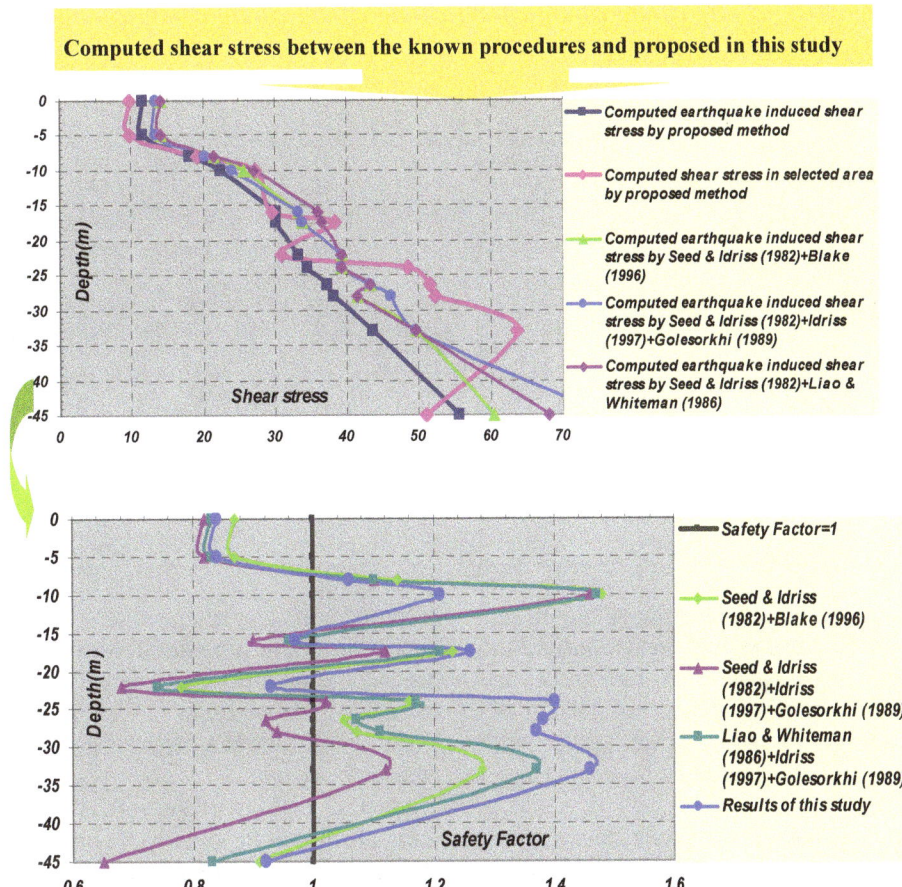

Figure 11. Comparison between safety factors of the proposed method by known procedures.

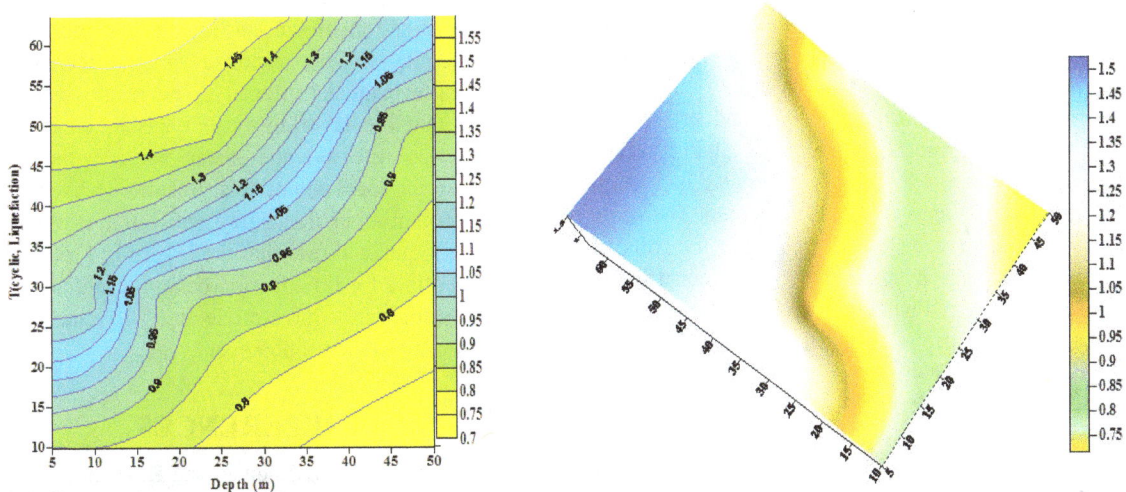

Figure 12. Contour map and 3D view of safety factor for selected area.

Table 4. Numerical results of the study.

Parameter	Condition	Maximum value at...	Parameter	Condition	Maximum value at...
Input motion	Elastic	0.1231 g (98.2 s)	Computed surface	Elastic	0.2127 g (98.4 s)
	Rigid	0.1402 g (98.2 s)	motion	Rigid	−0.236 g (94.6 s)
Input response	Elastic	0.590 g (1.32 s)	Computed surface	Elastic	1.054 g (1.080 s)
	Rigid	0.631 g (1.32 s)	response	Rigid	1.288 g (1.080 s)
Input stress	Elastic	−0.312 (98.3 s)	Computed surface	Elastic	−0.228 (98.4 s)
	Rigid	−0.336 (98.3 s)	stress	Rigid	−0.235 (98.4 s)
Input strain	Elastic	−0.1349% (98.3 s)	Computed surface	Elastic	−0.01521% (98.4 s)
	Rigid	0.1717% (94.5 s)	strain	Rigid	0.01728% (94.6 s)
Amplification	Elastic	5.40 (1.66401 Hz)	Spectral	Elastic	1.193 g (0.67 s)
ratio	Rigid	43.5 (1.66401 Hz)	acceleration	Rigid	3.47 g (0.61 s)

Table 5. Computed values by "NLGSS_Shahri".

Depth (m)	γ	σ_v	$\sigma'v$	C_N	r_d	$N_1(60)$	τ_{cyc}	CSR_L	$\tau_{cyc,L}$	FS_L
5	15.7	78.5	78.5	1.104	0.975	10.33	11.641	0.125	9.846	0.845
8	16	126.5	126.5	0.87	0.941	13.78	18.105	0.152	19.288	1.06
10	16.87	160.1	160.1	0.773	0.924	18.36	22.50	0.171	27.377	1.21
16	16.1	256.7	253.04	0.612	0.775	11.016	30.259	0.117	29.60	0.97
17.45	17	281.35	277.94	0.585	0.708	14.32	30.297	0.138	38.355	1.26
22	16.7	357.335	356.05	0.517	0.612	7.07	33.262	0.087	30.979	0.93
24	17.2	391.735	390.38	0.494	0.579	12.45	34.498	0.124	48.407	1.4
26.45	16.9	433.14	432.45	0.470	0.556	10.82	37.288	0.119	51.46	1.38
28	17.6	460.42	459.05	0.456	0.545	11.16	38.166	0.114	52.33	1.37
33	17.9	549.92	544.89	0.419	0.521	14.78	43.577	0.118	63.752	1.46
45	17.2	756.32	750.90	0.357	0.482	9.51	55.447	0.0679	51.027	0.92
50	18.1	846.82	836.74	0.339	0.441	18.30	56.801	0.0651	54.471	0.958

tion of earthquakes was collected, the need to built dams that could withstand earthquakes was recognized. Earth embankment dams may be damaged by earthquakes in several ways including dam movement, liquefaction of fill in a dam, water waves caused by an earthquake over topping a dam, and direct damage caused by a dam being located on a fault.

In this study, after calibrating the input parameters, the constructed model was used together with employed several software packages to obtain the response of a layered soil profile. The main target of this phase of the study was to evaluate the "NLGSS_Shahri" capabilities in response of liquefiable soils in order to manipulate large amount of geotechnical data and to prepare a data input file for performing dynamic analyses.

At the present paper, a methodology and processing principles of $C^{\#}$ developed GUI "NLGSS_Shahri" is introduced, and its application to the Nemat Abad earth dam in Hamedan province of Iran is presented. The large amount of geological and geotechnical data for soils of the selected area have been loaded to the constructed dynamic soil database. Dynamic site response analyses are performed using proposed method and liquefaction analyses are performed with generated code using results of dynamic analyses. The results of this study indicate that the generated program is a reliable tool for site response analysis and the proposed method can be used for site response analysis as well as the other procedures because comparison of the site response analysis of the

proposed profile agreed good reasonable matching by the known applicable procedures. More that in this study, the dependence on local soil instability conditions related to mechanical characteristics of surface soils, such as the slope of soils and the depth of ground water table, was taken into account.

Spectral analysis of the results showed that the stiffness of the soil deposits had a significant effect on the characteristics of the input motions and the overall behavior of the structure. The peak surface acceleration measured by the proposed method was significantly amplified, especially for low amplitude base acceleration. The amplification of the earthquake shaking as well as the frequency of the response spectra decreased with increasing earthquake intensity. The results clearly demonstrate that the layering system has to be considered, and not just the average shear wave velocity, when evaluating the local site effects. Result of presented liquefaction potential in this study subjected to Avaj-Changureh earthquake shows that the layers 1, 4, 6, 11 and 12 are susceptible for liquefaction behavior and also showed that the studied area have moderately liquefaction potential regarding to mentioned event.

REFERENCES

[1] A. A. Shahri, B. Esfandiyari and H. Hamzeloo, "Evaluation of a Nonlinear Seismic Geotechnical Site Response Analysis Method Subjected to Earthquake Vibrations (Case

Study: Kerman Province, Iran)," *Arabian Journal of Geosciences*, Vol. 4, No. 7-8, 2011, pp. 1103-1116.

[2] A. A. Shahri, K. Behzadafshar and R. Rajablou, "A Case Study for Testing the Capability of an Intermediate Generated Geotechnical Based Computer Software on Seismic Site Response Analysis," *International Journal of the Physical Sciences*, Vol. 6, No. 2, 2011, pp. 280-293.

[3] M. H. Baziar and R. Dobry, "Residual Strength and Large Deformation Potential of Loose Silty Sands," *Journal of Geotechnical and Geoenvironmental Engineering*, Vol. 121, No. 12, 1995, pp. 896-906.

[4] C. E. Fear and P. K. Robertson, "Estimating the Undrained Strength of Sand: A Theoretical Framework," *Canadian Geotechnical Journal*, Vol. 32, No. 4, 1995, pp. 859-870.

[5] K. Ishihara, "Liquefaction and Flow Failure during Earthquakes," *Geotechnique*, Vol. 43, No. 3, 1993, pp. 351-415.

[6] J. A. Sladen and K. J. Hewitt, "Influence of Placement Method on the *in Situ* Density of Hydraulic Sand Fills," *Canadian Geotechnical Journal*, Vol. 26, No. 3, 1989, pp. 453-466.

[7] S. M. Olson and T. D. Stark, "Liquefied Strength Ratio from Liquefaction Flow Failure Case Histories," *Canadian Geotechnical Journal*, Vol. 39, No. 3, 2002, pp. 629-647.

[8] T. L. Youd, I. M. Idriss, R. D. Andrus, I. Arango, G. Castro, J. T. Christian, R. Dorby, W. D. L. Finn, L. F. Harder, M. E. Hynes, K. Ishihara, J. P. Koester, S. C. Laio, W. F. Marcuson, G. R. Martin, J. K. Mitchell, Y. Moriwaki, M. S. Power, P. K. Robertson, R. B. Seed and K. H. Stokoe, "Liquefaction Resistance of Soils: Summery Report from the 1996 NCEER and 1998 NCEER/NSF Workshop on Evaluation of Liquefaction Resistance of Soils," *Journal of Geotechnical and Geoenvironmental Engineering*, Vol. 127, No. 10, 2001, pp. 817-833.

[9] M. P. Byrne and M. SeidKarbasi, "Seismic Stability of Impoundments," 17*th Annual Symposium, Vancouver Geotechnical Society*, 2003.

[10] http://www.usgs.gov

[11] B. O. Hardin and V. P. Drnevich, "Shear Modulus and Damping in Soil: Design Equations and Curves," *Journal of the Soil Mechanics and Foundations Division*, Vol. 98, No. 7, 1972, pp. 667-692.

[12] H. B. Seed and I. M. Idriss, "Soil Moduli and Damping Factors for Dynamic Response Analyses," Report EERC

70-10, Earthquake Engineering Research Center, University of California, Berkeley, 1970.

[13] H. B. Seed and I. M. Idriss, "Simplified Procedure for Evaluating Soil Liquefaction Potential," *Journal of the Soil Mechanics and Foundations Division*, Vol. 97, No. SM9, 1971, pp. 1249-1273.

[14] T. Imai and K. Tonachi, "Correlation of N-Value with S-Wave Velocity and Shear Modulus," *Proceedings of 2nd European Symposium on Penetration Testing*, Amsterdam, 1982, pp. 57-72.

[15] Y. Ohta and N. Goto, "Estimation of S-Wave Velocity in Terms of Characteristic İndices of Soil," *Butsuri-Tanko*, Vol. 29, No. 4, 1976, pp. 34-41.

[16] Y. Ohta and N. Goto, "Estimation of S-Wave Velocity in Terms of Characteristic İndices of Soil," *Butsuri-Tanko*, Vol. 29, No. 4, 1976, pp. 34-41.

[17] T. F. Blake, "Formula (4)," In: T. L. Youd and I. M. Idriss, Eds., *Summary Report of Proceedings of the NCEER Workshop on Evaluation of Liquefaction Resistance of Soils*, Technical Report NCEER 97-0022, 1996.

[18] R. Golesorkhi, "Factors Influencing the Computational Determination of Earthquake-Induced Shear Stresses in Sandy Soils," PhD Dissertation, University of California at Berkeley, Berkeley, 1989.

[19] I. M. Idriss, "Evaluation of Liquefaction Potential and Consequences: Historical Perspective and Updated Procedures," Presentation Notes, 3rd Short Course on Evaluation and Mitigation of Earthquake Induced Liquefaction Hazards, March 13-14, San Francisco, 1997, p. 16.

[20] S. S. C. Liao and R. V. Whitman, "Catalogue of Liquefaction and Non-Liquefaction Occurrences during Earthquakes," Department of Civil Engineering, Massachusetts Institute of Technology, Cambridge, Mass, 1986.

[21] S. S. C. Liao, D. Veneziano and R. V. Whitman, "Regression Models for Evaluating Liquefaction Probability," *Journal of Geotechnical and Geoenvironmental Engineering*, Vol. 114, No. 4, 1988, pp. 389-411.

[22] H. B. Seed and I. M. Idriss, "Ground Motions and Soil Liquefaction during Earthquakes," Monograph Series, Earthquake Engineering Research Institute, Oakland, 1982, p. 134.

[23] T. D. Strak and S. M. Olsen, "Liquefaction Resistance Using CPT and Field Case Histories," *Journal of Engineering Geology*, Vol. 121, No. 12, 1995, pp. 856-869.

2

Explication of Emergency Dwellings in Turkey in the Context of Cultural Continuity: Fieldwork in Alagoz Village

Ferah Akinci, Kader Dağistanli
Faculty of Architecture, Yildiz Technical University, Istanbul, Turkey

ABSTRACT

The study is a fieldwork in Alagöz Village in Bingöl which is situated in the Eastern Anatolia Region of Turkey. Majority of the traditional houses in Alagöz Village were destroyed by earthquakes. Standard type houses were built within the scope of "Earthquake houses" project by the state under the initiative for closing the gap of housing. State adopted the resolution of demolishing the traditional houses and implemented such resolution as a consequence of standard type earthquake houses which do not fit with the cultural life of the region are not preferred by their users. Population who compulsorily started to live in "Earthquake Houses" began to utter their complaints. Even, they attempted to reconstruct and complete their partially demolished houses and returned to live in their traditional old houses. Forwhy, these houses fit with their social lives. Within this context the houses in the village compose of Traditional Houses, Earthquake Houses, houses they developed by being adversely affected from the Earthquake Houses. Each type of housing will be taken up within the scope of the study and their physical, social tissues will be analyzed and scrutinized, and as a result of this detailed study, conception of the region that is transmitted from past to today will be defined. In the study, it is emphasized that traditions which reached from past to today can not be ignored and should be taken into consideration by all means. Within this context, houses located in the region were analyzed as per their periods, opinions of the users were obtained and requests were highlighted in accordance with these findings and the houses which were developed in consequence of the requirements were scrutinized. Outputs were assessed on behalf of protecting the continuity in settlement and finalized with the clues of what are the obvious things to be done. For permanence of the settlement it is required to analyze the culture which preserved and protected its existence. Especially, in the countries such as Turkey, it is required to resolve the dynamics of the migration in consequence of various facts from rural areas to urban centers. It is required to provide approaches which may satisfy their needs and close their gaps by supporting the segment who is pleased with living in its village rather than creating new formations.

Keywords: Housing; Earthquake; Continuitibility; User Demand

1. Introduction

Generally speaking; besides being an indispensable need of human being, housing has reached the point today from cave dwellings, being consolidated by a concept of continuity and development. Due to the instinct of self-protection from wild animals, the first space that human being formed, had an upgraded expectation after the settled life, is now constituted by living units which are produced by the intersection of interdisciplinary professions. The functions of the dwellings which are open to the innovative and developing technology has been interwove Ned with the culture. According to the definition which was done by the specialists of UNESCO (Dictionary of Social Sciences, UNESCO, 1964), culture is: "the consciousness that the society has about its historical continuity; namely, this society shows resolution to continue its existence and development with reference to this consciousness." [1]. The physical environments, that are created by the sociocultural and economic structures of individuals who forms the society, present the awareness of their history to the new generations. The concrete formations (the physical environments); are one of the keys of sustainability. The essence is the sustainability of the structures, which has proved their permanence in time and their adaptation of development.

2. Study Purpose and Scope

Whatever their function is, the formations contradictory to the cultural continuity have made us question their permanence. This study, whose aim is to properly constitute a link between the past and the future, having clearness after the field study datum, emphasises the importance of the culture. Every architectural structure in the world is formed by referring positively or negatively the previous one. It is not suddenly occurred... In that respect; the study introduces the significance of the users in the permanence/life of the dwellings, therefore, by focusing on the dwelling scale, the reality that the culture of the life can not be ignored. In the societies; the spaces which are appropriate for the local conditions and show a functional harmony with the users and which are transferred through generations, are adopted as a traditional understanding.

In this article, the earthquake issue, which is one of the natural disasters that interrupt the continuity, is discussed. The studies after the earthquake should be carefully dealt with and well-organised. In this context, the experiences and the way of urban planning after the destroyable earthquakes in the world are important. In the article, the dwellings which are made due to the necessity after the earthquake are mentioned. The duty which underlines the need of a dwelling reformation adapted to the life style and the local culture even though they are destroyed by an earthquake, is sampled within Alagoz Village. It is analysed by means of the participants from Alagoz Village examining the dwellings done by the government for the injured people. The insertion of the typical emergency dwelling units in every region whenever needed is a very irritating approach. The spaces which exist for the users can not ignore the culture of the users.

3. Earthquakes in Turkey

There are several fault lines in Turkey. The most effective ones are the fault lines of Northern and the Eastern Anatolia.

When the magnitude and the effects of the earthquakes on earth are considered, two of the regions are very interesting seismic belts. The first one is Pacific seismic belt which surrounds the Pacific Ocean and has effects especially on Japan, the other one is Mediterranean-Himalaya seismic belt which spreads from Gibraltar to Indonesia and includes also Turkey. Owing to the fact that there are lots of small slabs among the bigger ones, a big portion of Turkey remains in the seismic belt (**Figure 1**). Turkey is under the impression of three big seismic slabs; Eurasia, Africa and Arabic slabs. The Anatolian slab, on which a great part of Anatolia sits, is a small part of the Eurasian slab [2]. The geography has a seismic character. Consequently, earthquake should be one of the primarily effective factors in the country planning. The post earthquake period should be dealt with in two stages and the urgent precautions and the later ones should be thoroughly evaluated from the scale of country to the region and to the unit structure. The fieldwork done in Turkey, province of Bingol, Village of Alagoz mentioned in this article will be a documentary for the precautions in the scale of rural area referring the unit structure.

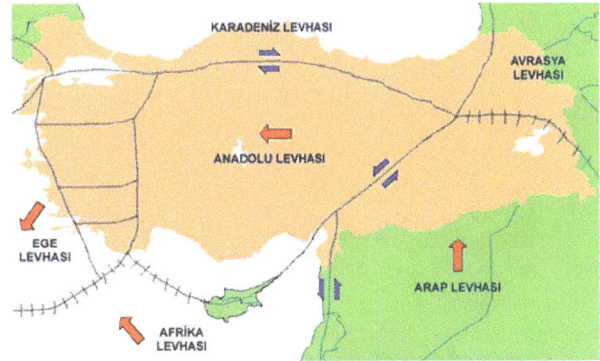

Figure 1. Different seismic slab positions in Turkey.

4. Emergency Dwellings in Turkey and around the World

The construction is generally risky through the country. As a matter of fact the earthquakes in the past have shown that lots of people have lost their lives in the wrongly constructed buildings. The realm of the earthquake was better understood after the earthquake of 1999 which was also strongly felt in one of the most important cities of Turkey, in Istanbul. The modifications in the regulations are a result of this fact. The injured people who were confronted with the jerry built tents, and then maintained their lives for a period with "temporary housing system". Besides being prefabrically produced and plastic-based, the units of this type were preferred due to being also economic and easily constructible. The prefabricate units are an intermediary step for the injured people to have the permanent dwellings.

For the accommodation of the sufferers in a better housing, in the temporary dwellings time is gained. However, it is analysed that there are several unfavourable examples for such kind of "better" housing units when their ability in making up for the necessities of the users are considered. The units which were produced just in one type were not designed according to the geographical data. So, there were cases where factors such as the orientation and the culture were not evaluated. On the other hand, the psychology of an event in which the sufferer almost collided with death and lost his relatives was not taken into account and he was put into a similar type of dwelling [3]. Moreover, in the areas thought for the new housing formations after the earthquake of 1999 in Marmara Region, earthquake resistant dwellings in tunnel formwork type were constructed.

The ways in replying the housing necessity after earthquake show variety in the world. Managua, the capital city of Nicaragua [4], witnessed an earthquake in 1972. The need of the new dwellings was already evident as a problem in the city due to the increase in the population because of the immigration. By the earthquake, this existing problem got bigger. After the disaster, to urgently solve the problem "cities of tents" were formed;

but they were not used owing to the inadaptability to the family life. The temporarily made wooden structures were not used either. Additionally, the igloos made by polyurethane foam were not a valid system due to the fact that they had the conditions to live in, after 5 months from the earthquake. The best solution for the victims was to stay near the relatives; or else to construct their own jerry built houses by using the materials collected from the wreckage. As a result of this a concept of "shanty house" was introduced. Some other part of the victims searched for a solution by staying in the cheap hostels. Hence, it was observed that the 95% of the quake victims accommodated in the houses of their relatives rather then in the emergency dwellings. What is more the following solutions for the permanent dwellings done in a more organised manner, did not also work. Blocks of houses from solid package concrete were executed. Another solution was the site which had a certain quality of living, but served to a limited number of habitants. However, the problems of the habitants came up and indeed the life style of the users sometimes contradicted the site's required understanding of life. It was observed that the dwellings were not adequate for the financial and cultural circumstances of the country. As a result of the fact that the urban land was not expropriated and the studies were not integrated with the urban plan, the governmental precautions remained deficient.

In 1963, Republic of Yugoslavia, the regional capital city of Macedonia, in Scopje [4], after the earthquake the tents were set up for the fist phase. The difficulties in planning the permanent dwellings were not seen because financial support arrived from a lot of country. The first planning approach after the earthquake was the ascertainment of damage in the buildings. The ones with repairable damage were again opened to use. Moreover, in the temporary dwelling stage no disadvantage was seen, because besides having a strong international spiritual and material support, the prefabrication system and the government being the land owner facilitated the period. Although the donations of the prefabricated units were reflecting the different architectural properties, in the whole concept they were harmoniously integrated. The works for not to stop the production in the country were also noteworthy. In the new dwellings, even though the national architectural references are valid, a discontinuity in the cultural sense is also evident. Nevertheless, generally speaking, Scopje had a successful transformation, due to an organised approach and a correct usage of the financial resources. Each construction exhibits the harmony in the whole.

5. Scope of Fieldwork and Kigi District in Cultural Continuity

Bingol is the city in which the fieldwork of the article,

Alagoz Village is found. It is located in the east of Turkey. It is in the region which remains between the fault lines of North Anatolia and the East Anatolia. The city has got 7 districts and 323 villages.

Bingol is an underdeveloped city which ranks the 71st in progress among other 81 cities of Turkey. According to the census of the year 2000, the 70% of its population is engaged in agriculture. Although the main means of livelihood is kettle breeding, there has been a reduction by half in the number of great and small cattle since 1991. Since 2002, the production of beehives has reached to 43,477 in number and therefore 796 tones of honey was obtained. Hence, beekeeping has become another important financial gain. Nevertheless there is unemployment in the city. The immigration rate from city to the countryside has increased. 2822 house of people returned to their villages up to the year 2002. In one hand the difficulty of earning money is the chief reason for immigration to the city, in the other hand there is a high rate of unemployment and not qualified labor force in the city [5].

The Alagoz village which is appurtenant to the Kigi district of Bingol presents an expedient dwelling character for this article. The cultural structure of the houses in the village which is transferred over generations is noteworthy. These houses were effected from the earthquake, confronted to an understanding of almost abandonment so as to find a solution to the disaster by the authorities. This region suffered from the earthquake and there is a problem throughout the country which stems from the culturally and functionally inappropriate applications. This problem needs to be solved.

The Alagoz village which is appurtenant to the Kigi district of Bingol city which is located in the eastern part of Turkey is selected as the fieldwork. On the west there is the city of Tunceli, on the North there are the cities of Erzurum and Erzincan, on the South there are the cities of Bingol, Elazig and Diyarbakir.

The district is 73 km far away from the city center of Bingol. But, since this said distance is valid for a gravel road, the destination to Bingol is provided by another road of 145 km which passes through Karakocan district of Elazig. The Alagoz village however is 27 km far from the district of Kigi. This road is alos a gravel road.

Being suffered by the disaster, lots of houses were damaged. The content of the study is, on one hand, these houses which are an expression of a cultural continuity and have been carried to today; on the other hand the dwellings which are constructed after the earthquake. Both types were estimated focusing on the user demands and were analysed referring to the system of their togetherness in the base of a common life style. In this analyse, the starting point was the idea that one of the products of the cultural accumulation which calls the societies into being must be "the house".

There is an accessibility problem in the village due to the heavy snow precipitation in winter; due to mud and landslides in the other seasons. After the earthquakes happened in 1968 and 2003, the houses were affected or totally destroyed, or survived partially damaged. The traditional houses have been a part of the local culture and the life style. The history of the district of Kigi, for instance, dates back to B.C. 3000. The Hittites, The Urartians and The Persians successively had dwelled in the region. The district had been under the domination of Seljukids, Akkoyunlus and Safevid Empire after 1071, and then Ottoman culture had been effective after 1514. The presence of a through going culture is one of unquestionable realities.

The region has a continental climate. The summer is cool; the winter is long and brisk. The showers are usually seen in the autumn. The general vegetation type in the region is steppe. However there are also forested areas where the water rate is high. The most prevalent tree species are oak, juniper, wild poplar, willow, ash-tree and black tree. Other than those, apple, pear, walnut and mulberry trees can be planted as fruit giving trees [6].

While in the census of 1990 the population of the village was 115, it was indicated as 66 in the census of 2000. This, therefore, demonstrates the change in the structure of the families of the village. 35% of the families has the children immigrated to the cities. Yet, there are a percentage of 30 in the village having a patriarchal family structure. In such kind of families, the father and (or) mother of the housefather, his wife, his unmarried children, his married male children, daughter-in-law and grand children share the same house. The family members of this kind of a family are 7 - 10 in number. The educational level in the village is quite low. 50% of the population is primary school graduated, 10% is elementary school graduate, and just 7% is high school graduated. There is no university graduate in the village. The relatives and the neighbors live together as a large family in the whole village. The tradition of working together for the community or one of its members, gathers the relatives and the neighbors together for various works, and consequently the relations intensively continue. For instance, the women of the village come together on the roofs or terraces of the houses and do assorted works. The 40% of the population is over the age 50. 65% of the families manage their lives by agriculture and cattle breeding, 25% by agriculture, cattle breeding and old-age salary, and the remaining 10% by only old-age salary [7].

The region has an appropriate structure for also bee-keeping. The honey production has a district wide sales market and in Istanbul and is among the qualified productions. But there is hardly anyone who does this job in a professional manner.

Because there are no shops in the village for daily needs, the shopping is done by going to the district. For that reason, 90% of the people are not satisfied with the shopping conditions.

6. Houses in Alagoz Settlement

Generally speaking, there are two kinds of houses in the village; houses constructed by the villagers or by a craftsman and houses constructed by authorities. When the typolgy is more specifically examined, the houses of different understandings are seen. These are:

1) The traditional houses built by the villagers, survived from the late 19th century in which the village was established;

2) The houses built again by the villagers after 1950;

3) Reinforced concrete houses built by the owner with the help of the family members and the neighbours, between the years 2005 and 2006, instead of the houses destroyed by the government;

4) The permanent emergency houses built by the government in 1986;

5) The permanent emergency houses built by the government between the years 2005 and 2006.

6.1. Traditional Houses Built by Villagers

The traditional houses built by the villagers survived from the late 19th century when the village was established. These are the houses which reflects the culture of the village, has a history and prooved their permanency through the life. The local stone was used in a very talented manner in the structure of these houses. They were built as solid masonry and timber and stone were used as material. While the stone was used in the foundation and in the walls, the timber was used as a lacing course in the walls, ceilings, windows and the doors. The rooves of the houses are flat and covered by waterproof hardpan. owing to the slope, some of the houses were built with terraces and can be regarded as two-storey. The connection between the storeys is provided by exterior circulation. On the upper floor, there are living spaces, whereas on the lower floor there are kitchen, cellar, store, woodshed and animal shed. For the light penetration, the kitchen is on the front side of the house, while the animal shed is underground. The roof of the lower floor functions as the entranceterrace of the house.

The food is stored in the summer time, due to the long and snowy winters. Homemade macaroni, boiled and pounded wheat, dried layers of fruit pulp, canned food, and such kind of food for winter. are prepared collectively on the rooves of the houses. Besides, fruits and vegetables, such as wheat, barley, apple, pear, apricot, tomato, pepper, etc. are left to dry under the sun on the roof.

Those terraces are, at the same time, a kind of special spaces on which the villagers play various games. In the village where a coffe house does not exist, the men, for example, play game of checkers by drawing lines on the roof and utilizing the wooden pieces and the natural stone as the game stones. Moreover the children also prefer these areas to play line game and tipcat. Some of the houses in the village were built in single storey. For those ones, the spaces of kitchen, store and animal shed are formed in the independent units.

Because of the collective settlement, the houses do not have courtyards.however, the villagers have gardens and land around the village in which they grow various fruit and vegetable. Most of the houses were seriously affected from the earthquakes and damaged. The damaged ones were deserted by the governmental authorities and destroyed. Now, mainly the upper floors are seem to be collepsed and the lower floors are used as animal sheds. Some of the families preferred their damaged house rather than the emergency house and repaired the upper floors as they were before, using the same material.

6.2. Houses Built Again by Villagers after 1950

These houses are similar to the traditional ones, but they are constructed with different materials. Indeed the structure is finished by an inclined roof. As a binding material between the masonry walls, concrete is partially used. All of them are single storey. With a different solution, the space for the store is met by the utilization of attic. The entrance of this store, instead of an inner connection, is provided outside, by using the appropriate level of the slope. Additionally, the animal shed is built separately, but near to the house.

6.3. Houses Built by Owners in 2005 and 2006

Reinforced concrete houses were built by the owners with the help of the family members and the neighbours in 2005 and 2006 instead of the destroyed houses on the same land. Concrete and brick were used as building materials since the owners were influenced by the emergency dwellings done in the region. For the windows, a plastic based material was used instead of timber frames. In these houses whose design schema is similar to the traditional terraced houses, the kitchen is interpereted as a separate space.there is a fireplace inside the kitchen. The walls of the kitchen is masonry and thicker than the other walls of the house. The orientation of the rooms, for instance, is same as the previous house. The inclination of the roof is 40% - 45% and the attic space is used as the store. The animal sheds are considered in the units close to the house.

6.4. Permanent Emergency Houses Built by Government in 1986

Therse are the houses which were constructed by the government. They were built in a permanent manner in 1986 for the families whose houses were damaged in the earthquake of 1968. the construction was done in an empty land close to the settlement border of the village. The emergeny dwellings were desigend by governmental architects and they all have the same spatial organization and plan. These projects were designed by Ministry of Public Works and Settlement without any consideration of local identity, user profile, culture, geographic conditions and orientation. They were propesed to the public as single houses and detached houses. The families were faced to back to back way of living by the introduction of such detached houses. However, the villages irrationally formed by the single houses were directed to a reluctant rational organization. Besides, the kitchen became a space in the house, whereas in the understanding of life the space for the kitchen was created indepentedntly or semi-independently to the house.

The traditional "Architectural Products Without the Architects", namely the emergeny dwellings with one type projects were not favorably used due to their irrelavant design to the traditions of the village. 70% of the users who lives in those houses because they have to, attached some additional spaces according to their traditions and needs. Kitchen was added to 65%, entrance terrace was added to 90%, toilet was added to 50% and courtyard was added to 80% of the used houses. Yet, 90% of the users ponited that they had willingly made those attachments. The attached kitchens also had the fireplace

6.5. Permanent Emergency Houses Built by Government in 2005 and 2006

The permanent emergency houses built by the government between the years 2005 and 2006 for the earthquake happened in Bingol in 2003. they were constructed by a method which is called "aid for the self-house builders". This method, however, did not worked in practice and the constructers mediated a settlement with the government on one project type and execceuted the "country style single storey emergency house" (2-roomed). Differently from the others, these were built on the previous site of the damged houses. The bathroom and the toilet were seperated and instead of gable roof, an icnlined roof was applied. The crafstman and the building materials are much more better than the other permanent emergency dwellings.

These houses which only have a combination with the local culture in the logic of sheltering does not present an example of the cultural continuity.

6.6. Evaluation of Houses in Alagoz

The natural environmental data of the geography and the socio-economic factors have a significant role in the formation of a building. In this context, the houses of Alagoz Village are evaluated in terms of natural environment and socio-economical factors. The houses are divided into two groups according to their construction phases. These are; the houses built by the villagers on their own or by the help of a craftsman (The traditional houses built by the villagers, survived from the late 19th century in which the village was established, The houses built again by the villagers after 1950, Reinforced concrete houses built by the owner with the help of the family members and the neighbours, between the years 2005 and 2006, instead of the houses destroyed by the government) and the houses built by government (The permanent emergency houses built by the government in 1986).

6.6.1. In Terms of Natural Environmental Factors

1) Traditional dwellings have been built considering the climatic data such as orientation, wind,etc. According to the climatic data, the houses have been oriented to the South or to the west as well, when the site is not appropriate in order to benefit from the sunlight inside the houses.

2) To provide the thermal insulation small scaled windows have rather been reduced in dimension on the North or comletely removed. Because of the long period of snowy weather in the region, to diminish the load on the roof, the inclination of the roof have been prefered in between 40% - 45%.

3) In typically planed emergency dwellings, instead, the climatic considerations have not been taken into account, as a result of the disoriented dwelling, the same spaces have been oriented to different directions. There are also houses whose entrance and two wide Windows oriented to the North.

4) It is observed that the traditional houses have been harmaonically settled on the topography , integrated with the site and the terraces have been formed in an accordance to the natural balance.

5) The inharmonious and environment-independent texture of the emergency houses which has a grid plan schema, instead, is structured differently from the organic traditional village dwelling and does not establish a relationship with topography.

6) In the tradional houses, the natural materilas are used which are directly related to the climate and vegetation and traditional construction methods are applied. In the buildings of the village the stone is mostly used and in various parts timber is prefered. The material is integrated with the environment due to its relation with the nature.

7) In the emergency houses, instead, contemporary industrial construction materilas are used. To provide the material economically is a current criteria. Because of the reinforcedconcrete and brick are the most available and the cheapest material, they have been prefered in these constructions like all over the country. In addition to that, owing to the capacity of heat absorbtion of reinforcedconcerete framework construction system and the hollowed brick wall is less then the traditional stone houses, they provide less thermal insulation; hence, whereas the house heats immediately in the summer, it can not heat in the winter and gets cold earlier.

6.6.2. In Terms of Socio-Cultural Environmental Factors

1) The traditional houses have been formed with respect to the life style and principle necessities.the spaces and the spatial relations have been determined in this frame kitchen, for insatance, decided seperately from the living spaces. Flat rooves have been occured which constitutes an open space for a lot of traditional work.

2) Contradictionally in the emergeny houses, a standard user profile is referred and their design is made typically all over the country. As a result of this, spaces were created which do not respond the needs of the users. Therefore, various attachments were made to those houses.

3) The family structure and the neighbourhood have been effective in the forming of the traditional houses. The distrubution of the spaces was affected from the family structure and multifunctionality was adopted as principle. The spaces were thought to gather and work collectivelly.

4) In the emergency houses, instead, the family structure was not taken into consideration and the multifunctionality of the spaces was not given importance, hence the rooms of private uses were designed. The organization and the scale of the rooms had a number of negative side.the spaces of common use were not also considered.

5) The traditional houses were generated under the influence of the relation between agricultural and cattlery production. Considering this situation, near or underneath the houses, animal sheds and the stores were built. Moreover, the houses were configured and the number of the rooms were determined according to the economic condition of the family .

6) In the emergency houses, instead, the construction finance was prior to the family structure and an economic solution was aimed at. the necessary spaces such as for animal sheltering, agricultural equipment and product storage were not even minded.

7) Politics and the law were not effective on the

reconstruction of the new traditional buildings, on the contrary did prevent them from living.

8) The excecuted emergency houses have indeed a lot of negative sides as a result of the various strategies of application.

When the user-built and government-built houses are compared, it is seen that the traditional houses are more superior in many aspects. These buildings are the most harmonious structures to either the natural effects and local culture. The houses built by government are not harmonious to the natural effects and do not reflect the local culture of the village in terms of visuality.

There is a great difference between the construction characters in terms of visuality. While the masonry of the region introduces a characteristic understanding to the region, in the new buildings this understanding have been abondoned.

In the context of the abovementioned issues, emergency dwellings are a product of inaccurate planning and design in terms of either cultural approach and the choice of physical material and its application. The first step to take in the village is to research the reasons of the damage in the traditional houses after the earthquake and if they are repairable and have the possibility of living inside again. In reconstructing the buildings harmonious to the local geography,climate and the culture instead should be formed without disturbing the existent texture of the village.the houses should be built suitable to the pyhsical and social values by integrating the modern life conditions and technology. The village should present a healthy way of living with either infrastructure and superstructure.

7. Recommendations for Future Projects

The climatic factors should be taken into consideration, the necessary comfort for human beings should be provided, in the construction the existent natural materials of the region should be predominantly used, the topographical data should be minded and the texture of the village should be continued in some degree. As a matter of fact, for a permananet housing approach in a small poppulation village, the harmony with the local culture should be prior to the easy construction methods.

The spatial organization should be created according to the dominant socio-cultural life in the village. The productivity dependent to agriculture and cattle breeding should be evaluated as an architectural data. Furthermore, since the sector of agriculture and cattle breeding are supported in the developement plan of the country, the villages such as Alagoz should be reconsidered by a

more powerful strategy. Therefore, the expectations of the families reimmigrating to their villages from Bingol will find an productive occupation. The conditional immigration concept will be avoided both to constitute a balance of the population in the country and not to detach the people of the villages from their identities. Throughout Turkey, the potential young population flowing to the big cities to find a job, becomes disappointed and do not have permanent living conditions neither in the city, nr in their villages. For that reason, the decisions taken for the villages will definetely provide benefit for the approach of the country planning. What is more, the bee keeping occured as a new job in Alagoz village should be promoted by the government specifically in the village it self. While the world is supporting the ecological formatios, it is conradictory to deny them. The emergency houses built for overcoming the all mentioned problems, cauesed more problematic situations after their realization, since the only issue regarded as criteria was the finance in the preconstruction period.

In this context, the values should be protected and the houses in the frame of continuity should be built with a consideration of "permanancy" in Turkey. The architectural modification in Alagoz Village is an example of a "cut off" in the cultural continuity and throughout Turkey, these typical problems are evident. To take a lesson from the experiences and orienting to the future will provide the permanency in the global world.

REFERENCES

[1] N. Köseoğlu, "National Culture and Identity," Ötüken Publications, İstanbul, 1992, p. 147, (in Turkish).

[2] Sanal Gazete Web Site (in Turkish). http://www.sanalgazete.com.tr/deprem/

[3] N. F. Akıncı, "The Aftermath of Disaster in Urban Areas: An Evaluation of the 1999 Earthquake in Turkey", *Cities*, Vol. 21, No. 6, 2004, pp. 527-535.

[4] I. Davis, "Housing Recommendations for Post Earthquakes," Translated and İnterpreted by F. Oymak, *Architecture Journal*, Vol. 14, No. 147, 1976, pp. 10-14, (in Turkish).

[5] F. Bayrak, Ş. Ünal and H. Taşçı, "Parliamentary Human Rights Investigation Commission Bingol Report," 2003, (in Turkish). http://insanhaklarimerkezi.bilgi.edu.tr

[6] Kigi District Governship Web Site, (in Turkish). www.kigi.gov.tr

[7] K. Dağıstanlı, "Architecture without Architect and a Study on Bingöl, Kığı, Alagöz Village," Masters Thesis, Architectural Design Program, Institute of Science, Yildiz Technical University, İstanbul, 2007, (in Turkish).

3

On the Relative Effect of Magnitude and Depth of Earthquakes in the Generation of Seismo-Ionospheric Perturbations at Middle Latitudes as Based on the Analysis of Subionospheric Propagation Data of JJY (40 kHz)-Kamchatka Path

Masashi Hayakaw[1,2], Alexander Rozhnoi[3], Maria Solovieva[3]

[1]Advanced Wireless Communications Research Center, University of Electro-Communications (UEC), Tokyo, Japan
[2]Hayakawa Institute of Seismo Electromagnetics, Co. Ltd., UEC Incubation Centre, Tokyo, Japan
[3]Institute of Physics of the Earth, Russian Academy of Sciences, Moscow, Russia

ABSTRACT

The relative importance of magnitude and depth of an earthquake (EQ) in the generation of seismo-ionospheric perturbations at middle latitudes is investigated by using the EQs near the propagation path from the Japanese LF transmitter, JJY (at Fukushima) to a receiving station at Petropavsk-Kamchatsky (PTK) in Russia during a three-year period of 2005-2007. It is then found that the depth (down to 100 km) is an extremely unimportant factor as compared with the magnitude in inducing seismo-ionospheric perturbations at middle latitudes. This result for sea EQs in the Izu-Bonin and Kurile-Kamchatka arcs is found to be in sharp contrast with our previous result for Japanese EQs mainly of the fault-type. We try to interpret this difference in the context of the lithosphere-atmosphere-ionosphere coupling mechanism.

Keywords: VLF/LF Subionospheric Propagation; Ionospheric Perturbation; Earthquake precursors; Earthquake Prediction

1. Introduction

There have been found different kinds of electromagnetic precursors of earthquakes (EQs), and the presence of those seismo-electromagnetic precursors is recently considered to be an irrefutable fact, though the physics of these phenomena remains poorly understood. (e.g., Molchanov and Hayakawa (2008) [1], Hayakawa (Ed.) (2009, 2012) [2,3], Uyeda et al. (2009) [4], and Hayakawa and Hobara (2010) [5]). Such electromagnetic EQ precursors can be customarily classified into the two types. The first type is the direct effects such as electromagnetic radiation from the lithosphere in a wide frequency range, while the second is the indirect effect such as seismo-atmospheric and ionospheric perturbations as detected by means of propagation anomalies of transmitter signals in different frequency ranges. When utilizing any one of those precursors for the practical short-term EQ predic-

tion, the most important issue is whether the precursor is statistically correlated with EQs or not, which can be only possible by analyzing a huge number of events based on the long-term observation.

Among many EQ precursors [1-5], the most promising one from the statistical point of view is the ionospheric perturbation, because the perturbations both in the lower ionosphere and upper ionosphere (F region) are found to be statistically significantly correlated with EQs. Hayakawa et al. (2010) [6,7] have obtained the significant statistical correlation between subionospheric VLF/LF propagation anomalies and EQs as for the lower ionospheric perturbation. On the other hand, Liu (2009) [8] has presented the result on the statistical correlation of the foF2 in the upper ionosphere with EQs. This paper deals with the lower ionospheric perturbations detected by subionospheric VLF/LF propagation anomalies. Hayakawa *et al.* (2010) [6] have found that a significant correlation is established for an EQ with significant magnitude (M) > 5.5 [9,10] and shallow depth (D) (D less than 40 - 50 km). These results are based on the observation in and around Japan (that is, mainly inland (or fault-type) EQs). So that, the influence of D is rather evident in generating seismo-ionospheric perturbations for EQs in Japan.

We would like to explore whether this conclusion is still valid for EQs in other geological areas or at different latitudes. By using the VLF/LF subionospheric propagation data from the JJY (Fukushima, Japan) to Petropavlovsk-Kamchatsky (PTK), Russia, we will study, in this paper, the relative importance of M and D of an EQ in generating seismo-ionospheric disturbances over this mid-latitude propagation path in order to explore whether there exists a significant difference in the characteristics of seismo-ionospheric perturbations of Japanese EQs and sea EQs. Finally, we discuss this difference in the context of the lithosphere-ionosphere coupling mechanism.

2. EQs and LF Propagation Data

In order to study the relative importance of M and D of an EQ in generating the seismo-ionospheric perturbation at middle latitudes, we will analyze the dependence of amplitude of the LF signal. For the analysis, we use the LF signal recorded at Petropavlovsk-Kamchatsky (PTK) station (geographic coordinates; 53°09'N, 158°55'E) in Russia from a Japanese transmitter with call sign of JJY (40 kHz) located at Fukushima [11]. The positions of the transmitter and receiving stations are illustrated in **Figure 1** and the distance between the transmitter and receiver is 2300 km. The data from three years (2005, 2006 and 2007) are used.

As a main characteristic of the LF signal, we estimate the difference between the current nighttime amplitude

and the monthly averaged amplitude calculated over night (this is called the conventional nighttime fluctuation method [11,10,6,7]).

The epicenters of EQs with M ≥ 4 from the catalogue of USGS (United States Geological Survey) are shown in **Figure 1**. We select the EQs with D less than 100 km and with epicenters close to the great-circle path between the transmitter and the receiver. One circle indicates an EQ, with its size reflecting the EQ M. As seen from this plot, many EQs are found to be located in the offsea of Hokkaido, Kurile islands and Kamchatka. At first we select the area around the wave path of JJY-PTK (like wave sensitive Fresnel zone), which is indicated by a black line in addition to the great-circle path of JJY-PTK in **Figure 1**. Then for every EQs in this area we calculate the radius of a zone for which the ionospheric precursor of an EQ may be found. According to Dobrovsky *et al.* (1979) [12], the preparation radius is given by $R = 10^{0.43 M}$. We also compute the distance (L) from the EQ epicenter to the great-circle path. In the following analysis we have included only EQs satisfying tentatively the ratio R/L ≥ 0.7. When we have several EQs on one particular day, we have selected the largest M EQ.

Figure 2 illustrates the distributions of EQs as a function of EQ M. The upper panel represents the total number of EQs on the basis of selection depending on D and distance from the great-circle path (R/L ≥ 0.7). The bottom panel illustrates the number of EQs after the further selection of the largest M on one day. In the latter case (bottom panel in **Figure 2**), the number of EQs with 4.5 ≤ M ≤ 5.5 is, on the contrary, more than that of EQs with 4.0 ≤ M ≤ 4.5. After this selection the total number of EQs is 543.

3. Analysis Results

Rozhnoi *et al.* (2004) [11] analyzed the data from the same propagation path of JJY-PTK during two years of

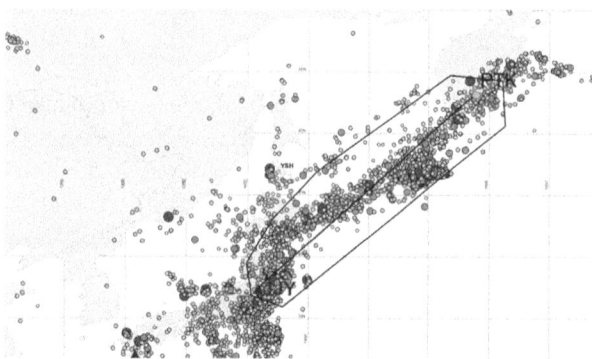

Figure 1. Relative location of the LF observing station in Petropavlovsk-Kamchatsky (PTK) in Russia, the LF transmitter JJY (40 kHz) in Fukushima, and epicenters of EQs with M > 4 (catalogue of USGS) for three years of 2005-2007. The area around the wave path of JJY-PTK is encircled by a black line.

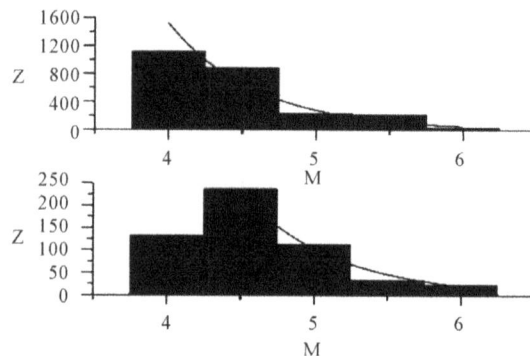

Figure 2. Distributions of EQs as a function of M. Upper panel shows the total number of EQs selected based on the D and the distance from the great-circle path (R/L ≥ 0.7) and bottom panel shows the number of EQs after the selection of the largest M in a day. Black curve is a regression curve.

On the Relative Effect of Magnitude and Depth of Earthquakes in the Generation of Seismo-Ionospheric Perturbations at Middle Latitudes as Based on the Analysis of Subionospheric Propagation Data of JJY (40 kHz)-Kamchatka Path

19

2001 and 2002. Then, in addition to the analyses of solar and geomagnetic effects in subionospheric LF data, they came to the conclusion that the seismic effect on this LF signal is observed only for EQs with M ≥ 5.5 and the anomaly appears a few to several days before an EQ. **Figure 3** illustrates the results of our analysis. The analysis was performed for a day when an EQ occurred (Day-0 in **Figure 3**) and during 5 days before the EQ (Day-1 to Day-5 in **Figure 3**). The ordinate (y axis) represents the amplitude of the LF signal averaged over night and normalized by the standard deviation (σ). While, the abscissa (x axis) is M/D (M: magnitude in Richter scale and D in km). If we would expect the similar effect of EQ D as in the results for Japanese EQs [6,7], we would anticipate that the signal amplitude would exhibit a significant increase with M/D. However, the correlation coefficients between the normalized amplitude and M/D in all plots in **Figure 3**, are found to be very weak of the order of 0.1, which means nearly no correlation between LF signal amplitude and M/D. So that, the regression equation for each scatter plot is not so meaningful. This means that the EQ D seems to be not so important as compared with M in inducing the seismo-ionospheric perturbation at middle latitudes in this paper. In any case, two groups of dots are seen in all plots of **Figure 3**, which corresponds to different Ds in the subduction area of our interest. One group includes EQs with 0 ≤ D ≤ 20 km, and another group of dots includes EQs with 20 < D < 100 km. The area of sensitivity along the wave path in **Figure 1** is seen to cover highly seismically active Izu-Bonin and Kurile-Kamchatka arcs. The epicentral zone can be divided into the different regions which are characterized by distinctly different seismic activity and focal zone depths. Maximal focal depths in this region are 600 - 650 km, and the upper mantle has a complicated mosaic-layered structure [11], composed of two characteristic layers as in **Figure 3**. The distributions of EQs depending on EQ M for two groups are shown in **Figure 4**. In the left the distribution of EQ M is observed for the first group of dots (deep EQs), in which we notice a nearly uniform temporal variation over the three years, with a large EQ on November 15, 2006. In the right panel of **Figure 4** we find that the temporal evolution of shallow EQs is very inhomogeneous over three years and that the period with increasing the number of EQs (shallow EQ, 0 < D < 20 km) is clearly observed during October 2006 to June 2007.

4. Concluding Remarks and Discussion

Based on the data analysis for three years (2005-2007) for the propagation path of JJY-PTK, we have investigated the relative importance of M and D in generating the seismo-ionospheric perturbations for sea EQs in the Kurile-Kamchatka region. Then we come to the follow

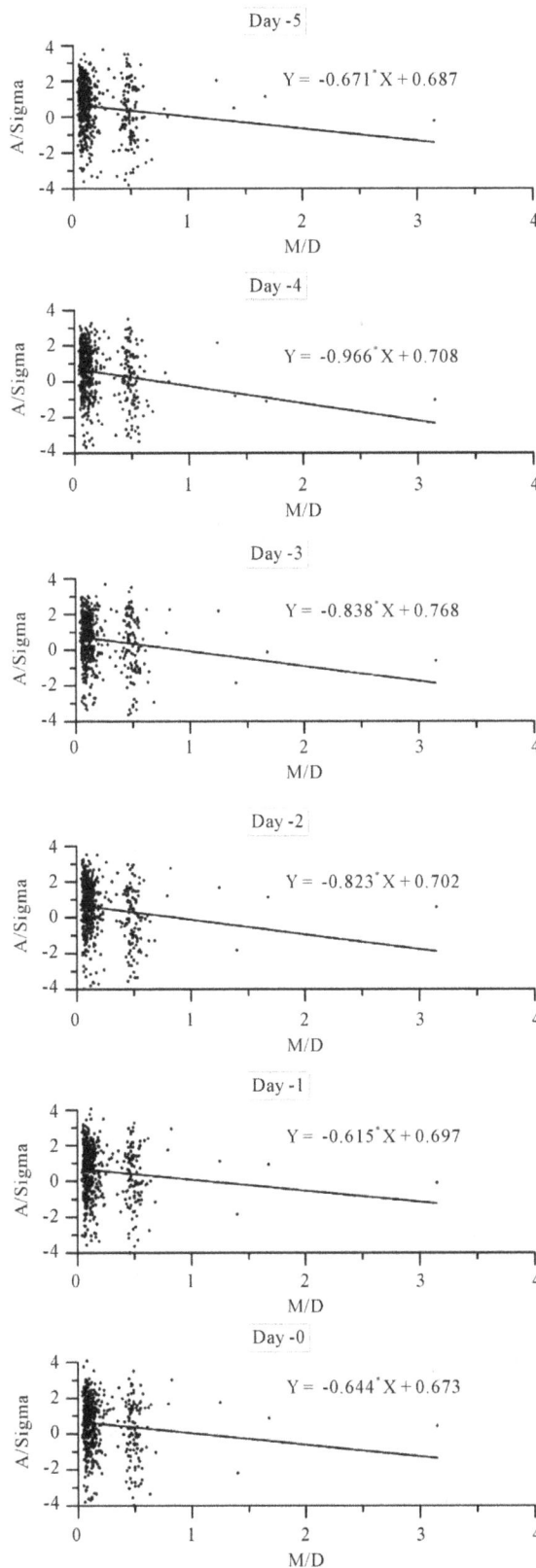

Figure 3. Dependences of amplitude of the LF signal on the parameter of M/D 1 - 5 days before an EQ (Day-1 - 5) and on the day of the EQ (Day-0). The ordinate is the LF signal intensity normalized by the standard deviation (σ).

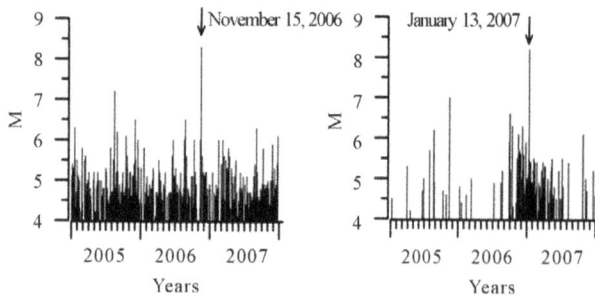

Figure 4. Temporal distributions of EQs as a function of M for two groups of dots. The left is for D larger than 20 km (deep EQs), and the right is for D smaller than 20 km (shallow EQs).

ing conclusion.

1) We have analyzed over 400 EQ events in order to study the correlation of LF signal intensity with a new parameter of M/D.

2) During several days before an EQ and on the day of the EQ, the LF signal intensity is found to exhibit nearly no correlation with M/D.

3) The above result is indicative that the EQ D is not so important as EQ M in the generation of seismo-ionospheric perturbations. Namely, EQ M is of the major importance in generating seismo-ionospheric perturbations at middle latitudes. This is the conclusion in the area of JJY-PTK path and for EQs with D less than 100 km.

The above result for sea EQs in the Kurile-Kamchatka region appears to be in sharp contrast with our previous statistical results for Japanese inland EQs. Hayakawa et al. (2010) [6] have analyzed the EQs mainly taken place inland; that is, their target EQs are of the fault-type, and they have found a significant effect of D. Namely, ionospheric perturbations are excited only by shallow (D ~40 - 50 km) EQs (with M ≥ 6.0), but it is difficult for us to find the ionospheric perturbation for deep (D > 50 km) EQs (with M ≥ 6.0). This point is consistent with a recent Japanese work by Kon et al. (2011) [13]. Of course, it may be possible that a very large (M > 7) EQ can induce the seismo-ionospheric perturbation even for D of the order of ~100 km.

When discussing the characteristics of VLF/LF perturbations at middle latitudes in this paper and in our previous paper [6], the first point for a possible difference is the latitudinal effect. Because Kamchatka is located at middle geomagnetic latitudes much higher than our Japanese low-latitudes [6,9], a majority of subionospheric VLF/LF perturbations are caused by solar-terrestrial effects such as solar flares, geomagnetic activities and only ~30% of the perturbations are related with EQs with M ≥ 5.5 [11]. On the other hand, the influence of solar-terrestrial effects on subionospheric VLF/LF propagation anomalies (though it does exist) is drastically

smaller at low latitudes such as Japan, and the "local" subionospheric VLF/LF propagation in Japan is highly likely to be related with EQs.

The clear discrepancy of the properties of seismo-ionospheric perturbation for inland and sea EQs is likely to be an important finding to be discussed in the context of lithosphere-atmosphere-ionosphere (LAI) coupling mechanism. There has been published a recent statistical study by Parrot (2012) [14], who has statistically studied the seismo-ionospheric perturbation (in the upper ionosphere) based on the about 5 years observation by the DEMETER satellite. His most fundamental point is that the ionospheric perturbation increases with increasing the EQ M. The second point is that more intense perturbation is observed for sea EQs than inland EQs; in other words, perturbations are not so important for deep (D > 40 km) EQs. In the conjunction with these statistical results by Parrot (2012) [14] we will try to reconsider our results in this paper. In the analysis of this paper dealing with EQs in the offshore of Hokkaido, Kurile Islands and Kamchatka, these EQs take place in the subduction region of Pacific plate, and these are believed to belong to sea EQs. Stronger perturbation is expected for sea EQs as shown in Parrot (2012) [14], which seems to be consistent with the insignificant effect of D (only down to 100 km) in our seismo-ionospheric perturbation for sea EQs in this paper. It is definite from our previous studies [9,11] and Parrot's result that the EQ M is the most essential parameter for seismo-ionospheric perturbations at any latitudes. It is easy for us to understand that the fact by Hayakawa et al. (2010) [6] that mainly shallow (D < 40 km) EQs are able to induce the ionospheric perturbation, seems to be consistent with Parrot (2012) [14] result which has indicated a small perturbation for inland EQs than for sea EQs.

There have been proposed two major hypotheses of LAI coupling; 1) electric field channel [15,16] and 2) atmospheric oscillation channel [e.g., 1,17,18]. In the model by Pulinets (2009) [15] radioactive radon is the main player, while Freund (2009) [16] assumes the appearance of positive holes. As for the second hypothesis, we have presented a lot of observational evidences in support to this channel [19] though there are very few observational results on the precursory physical processes near the ground surface [20]. Here we attempt to use the sea/land EQ asymmetry in the study of generation of seismo-ionospheric perturbation in order to elucidate which mechanism is more plausible.

In the case of second hypothesis, we have to take a fundamental premise that there must be present some kind of fluctuations on or near the ground surface (either ground motion, or any changes in atmospheric pressure (or temperature) or so). How to interpret the sea/land EQ asymmetry (stronger perturbation for sea EQs) in this hypothesis? For sea EQs, we assume ground motions in

On the Relative Effect of Magnitude and Depth of Earthquakes in the Generation of Seismo-Ionospheric Perturbations at
Middle Latitudes as Based on the Analysis of Subionospheric Propagation Data of JJY (40 kHz)-Kamchatka Path

21

the sea bed, which might lead to the sea surface waves with small amplitude but horizontally extended. This might form a great piston that pushes and pulls the air above the EQ hypocenter, which means that the sea surface is a good plough for exciting the atmospheric gravity waves and then leading to seismo-ionospheric perturbations. While, the ground EQ is considered to be a poor plough in the sense of launching the atmospheric gravity waves. On the other hand, when thinking of the 1st channel, there exists a difficulty to account for how to generate stronger electric field for the sea EQs. One point by Parrot (2012) [14] as related to the land/sea EQ asymmetry is that the electric conductivity above the sea due to the less contaminated condition than on the ground (land), which might be related with the 1st hypothesis. However it is not clear how this land/sea EQ asymmetry can be explained by electric field mechanisms.

REFERENCES

[1] O. A. Molchanov and M. Hayakawa, "Seismo Electromagnetics and Related Phenomena: History and Latest results," TERRAPUB, Tokyo, 2008, 189 p.

[2] M. Hayakawa, "Electromagnetic Phenomena Associated with Earthquakes," Transworld Research Network, Trivandrum, 2009, 279 p.

[3] M. Hayakawa, "The Frontier of Earthquake Prediction Studies," Nihon-Senmontosho-Shuppan, Tokyo, 2012, 794 p.

[4] S. Uyeda, T. Nagao and M. Kamogawa, "Short-Term Earthquake Prediction: Current State of Seismo-Electromagnetics," *Tectonophysics*, Vol. 470, No. 3-4, 2009, pp. 205-213.

[5] M. Hayakawa and Y. Hobara, "Current Status of Seismo-Electromagnetics for Short-Term Earthquake Prediction," *Geomatics, Natural Hazards and Risk*, Vol. 1, No. 2, 2010, pp. 115-155.

[6] M. Hayakawa, Y. Kasahara, T. Nakamura, F. Muto, T. Horie, S. Maekawa, Y. Hobara, A. A., Rozhnoi, M. Solivieva and O. A. Molchanov, "A Statistical Study on the Correlation between Lower Ionospheric Perturbations as Seen by Subionospheric VLF/LF Propagation and Earthquakes," *Journal of Geophysical Research: Space Physics*, Vol. 115, No. A9, 2010, Article ID: A09305.

[7] M. Hayakawa, Y. Kasahara, T. Nakamura, Y. Hobara, A. Rozhnoi, M. Solovieva and O. A. Molchanov, "On the Correlation between Ionospheric Perturbations as Detected by Subionospheric VLF/LF Signals and Earthquakes as Characterized by Seismic Intensity," *Journal of Atmospheric and Solar-Terrestrial Physics*, Vol. 72, No. 13, 2010, pp. 982-987.

[8] J. Y. Liu, "Earthquake Precursors Observed in the Ionospheric F-Region," In: M. Hayakawa, Ed., *Electromagnetic Phenomena Associated with Earthquakes*, Transworld Research Network, Trivandrum, 2009, pp. 187-204.

[9] S. Maekawa, T. Horie, T. Yamauchi, T. Sawaya, M. Ishikawa, M. Hayakawa and H. Sasaki, "A Statistical Study on the Effect of Earthquakes on the Ionosphere, Based on the Subionospheric LF Propagation Data in Japan," *Annales Geophysicae*, Vol. 24, 2006, pp. 2219-2225.

[10] Y. Kasahara, F. Muto, T. Horie, M. Yoshida, M. Hayakawa, K. Ohta, A. Rozhnoi, M. Solovieva and O. A. Molchanov, "On the Statistical Correlation between the Ionospheric Perturbations as Detected by Subionospheric VLF/LF Propagation Anomalies and Earthquakes," *Natural Hazards and Earth System Sciences*, Vol. 8, 2008, pp. 653-656.

[11] A. Rozhnoi, M. S. Solovieva, O. A. Molchanov and M. Hayakawa, "Middle Latitude LF (40 kHz) Phase Variations Associated with Earthquakes for Quiet and Disturbed Geomagnetic Conditions," *Physics and Chemistry of the Earth*, Vol. 29, No. 4-9, 2004, pp. 589-598.

[12] J. R. Dobrovolsky, S. I. Zubkov and V. I. Myachkin, "Estimation of the Size of Earthquake Propagation Zones," *Pure and Applied Geophysics*, Vol. 117, No. 5, 1979, pp. 1025-1044.

[13] S. Kon, M. Nishihashi and K. Hattori, "Ionospheric Anomalies Possibly Associated with M ≥ 6.0 Earthquakes in the Japan Area during 1998-2010: Case Studies and Statistical Study," *Journal of Asian Earth Science*, Vol. 41, 2011, pp. 410-420.

[14] M. Parrot, "Statistical Analysis of Automatically Detected Ion Density Variations Recorded by DEMETER and Their Relation to Seismic Activity," *Annales Geophysicae*, Vol. 55, 2012, pp. 149-155.

[15] S. Pulinets, "Lithosphere-Atmosphere-Ionosphere Coupling (LAIC) Model," In: M. Hayakawa, Ed., *Electromagnetic Phenomena Associated with Earthquakes*, Transworld Research Network, Trivandrum, 2009, pp. 235-253.

[16] F. Freund, "Stress-Activated Positive Hole Change Carriers in Rocks and the Generation of Pre-Earthquake Signals," In: M. Hayakawa, Ed., *Electromagnetic Phenomena Associated with Earthquakes*, Transworld Research Network, Trivandrum, 2009, pp. 41-96.

[17] O. A. Molchanov, "Lithosphere-Atmosphere-Ionosphere Coupling Due to Seismicity," In: M. Hayakawa, Ed., *Electromagnetic Phenomena Associated with Earthquakes*, Transworld Research Network, Trivandrum, 2009, pp. 255-279.

[18] V. Korepanov, M. Hayakawa, Y. Yampolski and G. Lizunov, "AGW as a Seismo-Ionospheric Coupling Responsible Agent," In: M. Hayakawa, J. Y. Liu, K. Hattori and L. Telesca, Eds., *Electromagnetic Phenomena Associated with Earthquakes and Volcanoes*, 2009, pp. 485-495.

[19] M. Hayakawa, Y. Kasahara, T. Nakamura, Y. Hobara, A. Rozhnoi, M. Solovieva, O. A. Molchanov and V. Korepanov, "Atmospheric Gravity Waves as a Possible Candidate for Seismo-Ionospheric Perturbations," *Journal of Atmospheric Electricity*, Vol. 31, No. 2, 2011, pp. 129-140.

[20] C. H. Chen, T. K. Yeh, J. Y. Liu, C. H. Wang, S. Wen, H. Y. Yen and S. H. Chang, "Surface Deformation and Seismic Rebound: Implications and Applications," *Surveys in Geophysics*, Vol. 32, No. 3, 2011, pp. 291-313.

4

Discuss the Properties of Structural Steel and Applications of Waste Concrete from Post-Earthquake Investigations

Xiaoshuang Shi[1*], Qingyuan Wang[1], Yunrong Luo[1,2]
[1]College of Architecture and Environment, Sichuan University, Chengdu, China
[2]College of Mechanical Engineering, Sichuan University of Science & Engineering, Zigong, China

ABSTRACT

This work proposes two aspects about construction materials abased on Wenchuan post-earthquake investigations. According to different feature failure modes in various damaged structures and the cause of the damage to the effects of the loading during the ground motion, the structural failures were found related to low cycle fatigue (LCF) properties of building steel. The hitherto research development is presented briefly. The characters of cycle response of the steels are tested and discussed. During the post-earthquake reconstruction process, the disposal of huge quantities of earthquake demolition waste brought great challenges. Utilizing the waste concrete taken from earthquake-stricken area as recycled coarse aggregate (RCA) in the new concrete is conducted. Furthermore, the application perspective of RCA is discussed.

Keywords: Earthquake; Failure Modes; Low Cycle Fatigue (LCF); Recycled Coarse Aggregate (RCA)

1. Introduction

The earthquake happens more frequently in recent years around places in the world. The damages of buildings and infrastructures cause huge loss of lives and economy. For example, the Wenchuan earthquake happened in 2008, which is measured at 8.0 Ms according to China Seismological Bureau, occurred on 12 May in Sichuan province of China. The total disaster zone covers an area of 440,000 square kilometers. The earthquake destroyed 5.3 million houses and damaged 21 million rooms, and left about 4.8 million people homeless. About 90,000 people were counted as dead or missing. More than 80% earthquake-related deaths were caused by the collapse of man-made structures. The building construction quality played a tremendous role in the death toll of the earthquake. The collapsed and damaged structures were estimated to produce about 300 million tons of waste concrete in this earthquake [1].

Earthquake damage depends on many parameters, including intensity, duration and frequency content of ground motion, geologic and soil condition, quality of construction, etc. Building design must ensure that the building has adequate strength, high ductility, and cyclic

load capacity in which would remain as one unit, even while subjected to very large deformation. Several factors of building construction determine the structural failure from the earthquake including, the type and age of building, foundation, materials used, seismic code used and construction quality, etc. Field investigation in Mianyang area revealed that 63% of raw-soil buildings, 36% of brick-wood structures, 25% of masonry structures, and 11% of frame structures were damaged during the earthquake (**Figure 1**). Obviously, the anti-earthquake ability of frame structure is the best due to its stronger materials and superior building type. The failure of the buildings charges upon the failure of inner materials. To improve the properties of construction materials, which are mainly concrete and steel, is an important way to avoid or alleviate damages caused by earthquakes.

On the other hand, facing with such large amount of wastes unprecedentedly is a tough challenge in post-earthquake reconstruction, especially the waste concrete takes up about 54.4% in all the types of demolition waste produced in this earthquake [2]. In general, we used to dispose the construction waste directly in landfills. However, this would take up large useful areas and result in secondary pollution, not to mention at one place in such a short time. Hence, one can see that the earthquake

---
*Corresponding author.

(a)

(b)

(c)

(d)

Figure 1. Earthquake damage to different buildings. (a) Raw-soil structure; (b) Brick-wood structure; (c) Masonry structure; (d) Frame structure.

disaster brought us two urgent issues to think about. One is how to improve the anti-seismic ability of construction materials. The other one is how to reuse the construction waste in post-earthquake reconstruction.

2. Low Cycle Fatigue Behavior of Structural Steel

As rapid development of high-rise buildings, steel as an indispensable construction material is now used more widely in the construction. The principal cause of earthquake-induced failure is ground shaking. As the earth vibrates, all buildings on the ground surface will respond to that vibration or cyclic loading. The earthquake loads acting on loading bearing elements are dynamic, in the form of high strain reversals, which can be simulated as single axis low cycle fatigue. Generally speaking, the strong shock lasts within one minute with the amplitude frequency among 1 Hz to 3 Hz. The damage to the buildings always happens during 100 to 200 cycles, which belongs to high strain low cycle (HSLC) fatigue problem [3]. Under the earthquake load, the HSLC property could be the control factor to the anti-seismic ability of structural steel. At present, many researches based on axial loading tests have been investigated [4-9]. The low cycle fatigue properties are studied according to Coffin-Manson empirical formula. The studies were more about ductility and energy dissipation effect of low yield steel, while the low cycle fatigue property of high-strength structural steels is barely reported.

The low cycle fatigue behavior of high-strength structural steel under asymmetrical total strain control can be investigated on cylindrical specimens with a gauge length of 20mm and central diameter of 10 mm, shown in **Figure 2**. The tests were carried out under uniaxial tension-compression loading with total strain control at a given strain amplitude of 0.4% - 1.0% and a strain ratio of −1. The triangular waveform was employed for all the fatigue tests. The total strain was measured using a dynamic extensometer with a span length of 12.5 mm which was attached to the specimen. The cyclic stress response curves of structural steel Q235 and Q345 under different strain amplitude are shown in **Figures 3** and **4**, respectively. The stress amplitude is the average value of the maximum tensile stress and compressive stress. It can be found that the feature of cyclic stress response curves is related to strain range. The materials are cyclic softening firstly when the strain amplitude is lower than 0.5%. After that, Q235 exhibited cyclic hardening and Q345 displayed cyclic stability until fracture. However, when the strain amplitude was upon 0.5%, both of the materials performed sharply cyclic hardening at the initial stage about 0.02 Nf, and then were followed by slightly cyclic hardening until fracture. Overall, cyclic hardening is the main character of the materials' cyclic feature, which is

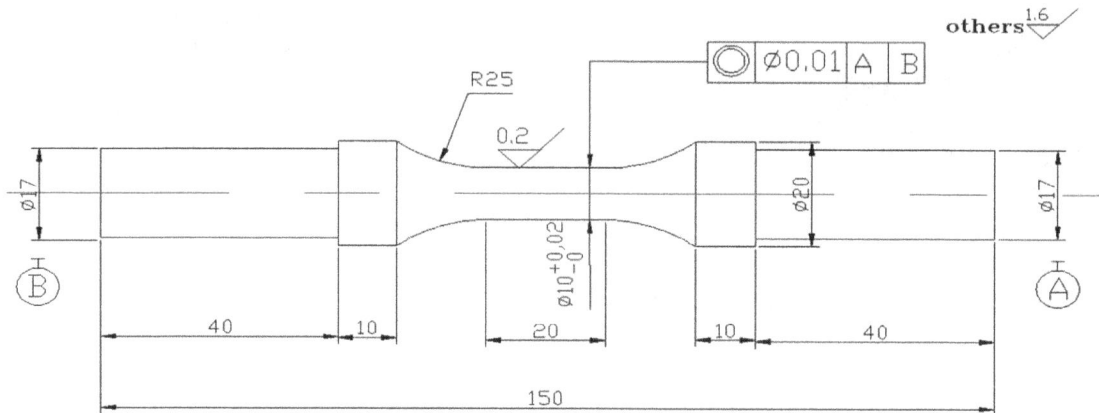

Figure 2. Geometry of specimen (Unit: mm).

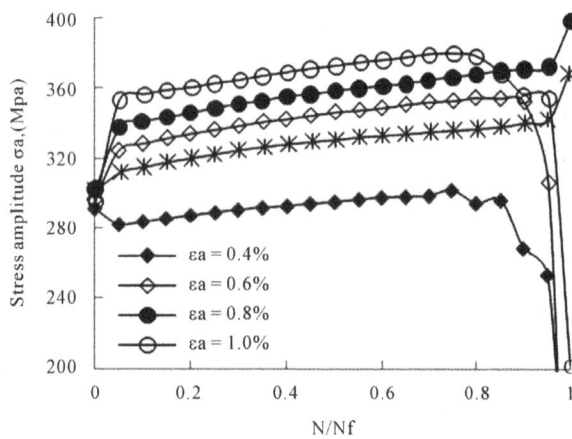

Figure 3. Cyclic stress response of Q235.

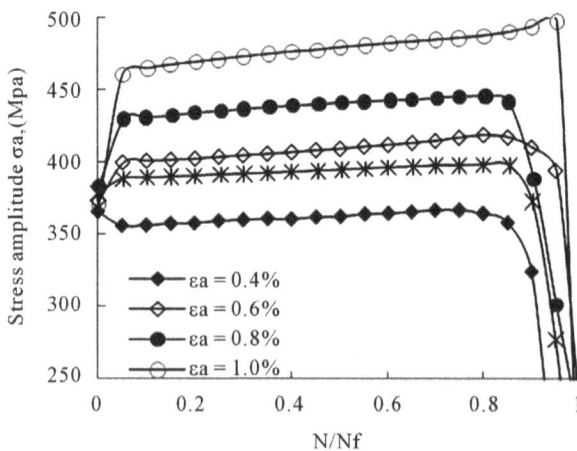

Figure 4. Cyclic stress response of Q345.

benefit for the material to avoid failure too early under large cyclic stress loading due to the decreased strength resulted from softening. Therefore, the structural steel equipped with good anti-seismic property should perform cyclic hardening or cyclic stability for the cycle response characters.

In recent years, some applications such as dampers

based on the use of stiffened plates made of special low-yield strength steel has become a trend of earthquake resistant design. The buildings are protected well by the devices which absorb a lot of seismic energy. Steels with low yield point have good ductility and plasticity, with yield strength as low as 110 MPa and elongation as high as 60%.

The cyclic stress response curve of the steel (at 3% strain amplitude) is shown in **Figure 5**. It can be found that the material exhibited cyclic hardening at the initial stage and kept cyclic stability until failed, which shows very good ability for energy absorbing under cyclic loading. So, this kind of structural steel shows great potential in earthquake design. The fracture surfaces of the samples are shown in **Figure 6**. The propagation of fatigue crack was mainly characterized by striations and dimples in **Figures 6(b)** and **(c)**. Finally, the fatigue cracks propagated resulting in ductile fracture with many dimples.

3. Using Recycled Aggregate in New Concrete

Recycled aggregate is made by process of crushing waste concrete into the diameter smaller than 40 mm. With the

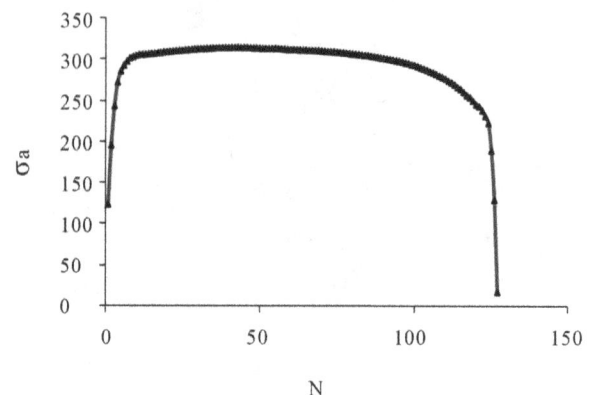

Figure 5. Cyclic stress response of special structure steel.

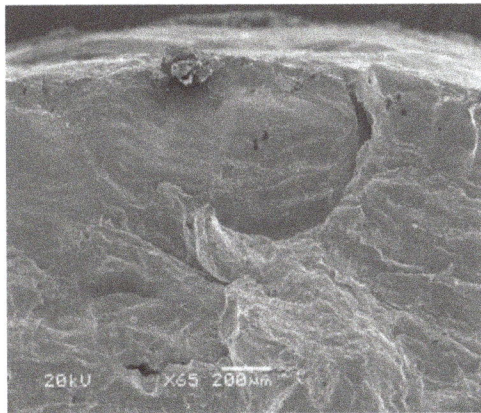

(a)

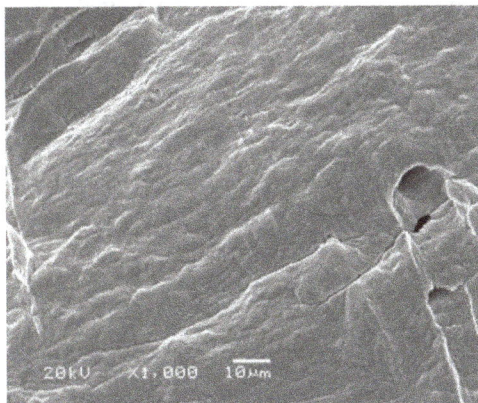

(b)

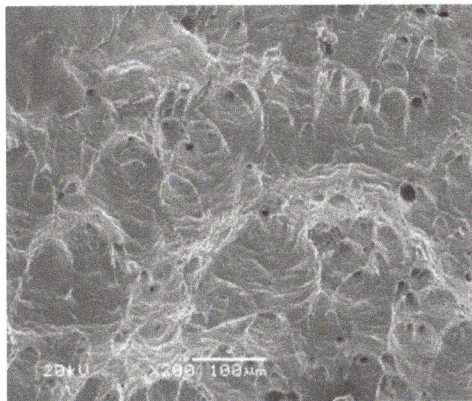

(c)

Figure 6. SEM images of fracture surfaces.

diameter among 5 mm to 40mm belongs to recycled coarse aggregate (RCA). Using RCA substituting natural aggregate partially or totally to manufacture new concrete is recycled aggregate concrete (RAC). The research about RAC is more and more popular in recent years, especially after Wenchuan earthquake [10-16]. As the waste concrete comparatively assembled at one place, this is an economic and effective way to solve the large amount of demolition waste in the earthquake-stricken area.

The properties of aggregate made from demolition waste in earthquake-stricken area Dujiangyan were investigated [17]. The test results show that the water absorption of RCA is the most different factor from natural aggregate, which is 3.5 times higher. The density of RCA is about 10% lower than that of natural aggregate. The 28-days compressive strength of RAC with 100% RCA replacement ratio decreased 16.7% compared with ordinary concrete, which is shown in **Figure 7**. The research studied by Hansen [18] indicated that the strength of RAC decrease about 5% to 32% compared with ordinary concrete. RCA contents and cement to water ratio are the key factors to influence the strength of the concrete.

Nowadays, the application of RCA in RAC has been accepted and developed into practical projects. This makes people have more confident on the utilization of RCA, so the research about novel application of RCA in various concretes has been heated up. From the sustainable development point of view, the research on RCA used in geopolymer concrete is studied. Six mixtures were made with different RCA replacement ratios (0%, 50% and 100%) to compare the properties of RAC and geopolymeric recycled concrete (GRC). The compressive strengths of different concretes shown in **Figure 8** [19] present that the influence of RCA to GRC is similar with that to RAC, and with appropriate mixture design the application of RCA in geopolymer concrete is feasible.

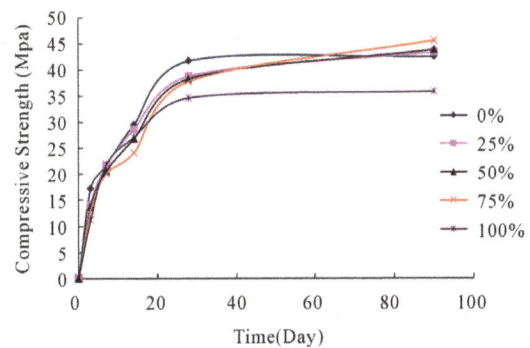

Figure 7. Compressive strength of RAC with different RA% vs. time [17].

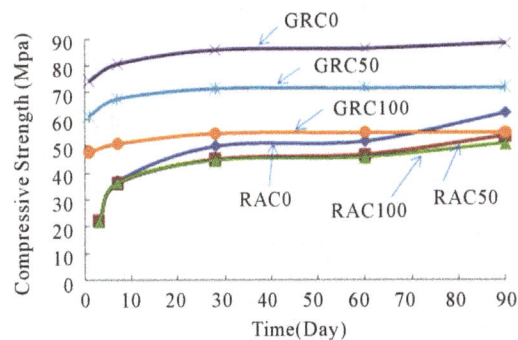

Figure 8. Compressive strength of RAC and GRC with different RA% vs. time [19].

Besides the study on the concrete with RCA, the mechanical properties of recycled concrete filled steel tube (RACFST) columns were also investigated in previous study [20]. According to the experimental results, the ultimate load capacity of RACFST is lower than ordinary concrete filled with steel tube columns (CFT) within 20%. The failure mode of RACFST is similar to those of conventional CFT columns with local buckling resulting in failure. The ultimate load of RACFST decreased with increasing RCA replacement ratio. Also, Yang and Han [21] reported the similar research work and results. Therefore, it could be concluded that RACFST column has slightly lower but comparable load capacity compared with conventional CFT column.

4. Concluding Remarks

The lessons from the deadly earthquake should be carefully studied, in order to improve the anti-seismic ability of the buildings. Overall, construction materials play a crucial role in earthquake and post-earthquake event. Structural steel as the major construction material used in the buildings is pointed out, and its LCF property is discussed in this paper which could demonstrate its anti-seismic ability. It is found that the material with cyclic hardening or cyclic stability characters benefits for anti-seismic ability. On the other hand, the disposal of demolition waste in earthquake-stricken area is also very important. The properties of RAC are comparable similar to ordinary concrete. Presently, the research and application about reusing waste concrete into RAC are approaching mature. However, more widely utilization of recycled aggregate in other kinds of concrete or novel materials should be explored and further studied.

5. Acknowledgements

Authors gratefully acknowledge the support from National Natural Science Foundation of China (NSFC 10925211 and NSFC 51208325), and Program for Changjiang Scholars and Innovative Research Team (IRT1027).

REFERENCES

[1] X. S. Shi, Q.-Y. Wang, C.-C. Qiu and X.-L. Zhao, "Recycling Construction and Demolition Waste as Sustainable Environmental Management in Post-Earthquake Reconstruction," *4th International Conference on Bioinformatics and Biomedical Engineering*, Chengdu, 18-20 June, 2010, pp. 18-20.

[2] J. Z. Xiao, H. Xie and C. Q. Wang, "Statistical Analysis on Building Waste in Wenchuan Earthquake-Hit Area," *Journal of Sichuan Univeristy* (*Engineering Science Edition*), Vol. 42, No. 41, 2009, pp. 188-194.

[3] S. H. Gong and G. M. Sheng, "Effects of Toughness of Steel on Seismic Performance of Building Structures," *Earthquake Resistant Engineering*, Vol. 6, No. 3, 2004, pp. 41-47.

[4] T. Nishimura and C. Miki, "Strain Controlled Low Cycle Fatigue Behavior of Structural Steels," *Proceedings of the Japan Society of Civil Engineers*, Vol. 279, No. 1, 1978, pp. 29-44.

[5] K. Shimada, J. Komotori and M. Shimizu, "The Application of the Manson-Coffin Law and Miner's Law to Extremely Low Cycle Fatigue," *Journal of Japan Society of Mechanical Engineers*, Vol. 53, No. 491, 1987, pp. 1178-1185.

[6] K. Masatoshi, "Extremely Low Cycle Fatigue Life Prediction Based on a New Cumulative Fatigue Damage Model," *International Journal of Fatigue*, Vol. 24, No. 6, 2001, pp. 699-703.

[7] J. B. Mander, F. D. Panthaki and A. Kasalanati, "Low-Cycle Fatigue Behavior of Reinforcing Steel," *Journal of Materials in Civil Engineering*, Vol. 6, No. 4, 1994, pp. 453-468.

[8] W. C. Liu, Z. Liang and G. C. Lee, "Low-Cycle Bending-Fatigue Strength of Steel Bars under Random Excitation. Part I: Bahavior," *Journal of Structural Engineering, ASCE*, Vol. 131, No. 6, 2005, pp. 913-918.

[9] K. A. S. Susantha, T. Aoki and T. Kumano, "Applicability of Low Yield Strength Steel for Ductility Improvement of Steel Bridge Piers," *Engineering Structures*, Vol. 27, No. 7, 2005, pp. 1064-1073.

[10] I. Gull, "Testing of Strength of Recycled Waste Concrete and Its Applicability," *Journal of Construction Engineering and Management-ACSE*, Vol. 137, No. 1, 2011, pp. 1-5.

[11] R. J. Wang and D. J. Yang, "Experimental Research on Compressive Strength and Elastic Modulus of Recycled Concrete," *International Conference on Civil Engineering*, Baoding, 19-20 July 2010, pp. 791-795.

[12] V. Corinaldesi, "Mechanical and Elastic Behaviour of Concretes Made of Recycled-Concrete Coarse Aggregates," *Construction and Building Materials*, Vol. 24, No. 9, 2010, pp. 1616-1620.

[13] J. Z. Xiao, J. D. Huang and M. Wei, "Study on Application of Recycled Aggregate Concrete in Sustainable Buildings," *Proceedings of International Conference of Technology of Architecture and Structure*, Shanghai, 15-17 October 2009, pp. 753-764.

[14] K. H. Obla and H. Kim, "Sustainable Concrete through Reuse of Crushed Returned Concrete," *Transportation Research Record*, Vol. 2113, No. 14, 2009, pp. 114-121.

[15] G. Durmus, O. Simsek and M. Dayi, "The Effects of Coarse Recycled Concrete Aggregates on Concrete Properties," *Journal of the Faculty of Engineering and Architecture of Gazi University*, Vol. 24, No. 1, 2009, pp. 183-189.

[16] Z. L. Cui, M. Kitatsuji and R. Tanaka, "The Experimental Research on Durability of Recycled Aggregate Concrete," *Proceeding of International Disaster and Risk Confer-

ence, Chengdu, 13-15 July 2009, pp. 449-452.

[17] X. S. Shi, Q.-Y. Wang, C.-C. Qiu and X.-L. Zhao, "Experimental Study on the Properties of Recycled Aggregate Concrete with Different Replacement Ratios from Earthquake-Stricken Area," *Sichuan Daxue Xuebao* (*Engineering Science Edition*), Vol. 42, Supp. 1, 2010, pp. 170-176.

[18] T. C. Hansen, "Recycled Aggregate Concrete Sencond State-Of-Art Report Developments 1945-1985," *Material and Structure*, Vol. 19, No. 3, 1986, pp. 201-246.

[19] X. S. Shi, F. G. Collins, X. L. zhao and Q. Y. Wang, "Experimental Study on Geopolymeric Recycled Concrete Using as Sustainable Construction Material," *Proceeding*

of International Conference on Advances in Construction Materials through Science and Engineering, Hong Kong SAR, 2011, pp. 186-194.

[20] X. S. Shi, Q. Y. Wang, C. C. Qiu and X. L. Zhao, "Mechanical Properties of Recycled Concrete Filled Steel Tubes and Double Skin Tubes," *2nd International Conference on Waste Engineering Management*, Shanghai, 13-15 October 2010, pp. 559-567.

[21] Y. F. Yang and L. H. Han, "Experimental Behaviour of Recycled Aggregate Concrete Filled Steel Tubular Columns," *Journal of Constructional Steel Research*, Vol. 62, No. 12, 2006, pp. 1310-1324.

5

Reliability Index of Tall Buildings in Earthquake Zones

Mohammed S. Al-Ansari

Department of Civil and Architectural Engineering, Qatar University, Doha, Qatar

ABSTRACT

The paper develops a reliability index approach to assess the reliability of tall buildings subjected to earthquake loading. The reliability index β model measures the level of reliability of tall buildings in earthquake zones based on their response to earthquake loading and according to their design code. The reliability index model is flexible and can be used for: 1) all types of concrete and steel buildings and 2) all local and international codes of design. Each design code has its unique reliability index β as a magnitude and the interaction chart corresponding to it. The interaction chart is a very useful tool in determining the building drift for the desired level of reliability during the preliminary design of the building members. The assessments obtained using the reliability index approach of simulated, tested, and actual buildings in earthquake zones were acceptable as indicators of the buildings reliability.

Keywords: Reliability Index; Earthquake; Interaction Chart

1. Introduction

The lateral displacement or drift of structural systems during an earthquake has an important impact on their potential failure. The probability of failure of structures is therefore reduced by limiting their lateral displacements or drifts. The reliability index β, which is typically used to measure the probability of failure of structural systems, allows structures to reach the desired reliability level through the assessment of the likelihood that their earthquake responses exceed predefined building drift values (roof displacement). The reliability index approach made it possible for the reduction of the probability of failure of building structures [1-6].

Drift limitations are currently imposed by seismic design codes, such as Uniform building Code (UBC) and International Building Code (IBC), in order to design safe buildings [7,8]. The acceptable range for the drift index of conventional structures lies between the values of 0.002 and 0.005 (that is approximately $\frac{1}{500}$ to $\frac{1}{200}$).

Excessive lateral displacements or drifts can cause failure in both structural and non-structural elements. Therefore, drifts at the final structural design stages must satisfy the desired reliability level and must not exceed the specified index limits [9-13]. While extensive research has been done in this important topic, no or limited studies addressed the use of reliability index to assess the building reliability in earthquake zones based on their lateral displacement and drift.

This paper develops a probabilistic model using the reliability index approach to assess the reliability of concrete and steel buildings subjected to earthquake loading in different zones and soil profiles based on their responses to earthquake loading, **Table 1**. The non-linear dynamic response of buildings was obtained using three simulated models of buildings, square, circular, and tube with different heights, **Figures 1-4**.

Other real buildings such as the full scale seven story reinforced concrete building that was tested statically and dynamically in Japan conducted under the US-Japan Cooperative Earthquake Engineering Research Program on the seismic performance of the building structure **Figure 5** [14].

Table 1. Soil profile and seismic factors.

Soil type (S)	Seismic factors (Z)
Hard Rock (S1)	0.075 gravitational acceleration (Z1)
Rock (S2)	0.150 gravitational acceleration (Z2)
Very dense soil and soft rock (S3)	0.20 gravitational acceleration (Z3)
Stiff soil (S4)	0.30 gravitational acceleration (Z4)
Soft soil (S5)	0.40 gravitational acceleration (Z5)

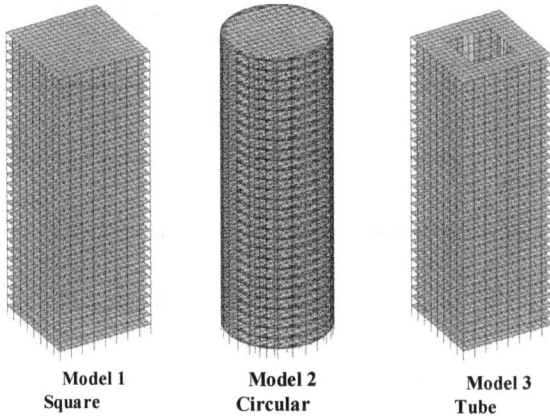

Model 1
Square

Model 2
Circular

Model 3
Tube

Figure 1. 3D Structural Building Models.

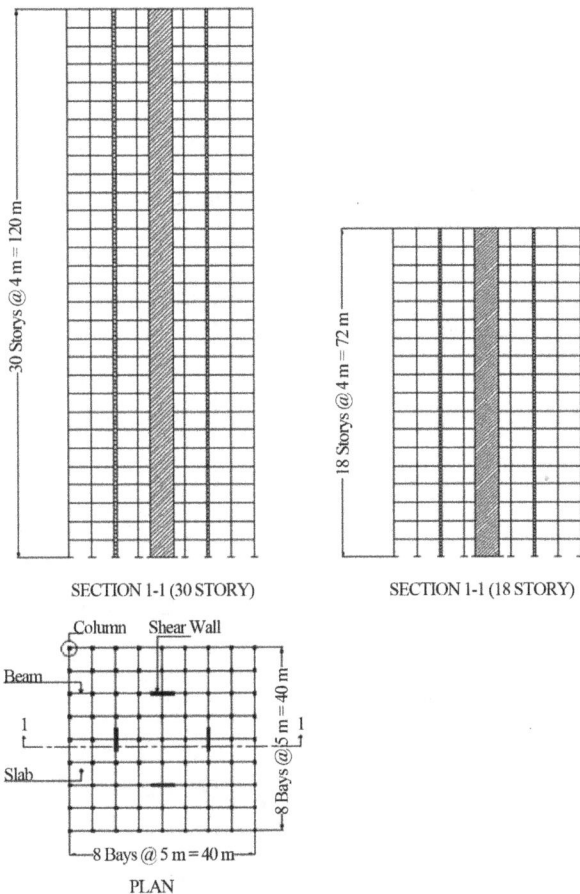

SECTION 1-1 (30 STORY)   SECTION 1-1 (18 STORY)

PLAN

Figure 2. Square Building Plan and Elevation.

Holiday Inn and the Bank of California in Los Angeles during the San Fernando earthquake 1971 were studied and analyzed, **Figures 6** and **7** [15].

2. Method Formulation

A structure fails when the actual building drift δ is higher than the maximum allowable drift Δ. The building margin of safety M is given by the following equation:

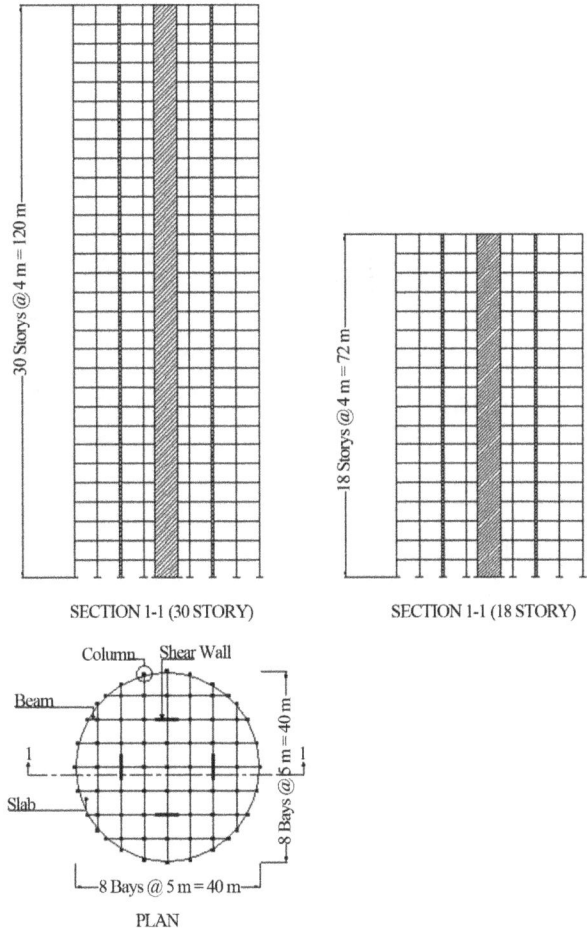

SECTION 1-1 (30 STORY)   SECTION 1-1 (18 STORY)

PLAN

Figure 3. Circular Building Plan and Elevation.

$$M = \Delta - \delta \qquad (1)$$

The allowable drift Δ is given by the following equation:

$$\Delta = H_T \cdot D_I \qquad (2)$$

where H_T = building total height, D_I = drift index (1/500 to 1/200), and δ = actual building roof displacement.

Hence the probability of failure (*pf*) of the building is given by the following equation:

$$pf = p\left(M \prec 0\right) = \varphi\left(\frac{0 - \mu_m}{\sigma_m}\right) \qquad (3)$$

where φ = standard normal cumulative probability distribution function, μ_m = mean value of M, and σ_m = standard deviation of M.

The parameter μ_m is given by the following equation:

$$\mu_m = \mu_\delta - \mu_\Delta \qquad (4)$$

The standard deviation of M is given by the following equation:

$$\sigma_m = \sqrt{\sigma_\Delta^2 + \sigma_\delta^2} \qquad (5)$$

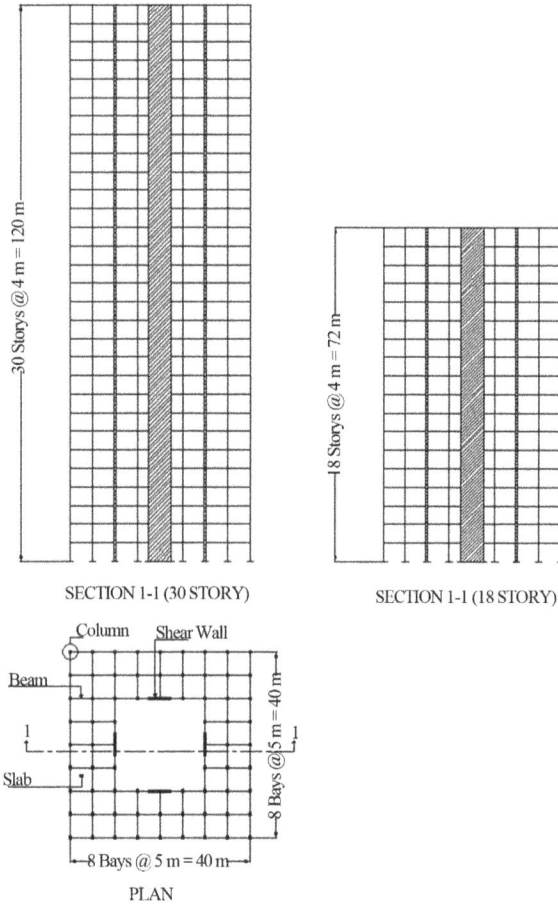

SECTION 1-1 (30 STORY)   SECTION 1-1 (18 STORY)

PLAN

Figure 4. Tube Building Plan and Elevation.

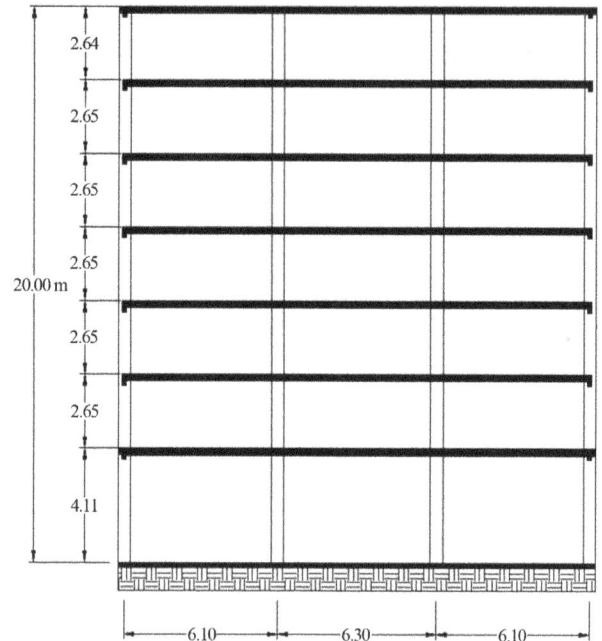

Figure 6. 7-Story Holiday Inn Building.

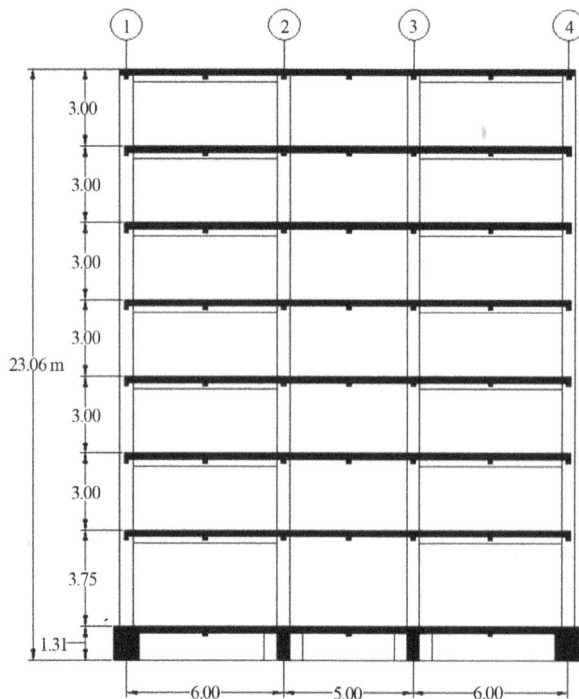

Figure 5. 7-Story Building (US-Japan Research Program).

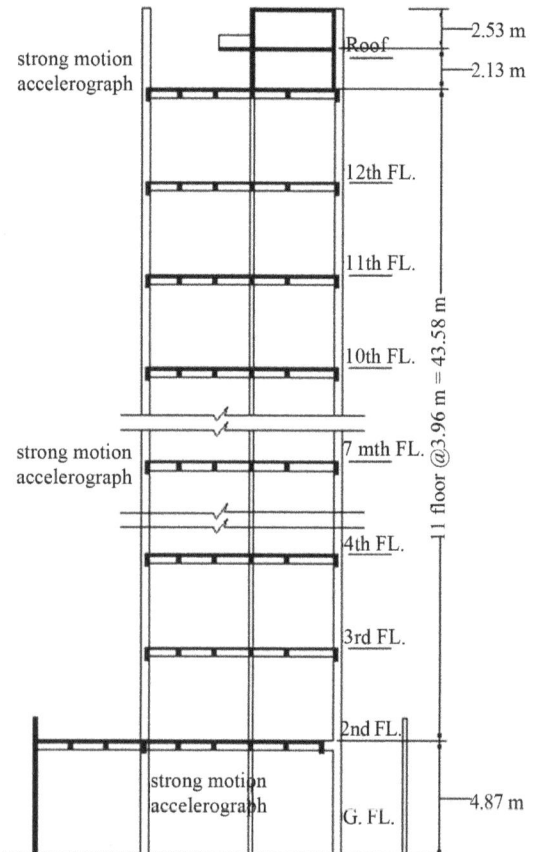

Figure 7. Bank of California Building Transverse Section.

Combining Equations (3)-(5) yields the following equation for the probability of failure (*pf*) of the building:

$$pf = \phi\left(\frac{\mu_\Delta - \mu_\delta}{\sqrt{\sigma_\Delta^2 + \sigma_\delta^2}}\right) \qquad (6)$$

Let us define the reliability index β using the following equation:

$$\beta = \frac{\mu_m}{\sigma_m} \qquad (7)$$

Combining Equations (4)-(7) yields the following equation for the probability of failure (pf) of the building:

$$pf = \phi(-\beta) \qquad (8)$$

Combining Equations (4)-(6) and (8) yields the following equation for the reliability index β:

$$\beta = \left(\frac{\mu_\Delta - \mu_\delta}{\sqrt{\sigma_\Delta^2 + \sigma_\delta^2}}\right) \qquad (9)$$

By setting the maximum allowable drift Δ equal to μ_Δ, the building earthquake response δ equal to μ_δ, and the standard deviation equal to the mean value times the coefficient of variation [16], Equation (9) can be re-written as follows.

$$\beta = \left(\frac{\Delta - \delta}{\sqrt{(C \cdot \Delta)^2 + (D \cdot \delta)^2}}\right) \qquad (10)$$

The parameter C is given by the following equation:

$$C = DLF * COV(DL) \qquad (11)$$

where DLF = dead load factor, which is equal to 1.2 as prescribed by ACI Code [17] and COV (DL) = coefficient of variation for dead load, which is equal to 0.13 as suggested by [16].

On the other hand, the parameter D is given by the following equation:

$$D = DLF * COV(DL) + LLF * COV(LL) \qquad (12)$$

where LLF = live load factor, which is equal to 1.6 as prescribed by ACI Code and COV (LL) = coefficient of variation for live load, which is equal to 0.37 as suggested by [17].

The displacement δ for a certain β is obtained by re-writing Equation (10) as follows.

$$\delta = \frac{-\beta}{2(D^2\beta^2 - 1)}\left[2D^2\beta \right. $$
$$\left. -2 \cdot \left[(-D^2)\beta^2 C^2 + C^2 + D^2\right]^{\frac{1}{2}}\right]\Delta + \Delta \qquad (13)$$

The maximum reliability index β_{max}, which is obtained by setting δ equal to zero in Equation (13), is given by the following equation:

$$\beta_{max} = \frac{1}{C} \qquad (14)$$

The formulation presented herein allows the estimation of the reliability of structures based on drift results when subjected to static equivalent earthquake loads defined by UBC and IBC codes of design for different seismic zones and soil profiles. The reliability index β is calculated for multi story buildings based on their top response to earthquake loading and the allowable drift (Δ) (i.e., on desired drift).The allowable and computed drifts are inserted into Equation (13) to determine the building reliability index β. Since the reliability index β is a function of the building height, the drift index, the roof displacement, and the dead and live load factors required by the code of design, the reliability index model can be used for all types of concrete and steel buildings. The reliability index β is in fact flexible because it can be used for all local and international design codes simply by changing the parameters magnitudes of the reliability index equation to the required parameters by the design code. Each design code has its unique reliability index β as a magnitude and the interaction chart that corresponds to it. The interaction chart is a very useful tool in determining the building drift for the desired level of reliability during the preliminary design of the building members.

3. Result and Discussion

Several steel and reinforced concrete buildings were simulated using STAAD PRO [18] as shown in **Figure 8**. The buildings, which have square, rectangular, and circular shapes and different heights, include reinforced concrete shear walls and slabs. The buildings were subjected to static equivalent earthquake loads, which are defined by UBC and IBC codes of design, for different seismic zones and soil profiles and their responses were obtained using the finite element software STAAD-PRO.

The reliability index β was computed for 18- and 30-story buildings in seismic zones 1, 2, 3 and 4 and for soil profiles S1, S2 and S3 for each zone. The results show that the reliability index β of the buildings with a drift limit index of 0.005 lied in the range of 3.5 and 6 for 18-story buildings and in the range of 2.3 and 5.8 for 30-story buildings. The buildings with a β value less or equal to 6 are considered as reliable while those with a β value less than 3.5 are considered as unreliable building (see **Table 2** and **Figures 9** and **10**).

Increasing the building strength will also increase its reliability. For example increasing the size of columns and shear wall will increase the level of reliability as shown in **Table 3**.

Figure 8. 30-Story Concrete Building Finite Element Model.

Table 2. Building Model Reliability Indexes.

Building Model	Height (meter)	Z^*/S^{**}	Concrete				Steel		
			δ (mm)	β		δ (mm)	β		
				1/200	1/500		1/200	1/500	
Square	72	Z1/S1	14.322	6.046	5.211	20.502	5.832	4.54	
		Z1/S2	19.084	5.883	4.693	27.328	5.566	3.841	
		Z2/S3	42.9	4.903	2.581	74.516	3.608	1.156	
		Z3/s2	119.804	3.706	1.238	118.624	3.732	1.26	
	120	Z4/S3	197.665	2.299	0.278	195.713	2.327	0.293	
		Z3/S1	95.85	4.276	1.782	94.905	4.3	1.808	
Circular	72	Z1/S1	16.988	5.957	4.921	20.422	5.835	4.549	
		Z1/S2	22.640	5.751	4.314	27.219	5.571	3.852	
		Z2/S3	50.912	4.555	2.105	74.285	3.617	1.163	
		Z3/s2	135.778	3.361	0.963	139.976	3.276	0.9	
	120	Z4/S3	224.012	1.959	0.093	230.945	1.878	0.051	
		Z3/S1	108.63	3.964	1.468	111.985	3.885	1.395	
Tube	72	Z1/S1	18.146	5.917	4.795	30.959	5.417	3.504	
		Z1/S2	24.124	5.694	4.16	41.258	4.87	2.692	
		Z2/S3	54.104	4.419	1.942	125.892	2.135	0.187	
		Z3/s2	154.706	2.992	0.701	151.665	3.048	0.739	
	120	Z4/S3	255.195	1.622	−0.078	250.029	1.673	−0.053	
		Z3/S1	123.787	3.617	1.164	121.399	3.67	1.207	

*Z, seismic zone factor; **S, soil profile.

4. Case Study

A case study dealing with the performance of actual building sunder seismic load is considered herein to further validate the use of their liability index β approach. The reliability index β was computed using Equation (10) to assess the building reliability in each seismic test. The first building is the 23.06-meter-high seven-story reinforced concrete building in Japan, which was tested statically and dynamically by the US-Japan Research Panel.

The building was subjected to a static test SL-3 and three dynamic earthquake tests SPD-1, SPD-3, and SPD-4. For each test, the reliability index β of the building were computed using Equation (10) for two drift indexes, namely, (1/200) and (1/500). The top displacement used in Equation (10) was the building actual top displacement recorded during each test. **Table 4** summarizes the building reliability index results obtained for the static and dynamic tests. The results shows that the reliability indexes β of the building during the static test for both drift limits are equal to −0.862 and −1.147, respectively. This indicates that the building is not reliable for a top displacement of 326 mm. This finding is in agreement with the actual status of the building after the test. It has been reported that the building actually failed and the main reinforcing bars of the boundary columns were fractured and the concrete was crushed along the full span of the shear wall. The values of the building reliability index β of the building for the dynamic earthquake tests SPD-1 were high indicating a reliable building. This finding agreed with the real building status during the test. Both SPD-3 and SPD-4 tests yielded low values of the reliability index indicating an unreliable building. These two findings are in agreement with the actual status of the building after the tests.

The reliability indexes of the Holiday Inn and the Bank of California buildings in Los Angeles during the San Fernando earthquake 1971 were computed to assess the reliability of the building. The Holiday Inn building is located approximately 8 miles from the San Fernando 1971 earthquake center. The seven story reinforced concrete frame building, which extends 20 meters above grade, suffered a considerable damage, which is in agreement with the computed low value of its reliability index (see **Table 5**).

The Bank of California building is a 12-story reinforced concrete frame building located approximately 23 kilometers from the center of San Fernando earthquake. The building, which has a height of 52.727 m, suffered a considerable damage, which is in agreement with the computed low value of its reliability index (see **Table 5**).

The displacement formula (Equation (13)), which is based on a simple iteration calculation, can be used to

Table 3. Member Sizes and Reliability Indexes.

Building Model	Height (meter)	*Z/S	Shear Wall m × m	Column m × m	Concrete		
						β	
					δ (mm)	1/200	1/500
Square	72	Z1/S3	0.3 × 10	0.3 × 0.6	21.465	5.796	4.438
		Z1/S3	0.2 × 10	0.3 × 0.3	51.612	4.544	2.068
		Z2/S2	0.3 × 10	0.3 × 0.6	58.889	4.219	1.716
		Z2/S2	0.1 × 10	0.2 × 0.2	212.184	0.878	−0.425
	120	Z2/S3	0.3 × 10	0.6 × 0.6	128.9	3.506	1.074
		Z2/S3	0.3 × 10	0.5 × 0.5	145	3.176	0.828
		Z3/S2	0.3 × 10	0.6 × 0.6	143	3.215	0.856
		Z3/S2	0.3 × 10	0.4 × 0.4	221.7	1.987	0.108
Tube	72	Z2/S2	0.3 × 10	0.3 × 0.6	44	4.855	2.51
		Z2/S2	0.3 × 10	0.3 × 0.5	52.5	4.487	2.022
		Z3/S3	0.3 × 10	0.3 × 0.6	76.9	3.522	1.087
		Z3/S3	0.3 × 10	0.3 × 0.5	88.256	3.135	0.799
	120	Z1/S2	0.3 × 10	0.6 × 0.6	61.807	5.155	2.995
		Z1/S2	0.3 × 10	0.4 × 0.4	97.172	4.243	1.747
		Z2/S3	0.3 × 10	0.6 × 0.6	139.04	3.294	0.913
		Z2/S3	0.3 × 10	0.4 × 0.4	218.590	2.024	0.128

*Seismic zone factor and soil profile.

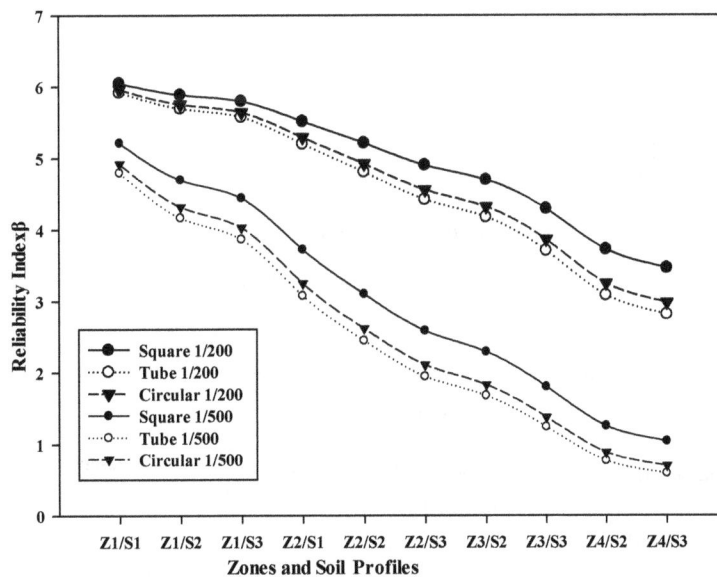

Figure 9. 18-Story Concrete Building Model Reliability Index.

compute the top building drift. An interaction chart that relates the maximum allowable drift (Δ) to the actual or computed building top drift (δ) is provided, **Figure 11**.

For example, a building with a total height of 120 meters, a drift index of $\frac{1}{200}$, and a desired reliability in-

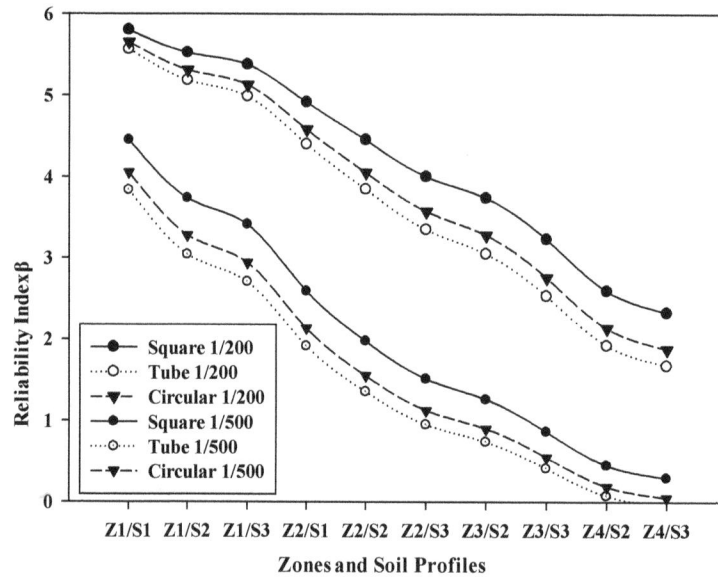

Figure 10. 30-Story Steel Building Model Reliability Index.

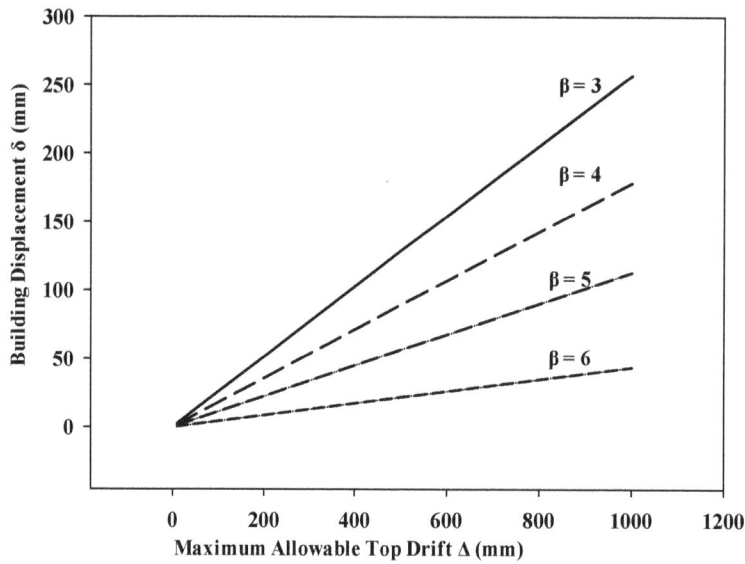

Figure 11. Interaction Chart.

dex of 3 will have 156 mm top drift (roof displacement) based on the interaction chart (see **Figure 11** and **Table 6**). The interaction chart is a very useful tool in determining the required top drift for the desired building reliability. In other words the building should be designed to have a roof displacement less or equal to the top drift δ obtained using the interaction chart.

5. Conclusion

The paper developed a reliability index approach to assess the reliability of concrete and steel buildings subjected to earthquake loading. The assessments obtained using the reliability index of the simulated, tested, and

Table 4. US-Japan Research Program.

Height (meter)	TEST			B	
	No.	Description	δ (mm)	1/200	1/500
	SL-3	Static	326	−0.862	−1.147
	SPD-1	Dynamic Miyagi Earthquake 1978	2.52	6.236	5.862
23.06	SPD-3	Dynamic Tehachapi Earthquake 1952	238	−0.686	−1.077
	SPD-4	Dynamic Tokachi Earthquake 1968	342	−0.884	−1.156

Table 5. Holiday Inn and Bank of California Buildings.

Building	Height (meter)	δ (mm)	β	
			1/200	1/500
Holiday Inn	20.00	241	−0.779	−1.114
Bank of California	52.727	279	−0.072	−0.829

Table 6. Top Drift Interaction Chart.

Building Height (meter)	β	Δ (mm)		δ (mm)	
		$\dfrac{1}{200}$	$\dfrac{1}{500}$	$\dfrac{1}{200}$	$\dfrac{1}{500}$
120	3	600	240	156	62
72	6	360	144	17	7
23	4	115	46	21	9
53	5	265	106	30	12

actual buildings in earthquake zones were acceptable as indicators of the buildings reliability. The reliability index model is flexible and can be used for: 1) all types of buildings and 2) all local and international design codes simply by changing the parameters magnitudes of the reliability index equation to the required parameters by the code of design. Each design code will have its unique reliability index β as a magnitude and the interaction chart that corresponds to it. The interaction chart is a very useful tool in determining the building drift for the desired level of reliability during the preliminary design of the building members.

6. Acknowledgments

This research work is supported by Qatar University Internal Grant QUUG-CENG-CA-12/10-2.

REFERENCES

[1] S. G. Buonopane and B. W. Schafer, "Reliability of Steel Frames Designed with Advanced Analysis," *Journal of Structural Engineering*, Vol. 132, No. 2, 2006, pp. 267-276.

[2] A. El Ghoulbzouri, A. Khamlichi, M. Bezzazi and A. Lopez, "Reliability Analysis for Seismic Performance Assessment of Concrete Reinforced Buildings," *Australian Journal of Basic and Applied Science*, Vol. 3, No. 4, 2009, pp. 4484-4489.

[3] R. Ranganathan and C. Deshpande, "Reliabilty Analysis of Reinforced Concrete Frames," *Journal of Structural Engineering*, Vol. 113, No. 6, 1987, pp. 1315-1328.

[4] M. S. Roufaiel and C. Meyer, "Reliability of Concrete Frames Damaged by Earthquakes," *Journal of Structural Engineering*, Vol. 113, No. 3, 1987, pp. 445-457.

[5] S. Terada and T. Takahash, "Failure-Conditioned Reliability Index," *Journal of Structural Engineering*, Vol. 114, No. 4, 1988, pp. 942-953.

[6] P. H. Waarts, "Structural Reliability Using Finite Element Methods," Delft University Press, Netherlands, 2000.

[7] International Conference of Building Officials, "Uniform Building Code," 1997.

[8] International Code Council, "International Building Code," 2009.

[9] D. C. Epaarachchi, M. G. Stewart and D. V. Rosowsky, "Structural Reliability of Multistory Buildings during Construction," *Journal of Structural Engineering*, Vol. 128, No. 2, 2002, pp. 205-213.

[10] M. S. Al-Ansari, "Drift Optimization of High-Rise Buildings in Earthquake Zones," *Journal of Tall and Special Building*, Vol. 2, 2009, pp. 291-307.

[11] J. W. Lindt and G. Goh, "Earthquake Duration Effect on Structural Reliability," *Journal of Structural Engineering*, Vol. 130, No. 5, 2004, pp. 821-826.

[12] J. W. Lindt, "Damage-Based Seismic Reliability Concept for Woodframe Structures," *Journal of Structural Engineering*, Vol. 131, No. 4, 2005, pp. 668-675.

[13] R. E. Melchers, "Structural Reliability Analysis and Prediction," Wiley, New York, 1999.

[14] J. K. Wight, "Earthquake Effects on Reinforced Concrete Structures, U.S.—Japan Research, SP84," American Concrete Institute, Detroit, 1985.

[15] D. A. Foutch, G. W. Housner and P. C. Jennings, "Dynamic Responses of Six Multistory Buildings during the San Fernando Earthquake," Earthquake Engineering Research Laboratory, California, 1975.

[16] B. Ellingwood, T. V. Galambos, J. G. MacgGregor and C. A. Cornell, "Development of a Probability Based Load Criterion for American Standard A58: Building Code Requirements for Minimum Design Loads in Buildings and Other," 1980.

[17] American Concrete Institute (ACI), "Building Code and Commentary," 2008.

[18] Bentley System Inc., "STAAD PRO V8i. Three Dimensional Static and Dynamic Finite Element Analysis and Design of Structures," 2009.

6

Shockproof Experimental Study of Automated Stocker System in the High-Tech Factory

Jyh-Chau Wang[1], Jenn-Shin Hwang[2], Wei-Jen Lin[1], Fan-Ru Lin[2], Chin-Lian Tsai[1*], Pin-Hong Chen[1]
[1]Innolux Corporation, Chinese Taipei
[2]National Center for Research on Earthquake Engineering, Chinese Taipei

ABSTRACT

This study, with the shake table experiments in the National Earthquake Engineering Research Center, investigated the seismic behavior of automation stocker system. The automation stocker system provided a fast and effective method for stocking products in the high-tech facilities. Firstly, the original machine tests for testing common stocker system were to investigate the seismic capacity and failing modes. Secondly, the team completed 2 kinds of reinforcement tests to investigate the seismic behavior: 1) the installation of bracing to improve overall stiffness; 2) the installation of the viscous dampers to improve the overall damping ratio. In comparing the results and performance of the three experiments, we learned from the results of the top-layer acceleration: the installation of bracing had the largest acceleration value, the original machine the second acceleration value and the damper the lowest acceleration value; the best effect was the installation of the damper. The result of the comparison of the top floor displacement meter showed that the highest data was the original machine; the second data was with the damper, the lowest data was the installation of the bracing. Based on the preliminary assessment on the best seismic retrofit ways of the storage system, we further examined the feasibility and applicability of automatic storage seismic retrofit in the high-tech factory.

Keywords: High-Tech Facilities; Automation Stocker System; Shake Table Test; Viscous Dampers

1. Introduction

1.1. Background and Motivation of the Study

The Science Park is the country's economic arteries (Hsiao, Pei-Fang, 2005) [1]. Because Taiwan is located on circum-Pacific seismic belt, one of the highly frequent seismic areas, the amount of investment in high-tech industrial equipments is usually really extremely; a single process equipment may cost over 3 million US dollars. Therefore, in the high-tech industries, from the stand points on operating and cost in business, equipment for seismic assessment should be given high priority [2]. To prevent potential serious damage caused by earthquake, the main process equipment and production lines are required to improve seismic design in the structure of the system. In order to ensure that Taiwan's high-tech Industry obtain satisfactory shockproof process equipment in dealing with earthquakes, the plants are requested to research and develop sustainable technology [3]. Therefore, the various high-tech industries are actively involved in issues of shockproof.

1.2. Literature Review

Different from traditional architecture, the special demand for the constructional structure of the high-tech factories not only requires stringent structural seismic safety, but also provides substantial prevention from micro vibration harm in the clean room of manufacturing environment [4].

With the high-tech industry continuing to rise, the increasing demands in manufacturing process are also facing new challenges in display industry. The most obvious changes are the increasing size of the glass panel, a single equipment size increasingly became larger and its function more complex. The facts in various equipment manufacturing process, complex construction and enormous numbers can cause impact on the baseline assessments of the equipment seismic demand (Chen, Chang-Liang, 2008) [5].

The Automated Material Handling System, AMHS, is the central system to reach manufacture process automation. The clean room stocker is one of important components in display factory. The automation Stocker System was shown in **Figure 1**, Lin, Kun-Bin (2007) [6].

The system is able to use stocker space efficiently by using computer online system to realize automated control manufacture, and continue to check the expired or overdue inventorial products. It can prevent poor inventory and improve management efficiency, as well, shorten producing time and reduce the operating cost. The AMHS application has become the main-stream in display factory.

Due to the nature of aluminum extrusion materials which is light weight, high stiffness, and simplicity to assemble, the aluminum extrusion materials have been generally used on cover and the internal structure for the AMHS in the clean room. This experiment aims to explore what the stocker's situation of the overall structure is, using aluminum extrusion materials, when earthquakes occur.

Following generations of evolution, the size of glass panel and the storage system are relatively enlarged. In the large stocker system, a full load of Cassettes weights about 600 kg; its length and width are 4 m, and height up to 7 m.

When an earthquake occurs, due to the overall cassette mass ratio is more than the shed bit, the whole stocker with low stiffness tends to have shaken violently which causes fraction of the glass panel as cassette edges slide off and collide. Another issue is that the automated stocker system is high precise equipment. If the shed bit is excessively deformed, it will affect the accuracy of the operation of the mechanical arm. When the situation is critical, the system is required to stop and be examined. When the duration is extended, the problems in production line are consequently followed.

The stocker overall structure is low stiffness; therefore, it tends to shake violently. Consequently, seismic force convey upwards to the top of the stocker structure and amplifies its effects. This study will examine and test the reinforcement measures: method 1, the team installed the bracing to improve the overall stiffness of the Stocker which will inhibit displacement issue, but it can be expected that acceleration will increase and enlarge. Method 2, the team installed a damper, the function of the damper is to increase the damping (or impedance) of the structure, and to reduce the vibration of external forces by earthquake or wind.

The main principle of the damper is to use the characteristics of the material itself to achieve the effect of energy conversion. The dampers in the market include fluid viscous damper (FVD), viscoelastic dampers (VE Damper) and Viscous Elastic Material (VEM), and the characteristics and speed of the force have proportional relationship. Usually, it will not excessively increase the overall stiffness of the structure. Therefore, the low stiffness Stocker sample with installation of the damper is expected to reduce the absolute acceleration and displacement of the stocker at the same time.

1.3. Study Aim

In this study, the AMHS took the high-tech factory as the research background; the main purposes are to explore the original machine tests because there is no reliable information for the equipment for high-tech industry. We need to understand both how to improve and reinforce the seismic strategies and technical content, in order to modify the failing mechanism in equipment during earthquake. To strengthen seismic retrofit to the failing mechanism, the study tested common stocker system to investigate the seismic capacity and failing modes. The stocker was designed by original manufacturer and was used in the high-tech factory.

The team completed 2 kinds of reinforcement tests to investigate the seismic behavior: 1) the installation of bracing to improve the overall stiffness; 2) the installation of the viscous dampers to improve the overall damping ratio. This pilot study used the stocker with the shake table experiments in the National Earthquake Engineering Research Center. It is expected to enhance the seismic capacity of the original designed stocker. The experimental results can be discussed to give concrete proposals for the best reinforcement and to upgrade the seismic capacity.

2. The Original Machine Tests of the Stocker System

This pilot study used the D6000 type stocker with the shake table experiments in the National Earthquake Engineering Research Center. The original machine tests for testing single stocker system, D6000 type, were to investigate the seismic capacity and failing modes. The

Figure 1. Automation stocker system.

D6000 type stocker was designed and manufactured in Japan and was used in the high-tech factories. The stocker for storing is an important equipment for storing a large number of both semi-finished and finished products.

2.1. Micro-Vibration Measurement for the High-Tech Factory

In order to accurately simulate the effect upon Stocker by seismic forces at its location, the National Center for Research Earthquake Engineering was commissioned to conduct the micro vibration measurement. The micro vibration measurement sensors were Model VSE15D speed type micro vibration meter, made by Tokyo SOKUSHIN Co., LTD. The micro vibration meter has both the horizontal and vertical radial velocity measurement; the duration of measurement was 250 seconds, with a sampling frequency of 200 Hz. The layout of micro-vibration sensors in the high-tech factory was shown in **Figure 2**. The sensors installed on each floor, e.g. placed at escape ladders, and are distant from the process equipments to avoid the influence of vibration by the process equipments themselves. The micro-vibration sensors in the stocker installed at top-layer. It helps to understand the factory structure and floor frequency by

the measurement. Then, the team used the measurement result to translate and meet the panel site's potential designed response spectra.

The result of micro-vibration measurement for this high-tech factory, showed that the structure's frequency of X direction is 1.22 Hz; the structure's Y direction is 1.04 Hz (as **Figure 2**). For the stocker system with cassettes, the natural frequency of X direction is 4.53 Hz; the natural frequency of Y direction is 4.9 Hz. For the high-tech factory, the stick model can be built based on the micro-vibration measurement results. The team chose the closest stations to the panel site that is the 0304 seismic wave, Chiahsien earthquake in Taiwan, from CHY099 Shanhua elementary measure station.

Figure 3 shows the inputted signals following the research of all the shake table tests, respectively, the X is shock waves GX (NS), Y is shock waves GY (EW), and Z is axial shock waves GZ (V). The time history showed at **Figure 3**. For simulation of the ground surface acceleration, the team established the response spectrum of artificial acceleration, the tri-axial duration history shown in **Figure 4**. The response spectrum met the Taiwanese official code Seismic Design Specifications and Commentary of Buildings. Furthermore, the team simulated the seismic waves of actual floor to investigate the seismic behavior of equipment in the high-tech factory.

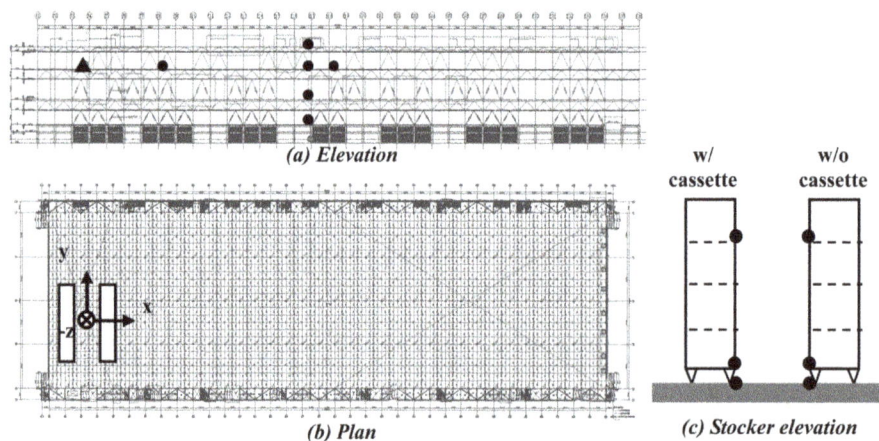

Figure 2. The locations of accelerometer in the factory and stocker are shown by the dots.

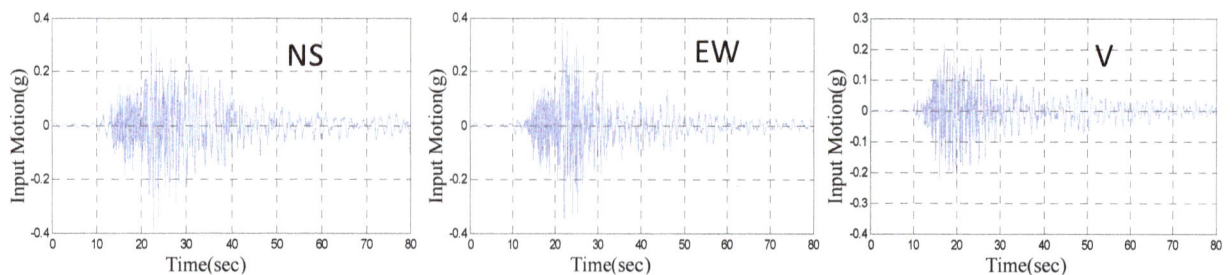

Figure 3. The signal, the X shock waves GX (NS), Y shock waves GY (EW), and Z axial shock waves GZ (V) were inputted respectively for the shake table test.

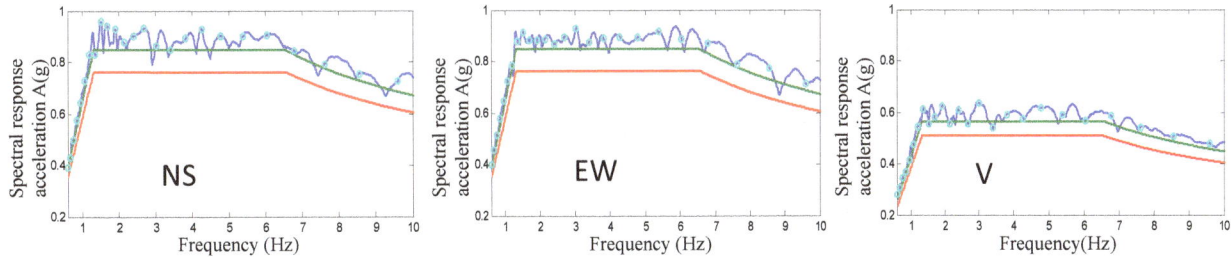

Figure 4. The tri-axial duration history: GX (NS), GY (EW), GZ (V). The 0304 seismic wave, Chiahsien Earthquake in Taiwan, from CHY099 the Shanhua Elementary Measure stations. The response spectrum met the Taiwan Official Code Building Seismic Design Specifications and Commentary.

2.2. Experimental Equipment and Establishing

The shake table tests were held in the National Earthquake Engineering Research Center; the specification of the shake table was 5.1 m × 5.1 m. The sample size of stocker-D6000 type was 4000 × 3050 × 7680 mm, and the total weight was 5650 kg. The stocker includes an empty scaffold with eight cassettes (Cassette, referred as CST), and a single CST weights 600 kg. The configuration of overall experimental stocker structure was shown in **Figure 5**. The fixed form of the stocker met the actual situation which was 12 feet locked solid on the base; and the base was fixed by expansion bolts to the concrete frame.

The two forms of MC-Nylon stopper were installed in Cassette shown in **Figure 6**. The right side of MC-Nylon CST refers to RCST. That was the original condition of the MC-Nylon stopper in the stocker. The left side of MC nylon stopper CST refers to LCST. The LCTS was mounted metal stopper and using rubber cushioned material. The purpose of the MC-Nylon stopper installation was to limit Cassette moving range when earthquakes occur.

The accelerometer and displacement meters of the installed position was shown in **Figure 7**. Both sensors are installed on the diagonal corners of the top layer, the first layer and the bottom layer of the stocker. The accelerometers measured the three axes—X, Y, and Z axes. The displacement meter measured for the X and Y axes.

To explore the CST seismic scenario, the three-axial accelerometer was equipped at the top layer of the two CST. The two CST, right side and left side, were at the top layer of stocker. For observing and simulating the glass sliding condition inside the CST during earthquakes, the team selected PVC board. In addiction, in order to avoid the damage of the shake table or personal injuries as the experiment of the shack table may result in glass breakage or falling. Also, the friction coefficient of PVC board is similar to glass. The result of the experimental test showed the friction coefficient of the glass with the support material is approximately 0.51; the friction coef-

ficient of the PVC board with the support material is about 0.45.

2.3. Result

2.3.1. The Damage Situation of the Stocker

The first phase of the experimental results showed that when PA was 200 gal, the torsion and shaking of the storage space occurred slightly. The CST on the top layer also found little displacement, but it did not cause stopper collision. when PA were 300 gal and 400 gal, the storage spaces of stocker vibrated significantly; it caused large displacement and collision between stopper and CST on the top layer of CST. Few parts of the PVC boards slid away.

The second phase: the aim for the experiment was to understand the ultimate strength of stocker. When PA was 500 gal, rocking and shaking in the storage spaces of the stock occurred violently. The top-layer CST are displaced and hit the stopper and darted out of the PVC boards, and the screws were loosened on the first layer of left cassette's flange with trellis joint, shown in **Figure 8**.

When PA was 600 gal, rocking and shaking violently in the storage spaces of the stock occurred. The RCST and LCST displaced rearward and exceeded beyond the limitation of the MC-Nylon. The cassette was stuck on the edge of the MC-Nylon stopper and could not to move; a large number of PVC board had slid; therefore, the experiment ended. Finally, the team inspected the overall stocker structure and found that the screws loosened on the first layer of left cassette's flange with trellis joint, the condition was the same with experiment in PA for 500 gal. Further, due to severe squeeze, the metal stopper with rubber cushioned material was fractured at the join.

2.3.2. Analyzed the Frequency-Measuring with White Noise

Following the experiment of 400 gal and 600 gal, frequency-measuring with White Noise was analyzed; it was PA for 50 gal, and 1 - 50 Hz. The purpose of frequency-measuring with White Noise was to observe storage spaces after 400 gal and the 600 gal limit tests of

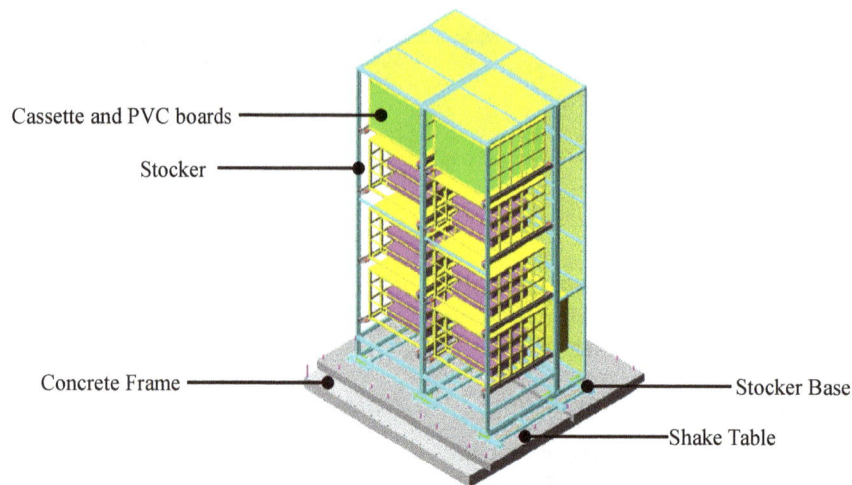

Figure 5. The configuration of overall experimental stocker structure.

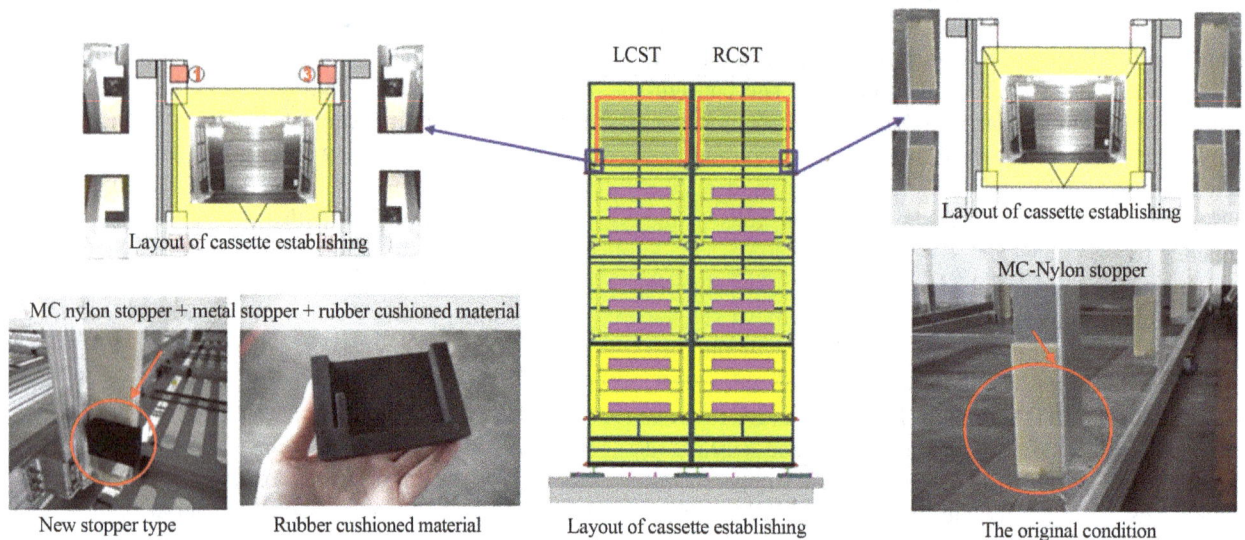

Figure 6. The two forms of MC-Nylon stopper were installed in cassette (*the right side CST referred to RCST, the left side CST referred to LCST*).

stocker to investigate whether or not the frequency of stocker structure was reduced. The method was used to check whether the internal material caused material damage or loosen at the joint by the experiment. It may also determine whether the experimental specimens damage or not.

The results showed that under the shock wave in PA 400 gal and 600 gal, comparing to the frequency of the specimen after and before the experiment, the frequency of length axle-Y axle was reduced approximately 20%.

The team speculated the reason to be the gravity position of the overall stocker's center was not located at the center of mass position. It may due to the large amount of FFU fans located behind the cassettes. The fans caused violently torsion in the course of the experiment which led to the Y axle of the structure destroyed.

2.4. Summary

The results of the experiment showed that low stiffness of Stocker storage spaces tends to be violently and structurally displaced. Consequently, it led to large num- ber of PVC boards darted out of the cassettes. In real condition of the high tech factory, the cassettes are used to store large semi-finished or finished glass. Therefore, if the glass was damaged by shocks, it was considered a type of destruction.

The result of frequency-measuring with White Noise analysis showed that after the shock wave in PA 400 gal, the natural frequency of length axle, Y axle, reduced approximately 20%. The possible reason is that the gravity position of the overall stocker's center was not located at the center of mass position. It caused violent torsion in the course of the experiment which led to the Y axle of

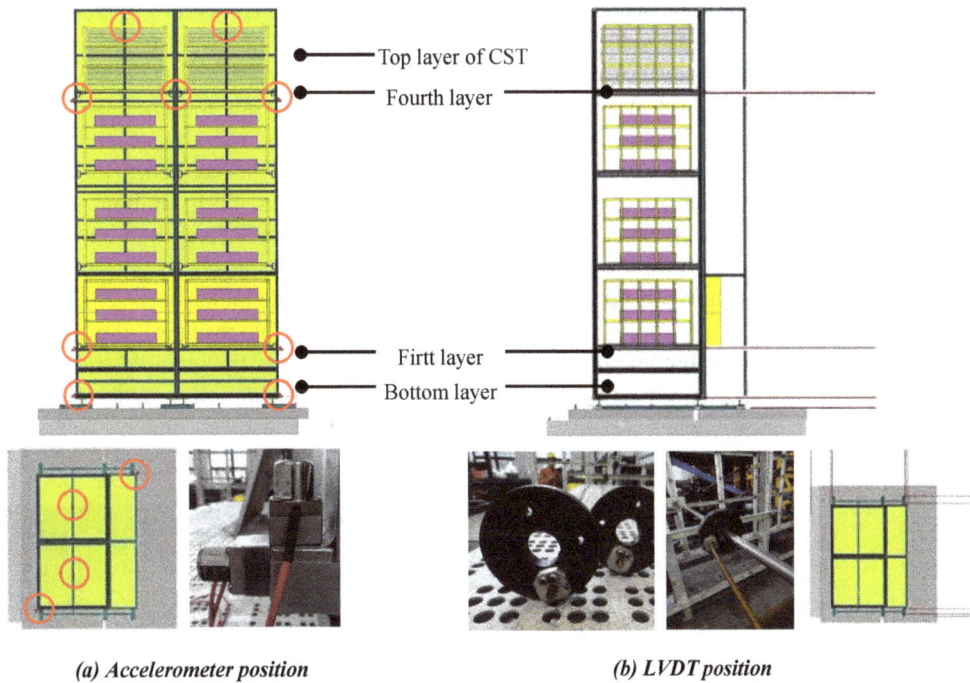

(a) Accelerometer position   (b) LVDT position

Figure 7. Accelerometer and displacement meter of the installed position.

Figure 8. The screw was loosened at the first layer of left cassettes flange with trellis joint.

the structure destroyed. The stocker feet and base were bolted by screws. After specimens removed for inspecttions, a slight deformation was found in the locking of aluminum extrusion materials.

3. The Seismic Tests for after Stocker System Reinforcement

From the results of the experiment of the original machine test, we learned the stiffness of the original Stocker framework design was too low. The structure tended to violently displaced which caused much unpredictable destructions. The team completed 2 kinds of reinforcement test to investigate the seismic behavior: the installation of bracing and the viscous dampers. We expected to

enhance the seismic capacity of the original design stocker. The preliminary assessment of experimental results could be discussed to give appropriate proposals for the best reinforcement and to upgrade the seismic capacity.

3.1. Method 1: The Installation of Bracing to Improve the Overall Stiffness

The installation of bracing directly provided the Stocker overall stiffness. It was expected to reduce the shake violently issue of Stocker, and reduce the amount of displacement between layers. The team observed the outcome after the improvement of Stocker, whether it could avoid distortion or skew of the Stocker framework, or cassette falling, and enhance the seismic capacity.

3.1.1. Experimental Equipment and Established

Method 1, the installation of bracing to improve the overall stiffness. The reinforcing braces installed on both sides of the stocker: the bracing section 50 × 100 mm of aluminum extrusion. The locking manner was the L-type angle iron block and T-fixed bolt to lock Stocker scaffold's aluminum extrusion groove. The installed bracing to improve the Stocker stiffness was shown in **Figure 9**.

3.1.2. The Result of the Experiment after Installing Bracing

The result of the experiment shown when stocker installed bracing, when PA was 200 gal, due to the force,

the link piece showed bending and deformation. The link piece: the L-type angle iron block at the bracing and aluminum extrusion framework groove. When PA was 300 gal, bracing joints loosened in a large number. The reason may due to the size of T-fixed bolt for L-type angle iron block locked being too small. Due to the excessive force, the T-fixed bolt was prone to cause destruction that the T-fixed bolt had been cut and stood idle and could not be removed. On the other hand, the un-loosened T-bolt joints buckled if the bracing loaded with the force. The result of the experiment after installing bracing was the main destruction was at the link piece on the bracing joints and framework. The destruction issue of installed bracing was shown at **Figure 10**.

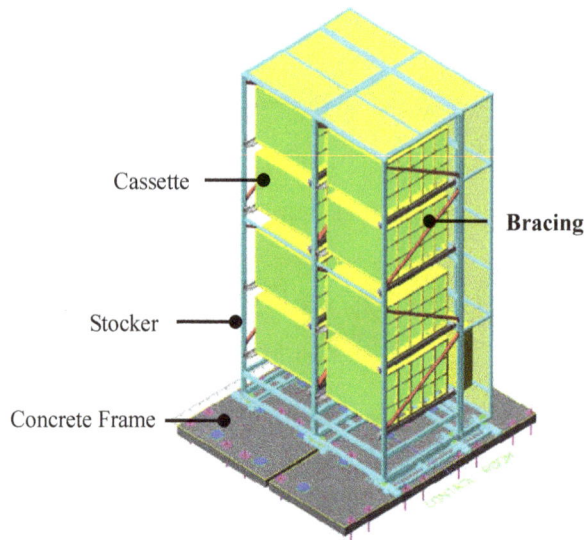

Figure 9. The installing bracing to improve the stocker stiffness.

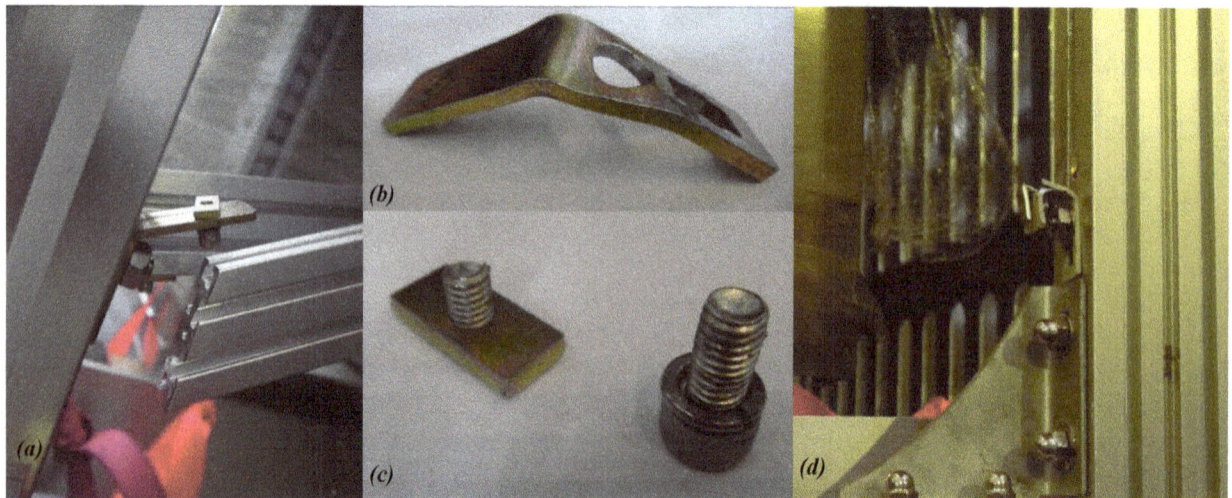

Figure 10. The destruction issue of installing bracing. (a) Bracing joint burst; (b) The link piece, L-type angle iron block, was bended and deformed; (c) T-fixed bolt was cut and stood idle; (d) The main destruction was link piece at the bracing joints and framework.

The possible reasons of destruction were summarized, 1) since the Stocker-D6000 size was bigger, the reinforcement test was relatively too slender; therefore, it could not provide the expected effect of the tensility and compression. 2) The bracing joints were fixed poorly. There were no beams on the top layer of the original Stocker design; damages are prone to occur easily at the line piece, when the bracing loaded force.

3.2. Method 2: The Installation of Viscous Dampers to Improve the Overall Damping Ratio

3.2.1. Experimental Equipment and Established

This pilot study of the seismic capacity was for the low stiffness Stocker system: install the viscous dampers to improve the overall damping ratio. It was expected to enhance seismic capacity and to give concrete proposals for the best reinforcement of upgrading the seismic capacity.

For the preliminary test, the National Earthquake Engineering Research Center provided the existed four sets of fluid viscous damper (FVD), and investigated whether the installation of a damper was a possible reinforcing method. According to the result of the original machine test, the maximum displacement of stocker was at the first layer. Therefore, the team focused on installing the dampers on the outside and inside of the first layer. The installation of the viscous damper was shown in **Figure 11**.

For installing the four sets of the fluid viscous dampers without increasing stiffness in Stocker; the team designed locking components on the dampers connector. They include: 1) transverse reinforcing beam: 50 (mm) × 50 (mm); 2) extensible square steel tube: 60 (mm) × 30 (mm) × 2070 (mm); 3) designed the damper connector; and 4) design extensible tubes and the joint of the framework.

For the damper only bear the axial force, the damper joint design is as follow: the damper extremity was hinged to the bolt groove. One side of the bolt groove was welded with extensible square tube; the other side of the bolt groove was jointed with 4 T-fixed bolts. The single damper installation condition was shown in **Figure 12**.

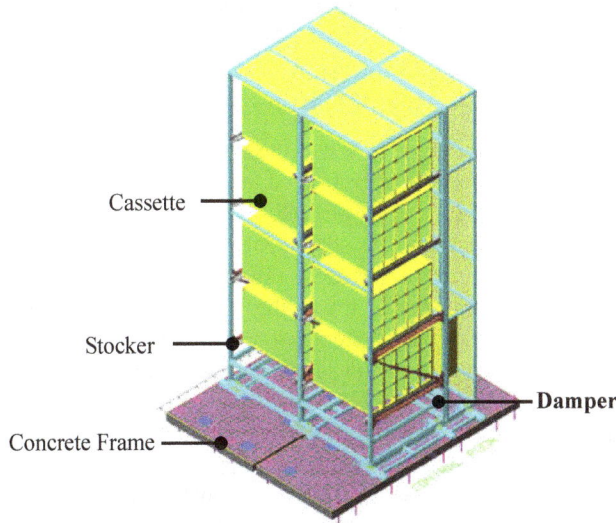

Figure 11. The team installed the dampers focus on the outside and inside of the first layer.

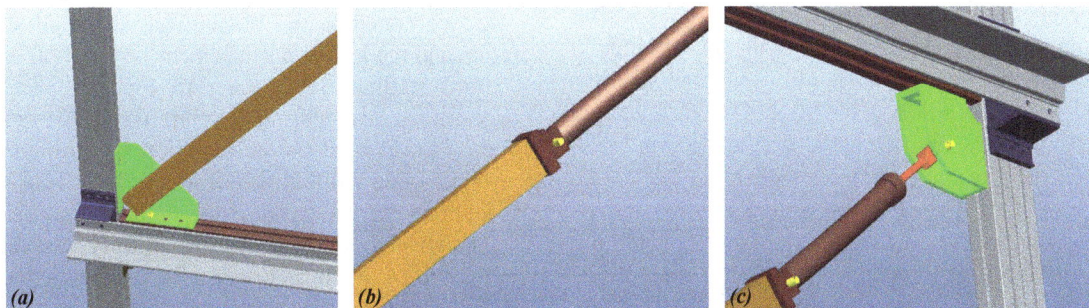

Figure 12. The installation condition of locking components for dampers connector. (a) Design of locking components for dampers connector; (b) Design of installing lower reinforcement beam; (c) Design of installing upper reinforcement beam.

In the experimental specimen, the original Stocker-D6000 type, the design between second and fourth layer is beamless. To avoid damage at the stocker specimen before experiment, it has not yet receiving force. The team installed a 50 (mm) × 50 (mm) transverse reinforcing beam. The beam was made of the aluminum extrusion material and installed at the damper connector in the stocker framework.

The objective of installing the lower reinforcement beams and upper reinforcement beams was to increase the strength of the Stocker, at the same time, to solve the weakness of the connector between the damper and the Stocker. The damper installation condition was shown in **Figure 12**.

3.2.2. The Result of Installing the Viscous Damper

To compare the effect before and after installing damper, the team compared the acceleration of the top layer of the stocker and the shock wave of the shake table in the original machine, and the result of the damping ratio of the original machine was calculated $\xi = 3.7\%$. After installing the damper, by observing the acceleration time history, the response peak of the acceleration was significantly attenuated and the energy demolished. The damping ratio of installed damper was $\xi = 6.01\%$. The top layer of the stocker and the shake table inputted the shake wave was shown in **Figure 13**.

The experimental result shown in PA for 400 gal, although the top layer cassette slid and collided with the stopper, the stocker structure was not seriously damaged. The preliminary results on increasing seismic capacity showed that the installation of damper was better than installation of bracing for low stiffness stocker. The experimental result shown in PA for 400 gal, although the top layer cassette slid and collided with the stopper, the stocker structure was not seriously damaged. The preliminary results on increasing seismic capacity showed that the installation of damper was better than installation of bracing for low stiffness stocker.

3.3. Summary

The experiment showed that due to enlarger stocker size, the reinforcement method for the installation of bracing, the bracing material or form should use higher stiffness material. The method bolting the joint of the damper also must be improved. The original L-type angle iron block was unable to bear the shear force. When the force from bracing passed to L-type angle iron block, the L-type angle iron block easily deformed. Therefore, it could not function stiffening effectively.

The preliminary observation of reinforcing method was the installation of damper had better effects. The damper was not the best damping ratio of the specimen,

so the seismic efficacy was limited. The direction for future research will continue to investigate on the installation of optimum damping ratio of the damper on the stocker.

4. Results of the Experiment and Comparison

The team investigated the seismic performance of stocker reinforcing measures and assessed the most appropriate reinforcement method.

4.1. Comparison of the Vibration Frequency: Original Machine, Bracing, and Damper

To compare the scanned of natural frequency, the original machine without any reinforcement of natural frequency was 0.98 Hz. After installing the damper, the natural frequency was 1.17 Hz. The damper without providing overall stiffness of the Stocker, the natural frequency was closer to the original machine. After installing the bracing, the natural frequency was 1.95 Hz, because the bracing upgraded stiffness of the Stocker. The result of the frequency comparison can preliminarily determine the reliability of the experiment.

4.2. Comparison of the Time History: Original Machine, Bracing, and Damper

In PA for 200 gal, the time history for Stocker at the top layer was compared and shown in **Figure 14**. From **Figure 14**, we see the time history of the absolute acceleration shown obviously that the installation of bracing, the acceleration response peak in the sample B-3.67 was the highest. As we can see, upgrading the stiffness of Stocker could increase the acceleration from bottom to top layer. The original machine, A-1.82, and damper, C-2.03, had no significant differences.

Comparison of the relative displacement in time history was shown in **Figure 15**. The displacement reaction issue, the original machine, sample A, was the highest, followed by installed damper, sample C.

Installed bracing, sample B, canceling out the acceleration response to inhibit the displacement reaction. Therefore, we conclude the bracing has the highest acceleration and the lowest displacement.

By installing the bracing to Stocker enhances the acceleration reaction and inhibits the displacement reaction. However, if the acceleration is too high, the top layer Cassette will outrush of the stocker. Thus, it is not recommended.

The absolute acceleration time history PA was shown in **Figure 16**. It shows the installation of bracing, sample B, has the highest acceleration response peak, followed by the original machine. The installed damper has the lowest acceleration response peak.

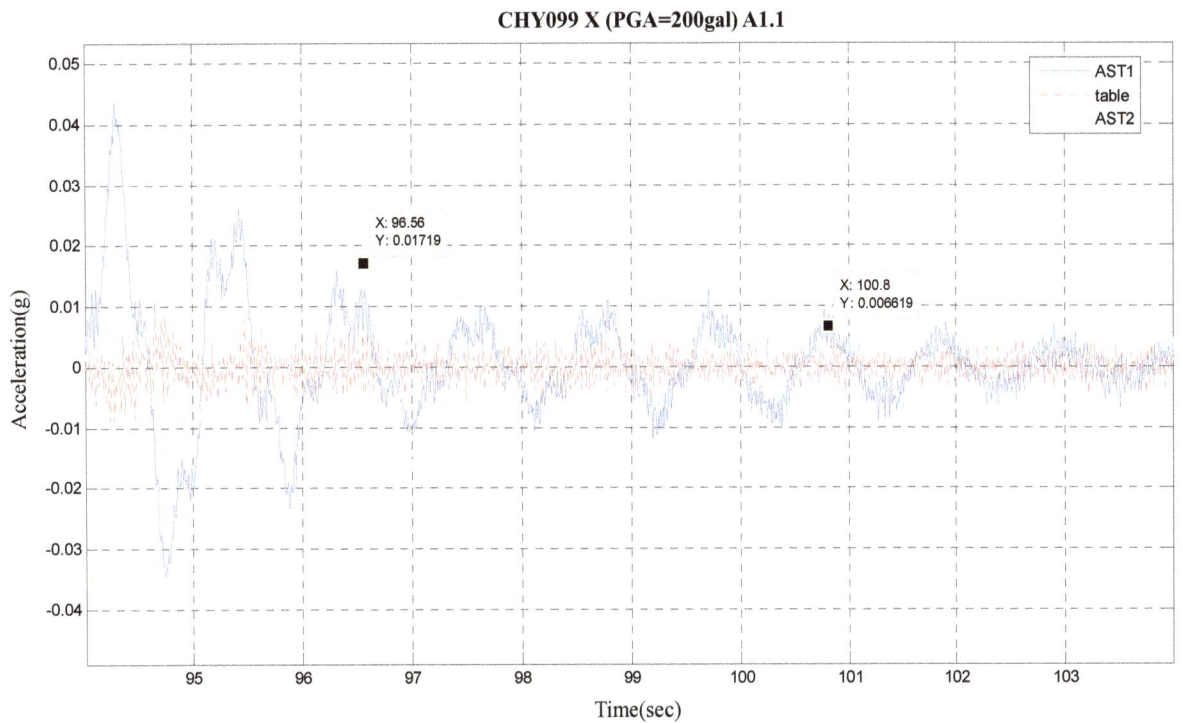

CHY099 X (PGA=200gal) A1.1

(a)

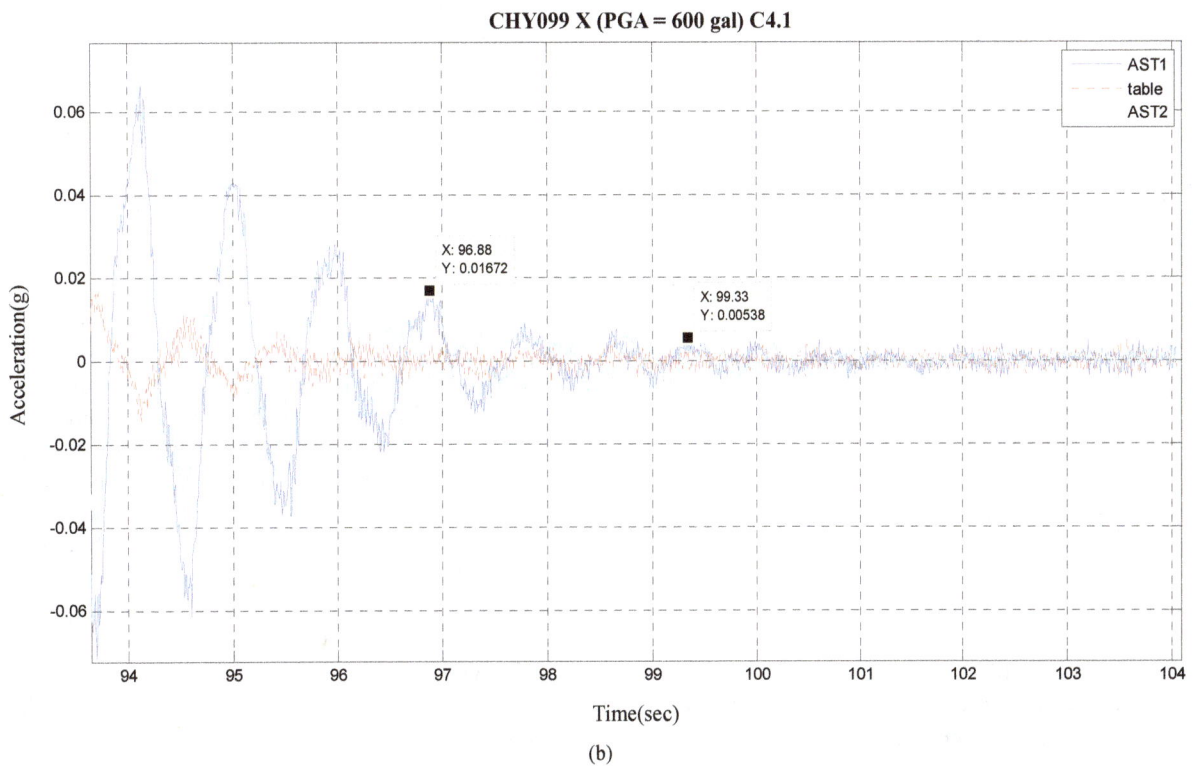

CHY099 X (PGA = 600 gal) C4.1

(b)

Figure 13. To compare the acceleration time history of the stocker top layer and shake table inputted the shake wave. (a) The damping ratio of original machine was $\xi = 3.7\%$; (b) The damping ratio of Installation of the damper was $\xi = 6.01\%$.

The preliminary observation of reinforcement method showed that the installed damper did work during earthquakes. However, the damper used in this experiment was not the best damping ratio of the stocker, so the seismic efficacy was limited. Future research will continue to investigate and analyze the stocker character, and

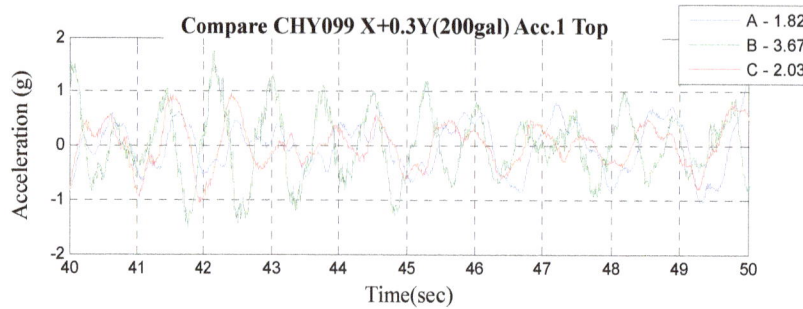

Figure 14. Acceleration time history in PA = 200 gal.

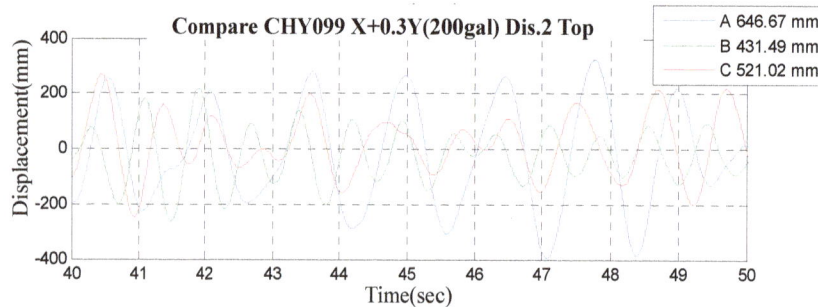

Figure 15. Displacement time history in PA = 200 gal.

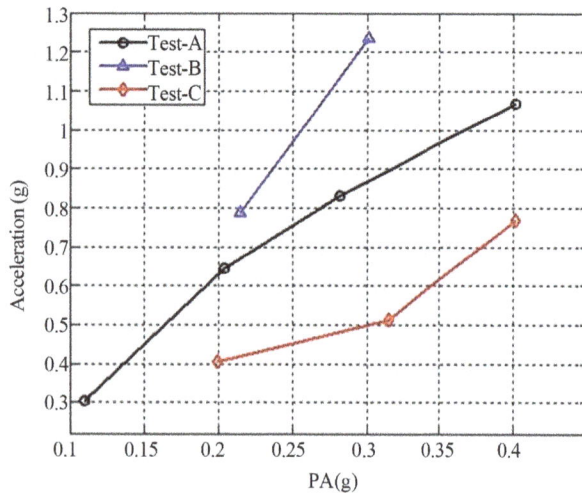

Figure 16. The absolute acceleration time history in the PA data comparison.

choose the optimum damping ratio of damper.

4.3. Comparison of the Maximum of Absolute Acceleration and the Maximum of Relative Displacement: Original Machine, Bracing, and Damper

The result of comparison of the maximum absolute acceleration of the shake table and the top layer of the stocker was shown in **Table 1**. In PA for 300 gal, acceleration of the top layer of the Stocker: the installed bracing (sample B) was the highest, followed by the

Table 1. Compare the maximum absolute acceleration of the shake table with the stocker top layer in PA = 300 gal.

Experimental equipment	Acceleration (g), Direction: X	
	Shake table	Stocker top layer
1) Original machine	0.297	0.832
2) Bracing	0.687	1.234
3) Damper	0.370	0.513

original specimen (sample A), and the installed damper (sample C) was the lowest.

The result of the comparison of the maximum relative displacement for the shake table and the top layer of the Stocker was shown in **Table 2**. In PA for 300 gal, the displacement reactions of northwest side (NW-Y) and the southwest side (SW-Y) were different. The reason may due to the large amount of FFU fans behind the cassettes. The FFU fans affected the overall stocker's position of center-gravity. It caused the overall stocker's center of gravity position to be not located at the center of mass position which led to the result of maximum relative displacement in the southwest side (SW-Y) being larger than it in the northwest side (NW-Y). The stocker's gravity position dislocated from the center of mass position was shown in **Figure 17**.

5. Conclusions

The experimental results of the original machine: Low stiffness of Stocker storage spaces easily tends to be

Table 2. Compare the maximum relative displacement of the shake table with the stocker top layer in PA = 300 gal.

Experimental equipment	Displacement (mm), Direction: X			Displacement (mm), Direction: Y		
	Shake table	Northeast (NE-X)	Northwest (NW-X)	Shake table	Northwest (NW-Y)	Southwest (SW-Y)
1) Original machine	62.1	167.4	165.7	32.6	24.2	85.9
2) Bracing	64.8	168.6	158.6	32.1	30.7	106.2
3) Damper	65.8	164.7	150.1	31.9	33.6	106.6

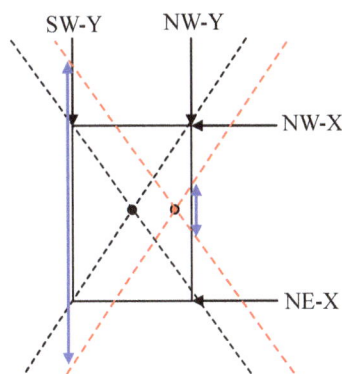

Figure 17. Stocker's gravity position dislocated from the center of mass position.

violently displaced which causes cassette violently displaced and large number of PVC boards (simulation of glass) slide.

The result of inspected frequency-measuring with White Noise shows that after the shock wave in PA 400 gal, the natural frequency of length axle, Y axle, reduced approximately 20%. The possible reason: the overall stocker's center of gravity position was not located at the center of mass position. It caused severe torsion in the course of the experiment, which led to the destruction of Y axle of the structure. The Stocker feet and base was anchored by screws. After specimens removed for inspections, we found that there was slight deformation in the locking of aluminum extrusion materials.

The installation of bracing: because the Stocker-D6000 size was bigger, the material used in bracing in the reinforcement test was relatively too slender. Therefore, it could not provide the expected effect of resisting the tensility and compression. The bracing joints were poorly fixed; there was no beams in the design on the top layer of the original Stocker. Failure at the line piece will occur easily when the bracing loaded with force.

Installing the viscous dampers: the preliminary observation of reinforcement method shown that the effect of installing of damper was better. The damper was indeed effective. However, in this experiment, the damper was not the best damping ratio of the stocker, so the seismic efficacy was limited. Future research will continue to

investigate and analyze characters of the stocker and to choose the optimum damping ratio of damper.

Through the comparison of relative displacement, we found that the Stocker's gravity position was not located at the center of mass position, and it caused severe torsion. In the future, installing tensional component and compressional component on the top layer to link the structure of the stocker can enhance the stability of the stocker system and reduce the effect of eccentric torsion.

The reinforcement method of stocker system: Installing tensional component and compressional component on the top layer to link the structure of the stocker can enhance the stability of the stocker system and reduce the effect of eccentric torsion. In addition, installing the speed type viscous dampers and designing the most appropriate damping coefficient of the damper can reduce acceleration and displacement response of the Stocker and avoid cassette sliding and collision, or glass panel sliding away from cassette.

6. Acknowledgements

In the completion of this study, we would like to express our sincere appreciation to those who have helped and accompanied us during the process of our study. Without their guidance, and support, we would not be able to complete this study.

Our highest appreciation goes to our advisor: 1) Yao George C, professor at department of architecture National Cheng Kung University; 2) Juin-Fu, Chai, National Center for Research on Earthquake Engineering; 3) Wang, Yen-Po, professor at National Chiao Tung University; and 4) Lyan-Ywan Lu, professor at National Kaohsiung First University of Science and Technology, who inspired us to develop the greatest interest in experiment of bracing and damper. Their professional knowledge and intellectual guidance helped us overcome the difficulties in conducting the research and shaped the final version of our study. Without their guidance and encouragement, we might have given it up.

Zhen-Yu, Lin and the members of Metal Industries Research and Development Centre, help us to complete the destruction experiment of Automated Stocker system.

In addition, we are also grateful to lecturer Zoe, who helps us to correct and complete this paper structure in English.

Finally, we are grateful to our team members' dedication to the study; they are Fu-Chun Huang, Chih-Hsien Lin, Lung-Ho Yeh, Chang Hsun, Wei-Jen Lin, Chih-Ming Chang, Yu-Ting Huang. With the shockproof expertise and practical experiments, they were consistent learning and their attitude to never giving up and overcoming all the difficulties have made this study happen successfully. We will also strive to work hard and study in the project for the equipment seismic in High-Tech Industry.

REFERENCES

[1]   P.-F. Hsiao, "Aseismatic Application of Fluid Dampers on High Tech. Fab. Structures," Master Thesis, National Cheng Kung University, Taiwan, 2005.

[2]   J.-C. Wang, C. Yao George, Y.-F. Lin, M.-J. Hsieh, C.-L. Tsai, P.-H. Chen, K.-C. Kuo and W.-T. Chen, "Seismic Experiments and SAP2000 Analytical Study on the Aseismic Footing of Precision Machinery in Hi-Tech Factories," 2012.

[3]   Chen, W.-C. "A Research on Evaluating Floor Vibration in Earthquakes for Non-Structural Elements Damage Estimate," National Cheng Kung University, Taiwan, 2008.

[4]   Y.-P. Wang, "Seismic Hazards Mitigation of the High-Tech Industries," Workshop on Protective System for Building Equipments and High-Tech Facilities, 2002.

[5]   C.-L. Chen, "A Study on Seismic Assessment and Vibration Reduction of High-Tech Facilities," Dissertation, National Kaohsiung First University of Science and Technology, Kaohsiung, 2008.

[6]   K.-B. Lin, "The study of In-Line Stocker Implementation in Large Size Panel Fabrications—A Case Study of TFT-LCD Company," Master Thesis, National Central University, Taiwan, 2007.

7

Control Parameters of Magnitude—Seismic Moment Correlation for the Crustal Earthquakes

Ernes Mamyrov
Institute of Seismology of the National Academy of Sciences of the Kyrgyz Republic, Bishkek, Kyrgyz Republic

ABSTRACT

In connection with conversion from energy class K_R ($K_R = \log_{10} E_R$, where E_R—seismic energy, J) to the universal magnitude estimation of the Tien Shan crustal earthquakes the development of the self-coordinated correlation of the magnitudes (m_b, M_L, M_s) and K_R with the seismic moment M_0 as the base scale became necessary. To this purpose, the first attempt to develop functional correlations in the magnitude—seismic moment system subject to the previous studies has been done. It is assumed that in the expression $M(m_b, M_L, M_s) = k_i + z_i \log_{10} M_0$, the coefficients k_i and z_i are controlled by the parameters of ratio $\log t_0 = a_t + b_t \log_{10} M_0$ (where $t_b = f_0^{-1}$; f_0—corner frequency, Brune, 1970, 1971; M_0, N·m). According to the new theoretical predictions common functional correlation of the advanced magnitudes M_m ($m_{bm} = m_b$, $M_{Lm} = M_L$, $M_{Sm} = M_S$) from $\log_{10} M_0$, $\log_{10} t_0$ and the elastic properties (C_i) can be presented as $M_m = d_i \log_{10} M_0 - 2 \log_{10} t_0 + C_i$, where $z_i = d_i - 2b_t$, and $k_i = C_i - 2a_t$, for the averaged elastic properties of the Earth's crust for the m_{bm} the coefficients $C_i = -11.30$ and $d_i = 1.0$, for M_{Lm}: $C_i = -14.12$, $d_i = 7/6$; for M_{Sm}: $C_i = -16.95$ and $d_i = 4/3$. For the Tien Shan earthquakes (1960-2012 years) it was obtained that $\log_{10} t_0 = 0.22 \log_{10} M_0 - 3.45$, and on the basis of the above expressions we received that $M_{Sm} = 1.59 m_{bm} - 3.06$. According to the instrumental data the correlation $M_s = 1.57 m_b - 3.05$ was determined. Some other examples of comparison of the calculated and observed magnitude—seismic moment ratios for earthquakes of California, the Kuril Islands, Japan, Sumatra and South America are presented.

Keywords: Magnitude; Seismic Moment; Energy Class; Earthquakes; Frequency

1. Introduction

In world practice, seismological research in assessing the scale of earthquakes magnitude scale of Gutenberg and Richter [1-3] is fundamental. In the countries of the former Soviet Union has been used scale independent energy class K_R, defined as the logarithm of the seismic energy E_R, highlighted by an earthquake, measured in joules ($K_R = \log_{10} E_R$, [4-6]).

For crustal earthquakes Tien Shan when considering the transition to magnitude scale was necessary to develop a self-consistent system of quantitative relationships that justify numerous empirical relationships body-wave magnitude m_b, local magnitude on surface waves M_L, surface wave magnitude for M_S and K_R from seismic moment M_0 (N·m), as the reference scale. In connection with the above purpose is to study the quantitative relationships m_b, M_L, M_S and energy of seismic radiation E_S c

M_0 based on the following findings:

1) proportional magnitudes and the maximum amplitude of seismic vibrations [1-3];

2) the statistical dependencies of the average magnitude of displacement along the fault u [7-12] and u functional relationship with the seismic moment, the shear modulus μ and the gap area S [13-14];

3) functional relationship corner period $t_s = f_0^{-1}$ s with M_0, the source radius r_0, speed S—wave v_S and static stress drop $\Delta\sigma$ [15,16], as well as the similarity of the angular frequency f_0 with a fundamental frequency of the acoustic Debye [17] f_D, depending on the amount of source and the elastic properties of the geophysical medium [18].

Our further quantitative construction is based on the following empirical relationship Gutenberg and Richter [3,12]:

$$\log_{10} E_{GR} = K_{GR} = 4.8 + 1.5 M_S \qquad (1)$$

$$M_S = 1.59 m_b - 3.97 \qquad (2)$$

$$\log_{10} t_0 = 0.32 M_L - 1.4 \qquad (3)$$

where E_{GR}—seismic energy according to Getenberg and Richter, J; t_0—fluctuations with a maximum duration of vibration speed A/T in the near field (A—amplitude, T—period), s.

Use the following generalization of Soviet seismologists, which were introduced scale energy class K_R [5], the magnitude of surface waves M_{LH} (IC device) and body waves m_{PV} on device SCM [4,9]:

$$\log_{10} E_R = K_R = 4.0 + 1.8 M_{LH} \qquad (4)$$

$$\log_{10} t_m = 0.35 M_{LH} - 1.4 \qquad (5)$$

$$m_b = 5.53 + 0.45(K_R - 14) \qquad (6)$$

$$m_{PV} = 0.35 \log_{10} M_0 - 2.75 \qquad (7)$$

where E_R—seismic energy according to [5], in J; $K_R = \log_{10} E_R$; t_m—increase the maximum duration of the seismic intensity in the near field, in sec.

The basis of the theoretical constructs are the following functional relations [10,13,15,16,19]:

$$M_0 = \mu \cdot S \cdot u = (16/7) \Delta\sigma \cdot r_0^3$$
$$= (16/7)(2.35/2\pi)^3 \Delta\sigma \cdot V_S^3 \cdot t_b \qquad (8)$$

$$E_{SK} = (\Delta\sigma/2\mu) M_0 \qquad (9)$$

$$r_0 = (2.34/2\pi) \cdot v_s \cdot t_b \qquad (10)$$

$$M_W = (2/36) \log_{10} M_0 - 6.07 \qquad (11)$$

where r_0—radius of the source, in м; $\Delta\sigma$—static seismic stress drop, in Pa; t_b—corner period, s; M_W—moment magnitude; (E_{SK}, in J; M_0, in N·m; u in m; v_S in m/s); for the constructions made $t_0 = t_b = t_m$.

Many generalizations proved that for a wide range of changes $\log_{10} M_0$ or M_W empirical correlations magnitude m_b, M_L and M_S from M_0 are non-linear, as in Equation (8), as a function of $M_0 \sim f(t_0^n)$ value of n varies from 3 to 6, and is increase $\Delta\sigma$ [7,12,20-24].

However, for individual intervals M_0 or M_W communication between magnitudes relationships and dependencies of the magnitude $\log_{10} M_0$ can be represented as linear relationships.

2. Justification Relations Magnitude—Seismic Moment

Based on the original definition of magnitude on Richter [25], under which the numerical value of the earthquake magnitude is proportional to the logarithm of the maximum oscillation decimal θ_m, expressed in microns

(10^{-6}м), it is assumed that an upgraded body-wave magnitude m_{bm} (equivalent m_b, m_{PV}) is (considering doubling θ_m on the ground at the focus):

$$m_{bm} = \log_{10} \theta_m + 6.3 = \log_{10} u + 6 \qquad (12)$$

If $S = \pi r_0^2$ in (8) on the basis Equations (9) and (10) and Equation (12) value m_{bm} equal (M_0, N·m; t_θ, s; μ, Pa; v_s, m/s):

$$m_{bm} = C_1 + \log_{10} M_0 - 2\log_{10} t_0 \qquad (13)$$

where $C_1 = \log_{10}\left[2\pi(2.34)^{-2} \cdot \mu^{-1} \cdot v_S^{-2}\right] + 6.3$, value C_1 determines the springiness of the geophysical environment at m_{bm}.

Based on generalizations Christensen [26,27] for the crust taken: average density
$\rho = 2830$ kg/m³, $v_S = 3600$ m/s and
$\mu = \rho \cdot v_s^2 = 36.7$ GPa in what follows, these quantities ρ, v_s and μ taken as the standard.

When these elastic parameters of the geophysical medium expression Equation (13) is transformed to the following form:

$$m_{bm} = \log_{10} M_0 - 2\log_{10} t_0 - 11.30$$
$$= 1/3 \log_{10} M_0 + 2/3 \log_{10} \Delta\sigma - 4.80 \qquad (14)$$

Seismic energy radiation E_{SK} by Kanamori [19], based on Equations (8) and (9) and Equation (13) is:

$$\log_{10} E_{SK} = K_{SK} = C_2 + 2\log_{10} M_0 - 3\log_{10} t_0$$
$$= C_2 - C_1 + 2m_{bm} + \log_{10} t_0 \qquad (15)$$

where $C_2 = \log_{10}\left[7 \cdot \pi_3 \cdot 4^{-1} \cdot (2.34)^{-3} \cdot \mu^{-1} \cdot v_S^{-3}\right]$.
Taken for the elastic parameters and subject [19]. $E_{SK}/M_0 = \Delta\sigma/2\mu = 5\times10^{-5}$ obtain: $\Delta\sigma = 3.67$ MPa and 36.7 bar and the expression Equation (8) can be rewritten in a simple form $\log_{10} t_0 = 1/3 \log_{10} M_0 - 5.43$, then Equation (15) simplifies to:

$$K_{SK} = 2\log_{10} M_0 - 3\log_{10} t_0 - 20.61$$
$$= 1.5 M_W + 4.8 = 3m_{bm} - 3 \qquad (16)$$

On the basis of Equations (13)-(16), reflecting the functional relationship of E_{SK} from M_0, t_0, m_{bm} and μ at $E_{GR} = E_{SK}$ introduced upgraded the magnitude of surface waves M_{Sm} (equivalent of M_S, M_W), while maintaining that the formula Equation (1) Gutenberg and Richter [2,3], with Equation (9), Equations (15) and (16) will be:

$$M_{Sm} = (4/3)\log_{10} M_0 - 2\log_{10} t_0 + C_S$$
$$= 2/3(\log_{10} M_0 + \log_{10} \Delta\sigma) - 10.45 = (2/3)K_{SK} - 3.2 \qquad (17)$$

where $C_S = (2/3)C_2 - 3.2$.

Taken for ρ and v_S C_S value in Equation (17) is equal to $C_S = -16.95$, and for the special case of $\Delta\sigma = 3.67$ MPa = const and $E_{SK}/M_0 = 5 \times 10^{-5}$ equality: $M_{Sm} = M_W$.

We also introduce a modernized local magnitude on

surface waves M_{Lm}—equivalent M_L [18,28], functionally interconnected with $\log_{10}M_0$, $\log t_0$, K_{SK}, m_{bm} and M_{Sm}:

$$M_{Lm} = 0.5\left(m_{bm} + M_{Sm}\right) = \left(7/6\right)\log_{10} M_0 - 2\log t_0 + C_L$$
$$= 0.5\log_{10} M_0 + 2/3\log_{10}\Delta\sigma - 7.62 \qquad (18)$$

where $C_L = 0.5\ (C_1 + C_S)$: for standard values ρ and v_S value C_L is equal: $C_L = -14.12$.

Accepted values for ρ and v_S by Equation (8) and Equation (9) the following relationship:

$$\log_{10}\Delta\sigma = \log_{10} M_0 - 3\log_{10} t_0 - 9.74 \qquad (19)$$

With the standard values ρ, v_S and $\Delta\sigma = 3.67$ MPa, based on Equation (14) and Equation (17) we obtain the following theoretical relation:

$$m_{bm} = 2.60 + 0.5 M_{Sm} \qquad (20)$$

which is within the accuracy of the definitions of the same magnitude satisfactory empirical relation refined body wave magnitude $\hat{m}_b$ of M_W for large earthquakes [19,29] ($m_b \geq 6$):

$$\hat{m}_b = 2.70 + 0.53\ M_W \qquad (21)$$

which were used $\hat{m}_b$ to calculate the true maximum oscillation amplitude A_g, taken from seismograms; $A_g \sim f\left(M_0^{0.35}\right)$.

Here it should be emphasized that at a constant value of $\Delta\sigma$ Equation (12) and Equation (14) the value of the maximum amplitude ε_m is proportional to $M_0^{1/3}$ or $\varepsilon_m \sim f\left(M_0^{0.33}\right)$, that closely coincides with $A_g \sim f\left(M_0^{0.35}\right)$ on [19,29].

In the sequel will be shown $m_{bm} \approx \hat{m}_b$.

Equation (20) agrees satisfactorily with other empirical relationship [9] ($m_{PV} = m_b + 0.18$):

$$m_{PV} = 2.86 + 0.525 M_W \qquad (22)$$

The above quantitative ratios indicate that between modernized magnitudes M_m (m_{bm}, M_{Lm}, M_{Sm}) and $\log_{10}M_0$ may exist linear functional relationship of the form:

$$M_m\left(m_{bm}, M_{Lm}, M_{Sm}\right) = k_i + z_i \log_{10} M_0 \qquad (23)$$

in which the coefficients k_i and z_i at the control parameter a_t and ε_t in the ratio:

$$\log_{10} t_0 = a_t + \varepsilon_t \log_{10} M_0 \qquad (24)$$

where $\Delta\sigma = \text{const} = 3.67$ MPa $\varepsilon_t = 1/3 = \text{const}$ and $a_t = -5.43$, but for other cases ε_t is not a constant.

In view of Equations (23) and (24) correlations Equations (14), (17) and Equation (18) for m_{bm}, M_{Sm} and M_{Lm} (standard values ρ and v_S) can be written as follows:

$$m_{bm} = \left(1 - 2\varepsilon_t\right)\log_{10} M_0 - 2a_t - 11.30 \qquad (25)$$

$$M_{Sm} = \left(4/3 - 2\varepsilon_t\right)\log_{10} M_0 - 2a_t - 16.95 \qquad (26)$$

$$M_{Lm} = \left(7/6 - 2\varepsilon_t\right)\log_{10} M_0 - 2a_t - 14.12 \qquad (27)$$

which provide a self-consistent system of semi empirical inter magnitude dependencies. For example, the dependence of $m_{\varepsilon m}$ from M_{Sm} based on Equations (25) and (26) can be expressed as:

$$m_{bm} = \frac{1 - 2\varepsilon_t}{4/3 - 2\varepsilon_t}\left(M_{Sm} + 2a_t + 16.95\right) - 2a_t - 11.30 \qquad (28)$$

which is $\varepsilon_t = 0.33$ and $a_t = -5.43$ ransformed into simple formula Equation (20).

3. Discussion of Empirical and Theoretical Relations Magnitude—Seismic Moment

Local magnitude—seismic moment. Since the value of the local magnitude is directly related to the maximum oscillation amplitude of the surface waves and the first inter magnitude connections [2,3] have been developed for California earthquakes, relations $M_L - \log t_0 M_0$ consider according to Thatcher and Hanks [30] in this region ($2 \leq M_L \leq 6.8$).

For this region, the authors have taken $\rho = 2700$ kg/m^3 and $v_S = 3200$ m/s, and by (13) and (17) a constant values will be: $C_1 = -11.09$, $C_S = -16.5$, $C_L = -13.72$. With known ρ, v_s, $\Delta\sigma/2\mu = 5 \times 10^{-5}$ between $\log_{10}t_0$ and $\log_{10}M_0$ would expect the following relationship: $\log_{10}t_0 = \left(1/3\right)\log_{10} M_0 - 5.34$, but the instrumental data obtained (**Figure 1**, N—the number of data, r—correlation coefficient):

$$\log_{10} t_0 = 0.25\left(\pm 0.03\right)\log_{10} M_0 - 3.90\left(\pm 0.43\right) \qquad (29)$$

i.e. in accordance with (19) with increasing values of $M_0 \log_{10}\Delta\sigma$ increases:

$\log_{10}\Delta\sigma = 0.25\log_{10} M_0 + 2.12$. Therefore, for the considered data characteristic dependence $M_0 \sim f\left(t_0^4\right)$, said Nuttli [12] for mid-plate earthquakes.

If true theoretical Equations (13), (17) and (18), then

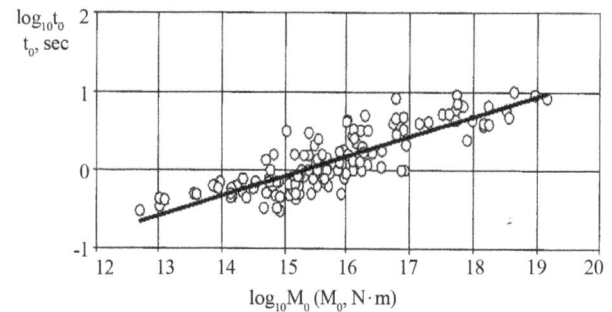

Figure 1. Correlation of $\log_{10}t_0$ from $\log_{10}M_0$ for Southern California earthquakes according to Thatcher and Hanks [30] ($\log_{10}t_0 = 0.25\left(\pm 0.03\right)\log_{10}M_0 - 3.90\left(\pm 0.43\right)$, $N = 138$, $r = 0.84$).

Equation (29) and the relationship between M_{Lm} and $\log_{10}t_0$ is given by:

$$\log_{10} t_0 = 0.37 M_{Lm} - 1.68 \qquad (30)$$

which is in good agreement with the expression (3) Gutenberg and Richter [2] and Equation (5) Soviet seismologists [31] which allows to consider $t_0 = t_s = t_m$.

In Figure 2 shows the correlation $\log_{10}t_0$ and M_L according to Thatcher [30], which also shows the relationship Equation (3) and Equation (30). The presented data show that the semi-empirical formula Equation (30) is in good agreement with generalizations instrumental data (**Figure 2**). It should also be noted that the $M_L = M_{Lm}$ based on Equation (3) Gutenberg and Richter [2], and

Equation (18) can be obtained

$$\log_{10} t_0 = 0.23 \log_{10} M_0 - 3.53 \qquad (31)$$

which is in satisfactory agreement with the expression (29).

In **Figure 3** in the range of $0.5 \le M_L \le 6.8$ shows the correlation ratio M_{Lm} of M_L for Southern California earthquakes [30], South-West Germany [32] and Central Japan [33]. In calculations M_{Lm} by Equation (18) for the earthquakes in these regions were considered elastic parameters of the geophysical medium according to these authors. The statistical data confirm the validity of our assumptions on the possible equality M_L and M_{Lm} (**Figure 3**).

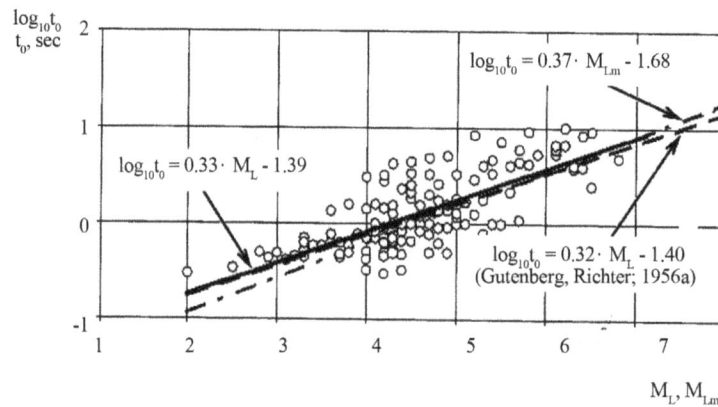

Figure 2. Correlation of $\log_{10}t_0$ from M_L for Southern California earthquakes according to Thatcher and Hanks [30], full line: $\log_{10}t_0 = 0.33(\pm0.05)M_L - 1.39(\pm0.23)$, $N = 138$, $r = 0.84$. Dashed line—$\log_{10}t_0 = 0.32M_L - 1.40$ by Gutenberg and Richter (1956a); dot-dash line—dependence $\log_{10}t_0$ from M_{Lm}, obtained from correlation $\log_{10}t_0$ from $\log_{10}M_0$ (Figure 1).

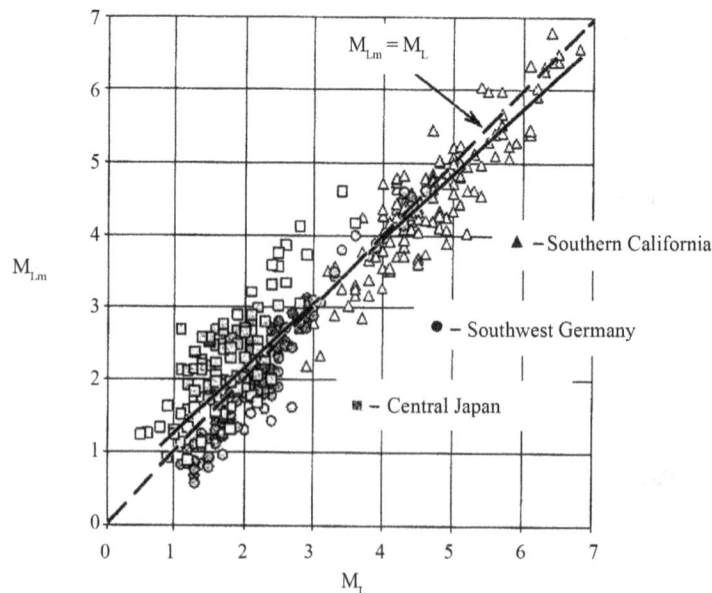

Figure 3. The ratio of calculated M_{Lm} and instrumental M_L for Southern California earthquakes by Thatcher and Hanks [30], South-West Germany (Scherbaum *et al.* 1983) and Cental Japan (Jin *et al.*, 2000). $M_{Lm} = 0.9(\pm0.03)M_L - 0.28(\pm0.05)$, $N = 384$, $r = 0.94$; dashed line $M_{Lm} = M_L$.

From numerous publications on nonlinear relations $\log_{10} M_0 - M_L$ acceptability of new assumptions considered on the basis of Hasegawa [34] for earthquakes in Eastern Canada. In the range $0 < M_L \leq 6.3$ are two of the interval $0 < M_L \leq 3.9$ and $3.9 \leq M_L \leq 6.3$, which have different dependencies on $\log_{10} t_0$ of M_L and $\log_{10} M_0$ from M_L [34].

For the first group of small earthquakes characterized by the following relationship ($10^5 < \Delta\sigma < 10^6$ Pa): $\log_{10} t_0 = 0.18 \log_{10} M_0 - 3.14$, but for another group ($10^6 \leq \Delta\sigma < 5 \times 10^6$ Pa): $\log_{10} t_0 = 0.28 \log_{10} M_0 - 4.54$.

On the basis of these empirical formulas for Equation (18) and Equation (24) with $C_L = -14.21$ ($\rho = 2800$ kg/m^3 and $v_s = 3800$ m/s) **Figures 4** and **5** shows the calculated dependences of $\log_{10} t_0$ from M_{Lm} and $\log_{10} M_0$ from M_{Lm}, which in satisfactory agreement with the relations $\log_{10} t_0 - M_L$ and $\log_{10} M_0 - M_L$ (**Figures 4** and **5**) by Hasegawa [34].

Finally, for the Southern California Earthquake Equation (18) and Equation (29) we can obtain the following relationship: $M_{Lm} = 0.67 \log_{10} M_0 - 5.92$, which coincides with the ratio of [30]:

$$M_L = 0.67 \log_{10} M_0 - 6.0 \qquad (32)$$

According to Equations (23) and (24) and Equation (27) if $\sigma_t = 0.25$ we get $z_i = 7/6 - 2\sigma_t = 0.67$, which indicates the acceptability of the proposed relations.

From Equation (32) it follows that $b_t = 0.25$ in Equation (24) the values of M_L and M_{Lm} magnitude M_W corresponds to Equation (11). Probably, the presence of the form Equation (29) between $\log_{10} t_0$ and $\log_{10} M_0$ explains equality $M_L = M_W$ for earthquakes with $M_W \leq 7.0$ North-West Europe [35], New Zealand [36], western Canada [37] and about Taiwan [38].

3.1. Ratio $m_b - \log_{10} M_0$: Design and Data Tools

As in the case of search based $M_L - \log_{10} M_0$, for body-wave magnitude m_b consider empirical relationships

According to Zapolsky [31], Gutenberg [1], specifically examining the relationship between the energy of focal radiation and earthquake magnitude according to

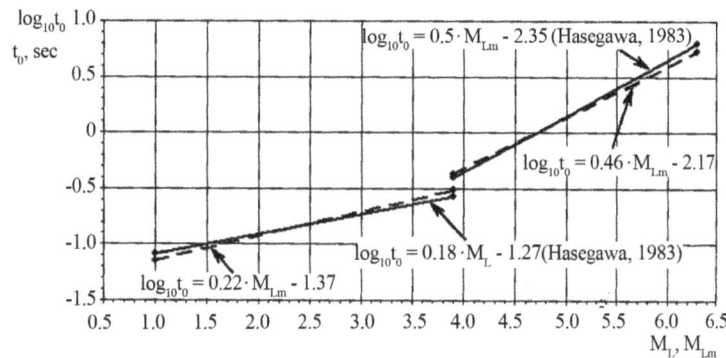

Figure 4. Correlation $\log_{10} t_0$ from M_L (full line—Hasegawa [34] and from calculated M_{Lm} (dashed line, see the text) for East Canada earthquakes.

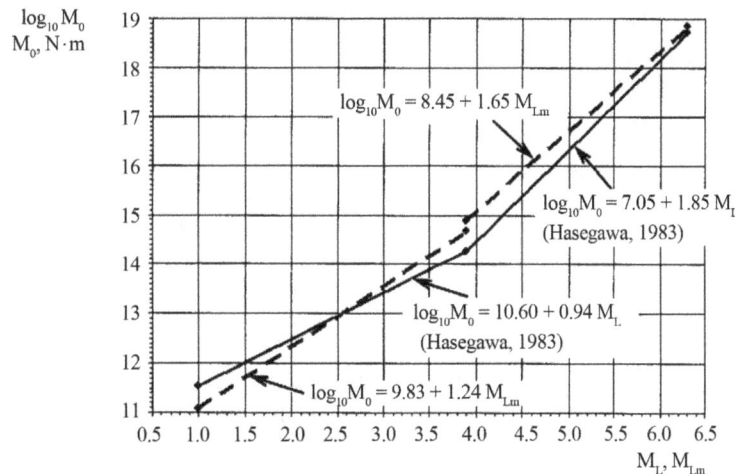

Figure 5. Correlation $\log_{10} M_0$ from M_L (full line—Hasegawa [34] and from calculated M_{Lm} (dashed line, see the text) for East Canada earthquakes.

the observations in the epicentral area, showed that the duration t_0, determine the energy of the oscillations with the maximum intensity depends strongly on the magnitude and 2.5-fold increases with increasing magnitude of m_b on unit [31].

$$\log_{10} t_0 = 0.4 m_b - 1.9 \tag{33}$$

A little-known empirical formula Equation (32) Gutenberg [1] is a key for further generalizations of our constructions on relations $m_b - \log_{10} M_0$ and $m_b - M_S$.

On the basis of (13) and (29) with $C_1 = -11.09$, we can get:

$$\log_{10} t_0 = 0.5 m_b - 2.26 \tag{34}$$

Substitution $\log_{10} t_0 = 0.23 \log_{10} M_0 - 3.53$ in Equation (31) into (13) leads to the following formula:

$$\log_{10} t_0 = 0.42 m_b - 1.84 \tag{35}$$

which is in good agreement with (33) provided $m_b = m_{bm}$.

Graphic expressions Equations (33)-(35) are shown in **Figure 6**, from which it can be assumed about the close convergence of these relations and the possible equality $m_b = m_{bm}$ (**Figure 6**). At equality $m_b = m_b m$—based Equations (13) and (33) for the standard ρ and v_S can obtain the expression:

$$\log_{10} t_0 = 0.22 \log_{10} M_0 - 3.57 \tag{36}$$

which is in good agreement with Equations (29) and (31), which may indicate the consistency of our constructions relating m_b, m_{bm}, M_L, M_{Lm} and $\log_{10} t_0$ with $\log_{10} M_0$ for earthquakes in California, despite the fact that the conclusions are based on statistical formulas in which the correlation coefficients are not equal to unity ($r = 0.75 - 0.90$)

If we use the Equation (36), on the basis of Equation (24) with $\varepsilon_t = 0.22$ and Equations (25) and (26) for the

standard values ρ and v_S, M_{Sm} dependence on m_{bm} can be expressed as:

$$M_{Sm} = 1.59 \; m_{bm} - 3.20 \tag{37}$$

which almost corresponds to the classical formula Equation (2) Gutenberg and Richter (1956в) and for which the equality $M_{Sm} = m_{bm}$ complied with $M_{sm} = 5.40$, which coincides closely with generalizations Chen [7], Gusev [9], Nuttli [12] and Utsu [24].

In **Figure 7** shows the correlation of $\log_{10} t_0$ from $\log_{10} M_0$ for earthquakes in the world (1981-1991) by the Catalogue Choy [39], for which the value of t_0 was taken from the Global CMT Catalogue. The ratio of $\log_{10} t_0$ from $\log_{10} M_0$ for these data is given by (**Figure 7**):

$$\log_0 t_0 = 0.30(\pm 0.01) \log_{10} M_0 - 4.88(\pm 0.01) \tag{38}$$

for which the range $17 \leq \log_{10} M_0 \leq 21$ value of $\log_{10} \Delta \sigma$ by Equation (19) increases from 6.60 to 7.10.

Substituting (38) in (13) leads to ($C_1 = -11.30$):

$$m_{bm} = 0.40 \log_{10} M_0 - 1.54 \tag{39}$$

which agrees closely with the empirical formula:

$$m_b = 0.22(\pm 0.02) \log_{10} M_0 + 1.85(\pm 0.02) \tag{40}$$

shown on **Figure 8**.

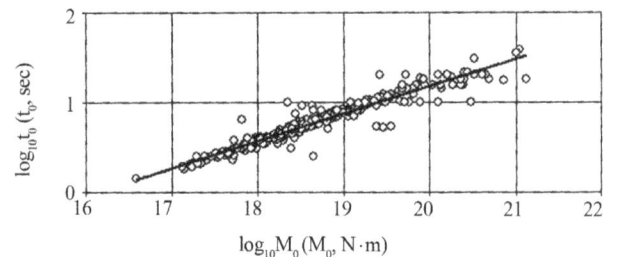

Figure 7. Correlation dependence $\log_{10} t_0$ from $\log_{10} M_0$ for major earthquakes of the world (1981-1991) by Choy's Catalogue [39]. $\log_{10} t_0 = 0.30(\pm 0.01) \log_{10} M_0 - 4.88(\pm 0.01)$, $N = 379$, $r = 0.96$.

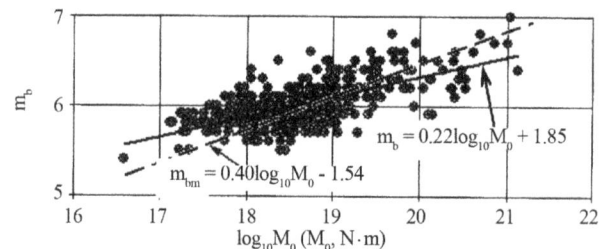

Figure 8. Correlation dependence m_b from $\log_{10} M_0$ (full line) for major earthquakes of the world (1981-1991) by Choy's Catalogue *et al.* (1995) for 1981-1991. $m_b = 0.22(\pm 0.02) \log_{10} M_0 - 1.85(\pm 0.02)$, $N = 362$, $r = 0.67$, dashed line—calculated dependence: $m_{bm} = 0.40 \log_{10} M_0 - 1.54$.

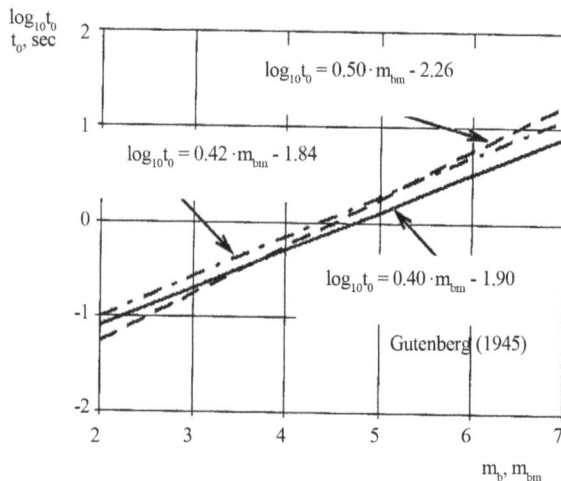

Figure 6. Correlation $\log_{10} t_0$ and m_b (full line), $\log_{10} t_0$ and m_{bm} (dashed line, see the text).

Equation (39) is in good agreement with the dependence on m_b from $\log_{10}M_0$ for Sumatra island earthquake ($\varphi = -10° + 10°$, $\lambda = +90° + 100°$) for 1993-2012 (**Figure 9**).

Table 1 shows a comparison of the magnitude $\hat{m}_b$ obtained by the true maximum amplitude [19,29,40] and the calculated value m_{bm} (**Table 1**) for a number of large earthquakes in 1960-1984. The presented data suggest that for most of the earthquakes characterized by the following inequality: $\hat{m}_b \geq m_{bm} > m_b$.

When $\log_{10}\Delta\sigma \geq 7.1$ value of $\hat{m}_b$ is close to the m_{bm} same as for Great Chilean earthquake $\hat{m}_b = 7.57$ and $m_{bm} = 7.71$, for Tangshan (1976) $\hat{m}_b = 6.9$ and $m_{bm} = 6.92$, Yanyuan (1976). $\hat{m}_b = 6.5$, $m_{bm} = 6.18$, and if $6.36 \leq \log_{10}\Delta\sigma < 7.0$ value of $\hat{m}_b$ more then m_{bm} (**Table 1**).

Table 2 presents a comparison of calculated m_{bm} and $\hat{m}_b$ (21) for 80 major earthquakes of the world for 2000-2012 for calculations m_{bm}, $\hat{m}_b$ and M_{Sm} used data from Global CMT Catalogue (**Table 2**). When comparing $\log_{10}\Delta\sigma$ from **Table 1** to **Table 2** shows that with increasing $\log_{10}M_0$ from 19.15 to 22.72 for the 2000-2012 earthquakes $\log_{10}\Delta\sigma$ value ranges from 6.75 - 7.58 with an average of 7.16, that is, much higher than for earthquakes 1960-1984 (**Tables 1** and **2**) and higher than the standard $\log\Delta\sigma = 6.56$.

For such high values $\Delta\sigma$ values m_{bm} closely coincide with the design $\hat{m}_b$, and for values M_{Sm} characterized by inequality: $M_{Sm} > M_W$ (**Table 2**) confirmed that conclusion is the relation $m_{bm} - \hat{m}_b$—for earthquakes in Japan and the Kuril Islands ($\varphi = 30° + 40°$, $\lambda = 140° + 150°$) for the 1993-2012 shown in **Figure 10**.

Thus for large earthquakes 1960-1984 and 1993-2012 at $\log\Delta\sigma > 7.1$ m_{bm} values coincide closely with the magnitude $\hat{m}_b$ calculated from the true maximum amplitude (A_g) of seismic vibrations, the magnitude of which is proportional to the seismic moment: $A_g \sim f\left(M_0^{0.35}\right)$ to Houston [29] and Kanamori [19]. Consequently, the m_{bm}

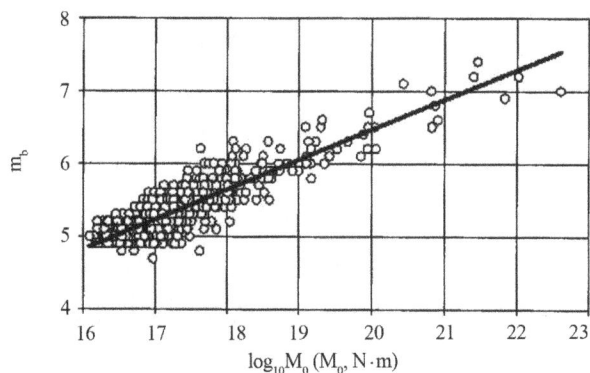

Figure 9. Correlation dependence m_b from $\log_{10}M_0$ for the earthquakes Sumatra region (1993-2012 years).
$m_b = 0.41(\pm0.02)\log_{10}M_0 - 1.76(\pm0.02)$, $N = 631$, $r = 0.88$.

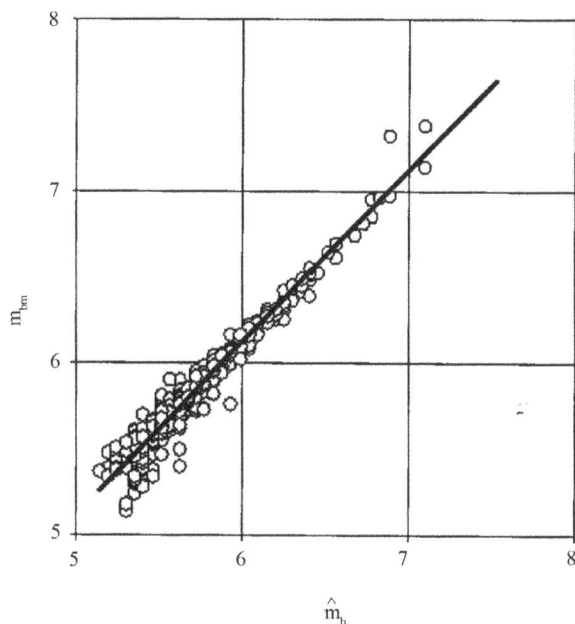

Figure 10. Correlation calculated magnitudes m_{bm} and $\hat{m}_b$ for the earthquakes in Japan and Kuril Islands for 1992-2012 years. $m_{bm} = 1.0(\pm0.02)\hat{m}_b + 0.14(\pm0.003)$, $N = 521$, $r = 0.97$. Values of $\hat{m}_b$ were calculated according to the formula (21).

value is proportional to the $\log_{10}A_g$.

The ratio of $M_S - \log_{10}M_0$. In Mamyrov's papers [18], [28] have shown that in the range of $16 \leq \log_{10}M_0 < 21.0$ if $\log_{10}\Delta\sigma \leq 7.0$ at the rated M_{Sm} closely coincides with M_S and M_W, and for high $\Delta\sigma \geq 10^7$ Pa following inequality $M_{Sm} > M_S$, as shown in **Table 2**.

In **Figure 11** shows the correlation of M_S from $\log_{10}M_0$ for earthquakes of the world for 1981-1991 according to the Catalog Chou *et al.* [39]:

$$M_S = 0.73(\pm0.03)\log_{10}M_0 - 7.47(\pm0.02) \quad (41)$$

which is in satisfactory agreement with the dependence $M_{Sm} = 0.73\log_{10}M_0 - 7.19$ (**Figure 11**, dashed line), derived from Equations (38) and (26). These relations with $M_S = M_{Sm}$ with $\log_{10}M_0$ are in good agreement with the generalization of Perez [41] for crustal earthquakes of the world for the years 1950-1997:
$\log_{10}M_0 = 1.33M_S + 10.22$.

In **Figure 12** shows the correlation M_S with $\log_{10}M_0$ (solid line) for the earthquakes in Japan and the Kuril Islands in 1993-2012:
$M_S = 0.77(\pm0.02)\log_{10}M_0 - 8.23(\pm0.02)$, here, we show the same relationship

M_{Sm} from $\log_{10}M_0$ (**Figure 12**, dashed line):
$M_{Sm} = 0.69\log_{10}M_0 - 6.09$, obtained with ($N = 521$, $r = 0.99$):

Table 1. A comparison of the magnitude $\hat{m}_b$ (Houston, Kanamori, 1986; Zhuo, Kanamori, 1987) and settlement m_{bm} for several major earthquakes of the world.

№№	Date	Time	φ	λ	Depth h, km	$\log_{10}M_0$, N·m	$\log_{10}t_0$ t_0, sec	m_b	$\hat{m}_b$	m_{bm}	M_S	M_W	M_{sm}	$\log_{10}\Delta\sigma$, $\Delta\sigma$, Pa	Region
	1	2	3	4	5	6	7	8	9	10	11	12	13	14	15
1	1960/5/22	19:11:17.5	−38.29	−73.05	35	23.35	2.17		7.57	7.71	8.5	9.6	9.76	7.10	Great Chilean
2	1963/10/13	5:17:55.1	44.76	149.80	26	21.85	1.88		7.23	6.79	8.1	8.5	8.35	6.47	Great Alaska
3	1964/3/28	3:36:12.7	61.02	−147.63	6	22.96	2.15		7.64	7.36	8.4	9.2	9.29	6.77	Rat Island
4	1965/2/4	5:01:21.7	51.21	178.50	29	22.15	1.94		7.19	6.97	8.2	8.7	8.63	6.59	Kurile Isl.
5	1967/7/22	16:56:55.3	40.63	30.74	4	20.20	1.32		6.38	6.26	7.1	7.4	7.28	6.50	Tyrkey
6	1968/5/16	0:49:0.4	40.90	143.35	26	21.45	1.71		7.18	6.73	8.1	8.3	8.16	6.58	Tokachi-oki
7	1969/8/11	21:27:37.6	43.48	147.82	46	21.34	1.71		6.90	6.62	7.8	8.2	8.01	6.47	Kurile Isl.
8	1971/2/9	14:00:41.0				19.08	0.85	6.2	6.41	6.08	6.7	6.6	6.73	6.79	San Fernando
9	1974/10/3	14:21:34.5	−12.25	−77.52	36	21.18	1.65		7.0	6.68	7.6	8.1	7.92	6.49	Haicheng, China
10	1975/2/4	11:36:7.1	40.67	122.65	16	19.61	1.05		6.76	6.21		7.0	7.03	6.72	Peru
11	1976/2/4	9:01:7.2	15.14	−89.78	16.3	20.31	1.14	6.2	6.66	6.73	7.5	7.5	7.78	7.15	Gua temala
12	1976/5/29	12:23:29.9	24.39	98.65	15	19.09	0.72	6.1	6.5	6.35	6.9	6.7	7.00	7.19	Longlin, China
13	1976/5/29	14:00:33.2	24.29	98.58	15	19.05	0.73	6.0	6.5	6.29	7.0	6.6	6.93	7.12	Longlin, China
14	1976/7/27	19:42:11.1	39.52	118.03	15	20.44	1.11	6.3	6.9	6.92	7.9	7.6	8.01	7.37	Tangshan, China
15	1976/7/28	10:45:45.9	39.75	118.78	15	19.55	0.88	6.3	6.7	6.49	7.4	7.0	7.29	7.47	Tangshan, China
16	1976/8/16	14:6:55.0	32.63	104.42	15	19.11	0.73	6.1	6.9	6.35	6.9	6.7	7.01	7.18	Songpan, China
17	1976/8/16	16:11:38.7	7.07	123.75	33	21.04	1.34	6.4	7.26	7.06	7.9	8.0	8.35	7.28	Mindanao
18	1976/8/21	21:49:57.8	32.37	104.29	15.3	18.50	0.52	6.1	6.7	6.16	6.4	6.3	6.61	7.20	Songpan China
19	1976/8/23	3:30:11.5	32.11	104.21	19.6	18.66	0.58	6.2	6.6	6.20	6.7	6.4	6.71	7.18	Songpan China
20	1976/11/6	18:4:16.0	27.50	101.40	22.7	18.56	0.54	5.8	6.5	6.18	6.5	6.3	6.65	7.20	Yanyuan, China
21	1976/11/15	13:53:7.2	39.45	117.71	15	18.63	0.56	6.0	6.3	6.21	6.3	6.4	6.71	7.21	Tangshan, China
22	1976/11/24	12:22:25.3	38.88	43.96	15	19.62	0.90	6.1	6.58	6.52	7.3	7.0	7.34	7.16	Sumbawa
23	1977/8/19	6:9:33.1	−11.14	118.23	23.3	21.55	1.48	7.0	7.47	7.29	7.9	8.3	8.75	7.37	Iran
24	1978/9/16	15:36:13.5	33.37	57.02	11	20.12	1.34	6.5	6.9	6.14	7.4	7.3	7.13	6.36	Oaxaca
25	1978/11/29	19:53:2.9	16.22	−96.56	16.1	20.72	1.36	6.4	6.87	6.70	7.7	7.7	7.89	6.90	Tyrkey
26	1979/3/14	11:7:31.1	17.78	−101.37	26.7	20.23	1.27	6.5	6.71	6.39	7.6	7.4	7.41	6.69	Petatlan
27	1979/10/15	23:17:0.8	32.62	−115.57	12	18.86	0.78	5.7	5.92	6.00	6.9	6.5	6.57	6.78	Imperial Yalley
28	1979/12/12	8:00:7.0	2.32	−78.81	19.7	21.23	1.35	6.4	6.91	7.23	7.7	8.1	8.58	7.44	Colymbia
29	1980/1/1	16:42:49.8	38.80	−27.74	10	19.45	1.00	6.0	6.3	6.15	6.7	6.9	6.92	6.71	Azores Isl.
30	1980/2/7	10:49:26.3	−54.29	158.43	15	19.36	1.01	6.1	6.2	6.04	6.5	6.8	6.79	6.59	Macguarie Isl.
31	1980/7/17	19:43:3.1	−12.44	165.94	34	20.68	1.24	5.8	6.79	6.90	7.9	7.8	8.07	7.22	Eureka
32	1980/10/10	12625:25.5	36.14	1.41	12	19.70	1.00	6.5	6.5	6.40	7.3	7.1	7.32	6.36	Santa Grus Isl.
33	1980/11/8	10:27:45.9	41.14	−124.36	15	20.05	1.00	6.2	6.7	6.75	7.2	7.3	7.72	7.31	El Asnam, Algeria

Continued

34	1981/1/23	21:13:55.6	30.86	101.35	10	18.86	0.78	5.7	6.5	6.00	6.8	6.5	6.57	6.78	Daofu, China
35	1981/2/24	20:53:49.2	38.07	23.04	10	18.95	0.78	5.9	6.6	6.09	6.7	6.6	6.69	6.97	Greece
36	1981/4/24	21:50:14.3	−13.51	166.43	44.4	19.55	0.90	6.1	6.2	6.45	6.9	6.8	7.25	7.11	Yanuati Isl.
37	1981/4/27	18:17:40.0	−57.58	147.86	10	18.91	0.78	5.7	6.1	6.05	6.5	6.5	6.64	6.83	Macguarie Isl.
38	1981/7/28	17:22:43.6	30.01	57.8	15.2	19.95	1.15	5.7	6.9	6.35	7.1	7.2	7.28	6.76	Kurile Isl.
39	1991/9/3	5:35:50.1	42.97	147.87	35.7	18.88	0.70	6.6	6.5	6.18	6.6	6.5	6.76	7.04	Papua, New Guinea
40	1981/11/6	16:47:51	−3.18	143.72	15	18.96	0.84	6.2	6.2	5.98	6.9	6.6	6.59	6.70	Loayltu Isl.
41	1981/11/24	23:30:41.9	−22.19	170.32	23.3	19.13	0.87	5.7	6.2	6.09	6.7	6.7	6.75	6.79	Kermades Isl.
42	1981/12/24	5:33:33.3	−29.81	−177.55	19.4	19.32	1.08	6.1	6.4	5.86	6.8	6.8	6.58	6.34	Iran
43	1982/8/5	20:33:2.0	−12.52	166.01	23.9	19.50	0.98	6.2	6.3	6.24	7.1	6.9	7.02	6.82	Santa Grus Isl.
44	1982/12/19	17:44:21.8	−24.31	175.0	29.2	20.30	1.33	6.0	6.4	6.34	7.7	7.5	7.39	6.57	Tonga Isl.
45	1983/4/3	2:50:26.4	8.85	−83.25	28	20.26	1.27	6.5	6.6	6.42	7.2	7.4	7.52	6.71	Panama
46	1983/5/2	23:43:44.7	36.42	−120.66	14.5	18.60	0.79	6.2	6.0	5.72	6.5	6.3	6.25	6.49	Akito Oki, Japan
47	1983/5/26	3:0:18.3	40.44	138.87	12.6	20.66	1.30	6.8	7.2	6.76	7.7	7.7	7.93	7.02	Coolinga
48	1983/10/4	18:52:37.8	−26.01	−70.56	38.7	20.53	1.46	6.4	6.8	6.31	7.3	7.6	7.50	6.41	Chili
49	1983/10/28	14:6:22.5	44.35	−113.98	13.7	19.49	1.00	6.2	6.6	6.19	7.3	6.9	6.97	6.75	Idaho
50	1983/11/16	16:13:5.9	19.40	−155.59	11	19.03	0.78	6.3	6.7	6.17	6.6	6.6	6.90	6.95	Hawaii
51	1983/11/30	17:46:28.9	−6.35	71.75	10	20.61	1.23	6.6	7.1	6.85	7.5	7.7	8.00	7.18	Chugos Arch
52	1984/2/7	21:33:36.1	−9.81	160.42	21.9	20.40	1.33	6.5	6.7	6.44	7.5	7.5	7.52	6.67	Solomon Isl.
53	1984/3/19	20:28:39	40.38	63.37	15	19.55	0.78	6.5	6.7	6.69	7.0	7.0	7.49	7.47	Uzbekistan

Notice: for the earthquakes N1—10 $\log_{10}t_0$ calculated according to $\Delta\sigma$ Kasachara (1984), Purcaru and Berkhemer (1982); for the rest earthquakes N11-52 all data have been taken according to Global CMT Cataloge, for the earthquakes N53—data have been taken from USSR's catalogue, 1984.

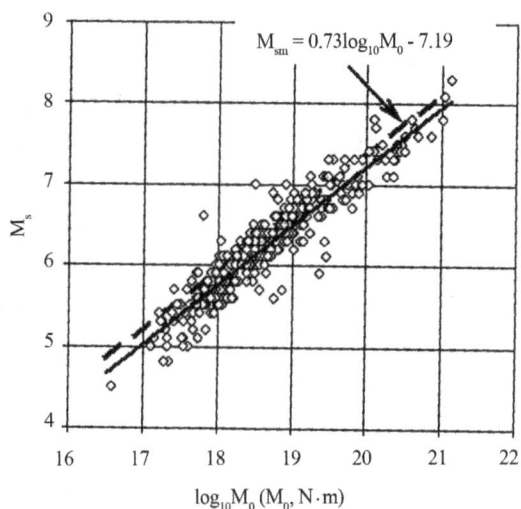

Figure 11. Correlation of magnitudes M_S and $\log_{10}M_0$ (full line) by Catalogue of major earthquakes of the world Choy [39]: $M_S = 0.73(\pm0.03)\log_{10}M_0 - 7.47(\pm0.02)$, $N = 372$, $r = 0.93$. Dashed line—calculated dependence. $M_S = 0.73\log_{10}M_0 - 7.19$.

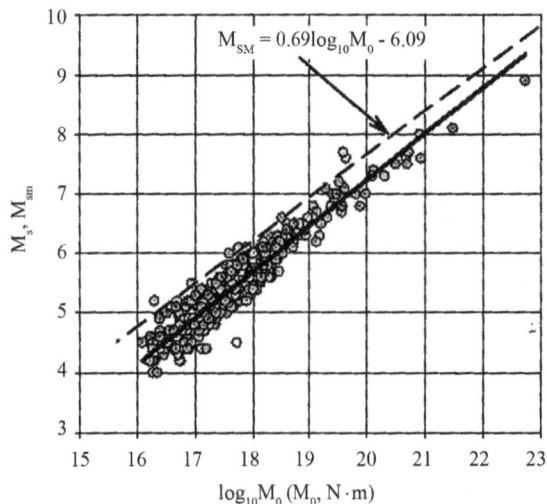

Figure 12. Correlation dependence of magnitude M_S from $\log_{10}M_0$ for the earthquakes in Japan and Kuril Islands for 1993-2012 years. $M_S = 0.77(\pm0.03)\log_{10}M_0 - 8.23(\pm0.02)$, $N = 514$, $r = 0.95$. Dashed line—calculated dependence. $M_{Sm} = 0.69\log_{10}M_0 - 6.09$.

Table 2. Comparison of calculated (m_{bm}, M_{Sm}) and instrumental (m_b, M_S, M_W) for a number of magnitude large earthquakes of the world 2000-2012 years.

№№	Date	Time	φ	λ	Depth H, км	$\log_{10}M_0$ M_0, Нм	$\log_{10}t_0$ t_0, с	m_b	$\hat{m}_b$	m_{bm}	M_S	M_W	M_{sm}	$\log_{10}\Delta\sigma,$ $\Delta\sigma,$ Па	Region
	1	2	3	4	5	6	7	8	9	10	11	12	13	14	15
1	2000/1/8	16:47:30.2	−16.84	−173.81	162.4	19.84	0.99	6.5	6.52	6.56	6.6	7.2	7.46	7.13	Tonga Isl.
2	2000/2/25	1:44:5.2	−19.55	174.17	16.8	19.70	0.95	6.1	6.46	6.50	7.1	7.1	7.35	7.11	Vanuatu Isl.
3	2000/5/4	4:21:33.4	−1.29	123.59	18.6	20.39	1.18	6.7	6.67	6.73	7.5	7.5	7.81	7.11	Sulawesi
4	2000/6/4	16:28:46.5	−4.73	101.94	43.9	20.87	1.46	6.8	6.83	6.65	8.0	7.8	7.89	6.75	Sumatra
5	2000/6/18	14:44:27.6	−13.47	97.17	15.0	20.90	1.41	6.8	6.89	6.78	7.8	7.9	8.03	6.93	Indian Ocean
6	2000/11/16	4:55:36.5	−4.56	152.79	24.0	21.09	1.34	6.0	6.94	7.11	8.2	8.0	8.42	7.33	New Ireland
7	2000/11/16	7:42:44.5	−5.03	153.17	31.2	20.81	1.36	6.2	6.83	6.79	7.8	7.8	8.01	6.99	New Ireland
8	2000/11/17	21:2:20.1	−5.26	152.34	17.0	20.75	1.33	6.2	6.83	6.79	8.0	7.8	7.99	7.02	New Britain
9	2000/12/6	17:11:14.7	39.60	54.87	33.0	19.59	0.94	6.7	6.41	6.41	7.5	7.0	7.22	7.03	Turkmenia
10	2001/1/1	6:57:24.0	6.73	127.07	44.0	20.24	1.08	6.4	6.62	6.78	7.2	7.4	7.81	7.26	Mindanao
11	2001/1/26	3:16:54.9	23.63	70.24	19.8	20.53	1.38	6.9	6.73	6.47	8.0	7.6	7.59	6.65	India
12	2001/6/23	20:34:23.3	−17.28	−72.71	29.6	21.67	1.63	6.7	7.15	7.11	8.2	8.4	8.61	7.04	Peru
13	2001/8/21	6:52:14.3	−36.70	179.08	59.0	19.71	1.01	6.4	6.41	6.39	7.1	7.0	7.24	6.94	New Zealand
14	2001/10/12	15:2:23.3	12.88	145.08	42.0	19.57	0.92	6.7	6.41	6.43	7.3	7.0	7.24	7.07	Mariana Isl.
15	2002/9/8	18:44:38.3	−3.27	143.38	19.5	20.47	1.26	6.5	6.73	6.65	7.8	7.6	7.75	6.95	Papua
16	2002/10/10	10:50:41.9	−1.79	134.30	15.0	20.41	1.18	6.5	6.67	6.75	7.7	7.5	7.83	7.13	Java Isl.
17	2002/11/02	1:26:25.9	2.65	95.99	23.0	19.95	1.09	6.2	6.52	6.47	7.6	7.2	7.40	6.94	Sumatra
18	2002/11/03	22:13:28.0	63.23	−144.89	15.0	20.87	1.37	7.0	6.83	6.83	8.5	7.8	8.07	7.02	Alaska
19	2003/03/17	16:36:26.6	51.33	177.58	27.0	19.62	0.89	5.9	6.41	6.54	6.7	7.0	7.36	7.21	Aleutian Isl.
20	2003/08/21	12:12:59.5	−45.01	166.87	31.8	19.87	0.98	6.6	6.52	6.61	7.5	7.2	7.52	7.19	New Zealand
21	2003/09/25	19:50:38.2	42.21	143.84	28.2	21.48	1.52	6.9	7.10	7.14	8.1	8.3	8.58	7.18	Hokkaido
22	2003/09/27	11:33:36.2	50.02	87.86	15.0	19.97	1.01	6.5	6.52	6.65	7.5	7.2	7.59	7.20	Siberia, Russia
23	2004/12/23	14:59:30.9	−49.91	161.25	27.5	21.21	1.43	6.5	7.00	7.05	7.7	8.1	8.40	7.58	Macquarie Isl.
24	2004/12/26	1:1:9.0	3.09	94.26	28.6	22.60	1.98	7.0	7.47	7.34	8.9	9.0	9.15	6.92	Sumatra
25	2005/03/28	16:10:31.5	1.67	97.07	25.8	22.02	1.69	7.2	7.26	7.34	8.4	8.6	8.96	7.21	Sumatra
26	2005/10/08	3:50:51.5	34.38	73.47	12.0	20.47	1.18	6.9	6.73	6.81	7.7	7.6	7.91	7.19	Pakistan
27	2005/11/14	21:38:59.3	38.22	144.97	18.0	19.57	0.86	6.7	6.41	6.55	6.8	7.0	7.36	7.25	Honshu
28	2006/02/22	22:19:15.0	−21.20	33.33	12.0	19.62	0.90	6.5	6.41	6.52	7.5	7.0	7.34	7.18	Mozambique
29	2006/04/20	23:25:17.6	60.89	167.05	12.0	20.48	1.18	6.8	6.73	6.82	7.6	7.6	7.93	7.20	Siberia, Russia
30	2006/05/03	15:27:3.7	−20.39	−173.47	67.8	21.05	1.37	7.2	6.94	7.01	7.9	8.0	8.31	7.20	Tonga Isl.
31	2006/11/15	11:15:8.0	46.71	154.33	13.5	21.54	1.54	6.6	7.10	7.16	8.3	8.3	8.62	7.18	Kuril Isl.
32	2007/01/13	4:23:48.1	46.17	154.80	12.0	21.25	1.44	7.3	7.00	7.07	8.2	8.1	8.43	7.19	Kuril Isl.
33	2007/04/01	20:40:38.9	−7.79	156.34	14.1	21.20	1.42	6.8	7.00	7.06	8.1	8.1	8.41	7.20	Solomon Isl.
34	2007/01/21	11:28:1.0	1.10	126.21	22.2	20.30	1.12	6.7	6.67	6.77	7.5	7.5	7.81	7.20	Molucca Sea
35	2007/08/15	23:41:57.9	−13.73	−77.04	33.8	21.05	1.37	6.7	6.94	7.01	8.0	8.0	8.31	7.20	Peru
36	2007/09/12	11:11:15.6	−3.78	100.99	24.4	21.83	1.63	6.9	7.20	7.27	8.5	8.5	8.82	7.20	Sumatra
37	2007/09/12	23:49:35.3	−2.46	100.13	43.1	20.91	1.32	6.6	6.89	6.97	8.1	7.9	8.22	7.21	Sumatra
38	2007/11/14	15:41:11.2	−22.64	−70.62	37.6	20.68	1.25	6.7	6.78	6.88	7.7	7.7	8.05	7.19	Chile

Continued

39	2008/10/05	15:53:1.1	39.50	73.64	12.0	19.15	0.73	6.4	6.25	6.59	6.9	6.7	7.06	7.22	Kyrgyzstan
40	2008/11/16	17:2:43.8	1.50	122.05	29.2	20.12	1.06	6.5	6.57	6.70	7.3	7.3	7.69	7.20	Sulavesi
41	2009/01/03	22:33:44.9	−0.58	133.48	18.2	20.15	1.07	6.7	6.62	6.71	7.4	7.4	7.71	7.20	Java Isl.
42	2009/01/15	17:49:48.3	46.97	155.39	45.2	20.18	1.08	6.9	6.62	6.72	7.5	7.4	7.73	7.20	Kuril Isl.
43	2009/05/28	8:25:4.8	16.50	−87.17	12.0	20.11	1.05	6.7	6.57	6.71	7.3	7.3	7.70	7.22	Honduras
44	2009/07/15	9:22:49.6	−45.85	166.26	23.5	20.76	1.28	6.5	6.83	6.90	7.8	7.8	8.10	7.18	New Zealand
45	2009/08/10	19:56:5.0	14.16	92.94	22.0	20.29	1.12	6.9	6.67	6.75	7.6	7.5	7.80	7.19	Andaman Isl.
46	2009/09/02	7:55:7.5	−8.12	107.33	53.2	19.56	0.87	6.8	6.41	6.52	7.0	7.0	7.32	7.21	Java Isl.
47	2009/09/29	17:48:26.8	−15.13	−171.97	12.0	21.22	1.43	7.1	6.99	7.06	8.1	8.1	8.41	7.19	Samoa Isl.
48	2009/09/30	10:16:17.4	−0.79	99.67	77.8	20.44	1.17	7.1	6.73	6.8	7.5	7.6	7.89	7.19	Sumatra
49	2009/10/07	22:3:28.9	−12.59	166.27	44.2	20.51	1.19	6.4	6.73	6.83	7.7	7.6	7.95	7.20	Vanuatu Isl.
50	2009/10/07	22:19:15.3	−11.86	166.01	41.7	20.82	1.30	6.4	6.83	6.92	7.9	7.8	8.14	7.18	Santa Crus Isl.
51	2009/10/07	23:14:0.6	−13.12	166.37	42.5	20.22	1.09	6.4	6.62	6.74	7.4	7.4	7.76	7.21	Vanuatu Isl.
52	2010/01/03	22:36:42.4	−8.88	157.21	12.0	19.76	0.94	6.4	6.46	6.58	7.1	7.1	7.45	7.20	Solomon Isl.
53	2010/02/27	6:35:14.5	−35.98	−73.15	23.2	22.27	1.78	7.2	7.36	7.41	8.3	8.8	9.11	7.19	Chile
54	2010/04/06	22:15:19.1	2.07	96.74	17.6	20.82	1.29	7.0	6.83	6.94	7.9	7.8	8.16	7.21	Sumatra
55	2010/08/10	5:23:53.9	−17.57	167.81	31.9	20.00	1.02	6.4	6.57	6.66	7.3	7.3	7.61	7.20	Vanuatu Isl.
56	2010/10/25	14:42:59.8	−3.71	99.32	12.0	20.83	1.30	6.5	6.83	6.93	7.8	7.8	8.15	7.19	Sumatra
57	2010/12/21	17:19:53.6	27.10	143.76	15.6	20.24	1.10	7.0	6.62	6.74	7.5	7.4	7.77	7.20	Bonin Isl.
58	2010/12/25	13:16:51.4	−19.67	168.04	16.6	20.05	1.04	6.8	6.57	6.67	7.4	7.3	7.64	7.19	Vanuatu Isl.
59	2011/01/02	20:20:26.6	−38.71	−73.84	19.4	19.80	0.95	6.6	6.46	6.60	7.1	7.1	7.48	7.21	Chile
60	2011/01/18	20:23:31.8	28.61	63.90	52.3	19.94	1.00	6.7	6.52	6.64	7.2	7.2	7.57	7.20	Pakistan
61	2011/03/09	2:45:32.0	38.56	142.78	14.1	20.09	1.05	6.4	6.53	6.69	7.3	7.3	7.67	7.20	Honshu Isl.
62	2011/03/11	5:47:32.8	37.52	143.05	20.0	22.72	1.84	7.2	7.52	7.74	8.9	9.1	9.59	7.46	Honshu Isl.
63	2011/03/11	6:15:58.7	35.92	141.38	29.0	20.93	1.33	6.8	6.89	6.97	6.8	7.9	8.23	7.20	Honshu Isl.
64	2011/03/11	6:26:12.6	38.27	144.63	21.1	20.49	1.19	7.1	6.73	6.81	7.5	7.6	7.95	7.18	Honshu Isl.
65	2011/04/07	14:32:50.6	38.82	141.85	53.3	19.77	0.94	6.9	6.46	6.59	7.1	7.1	7.49	7.21	Honshu Isl.
66	2011/07/06	19:3:32.5	−29.22	−175.83	22.3	20.47	1.18	7.0	6.73	6.81	7.8	7.6	7.98	7.19	Kermadec Isl.
67	2011/07/10	0:57:16.3	37.98	143.33	22.0	19.60	0.89	6.6	6.41	6.52	7.0	7.0	7.34	7.25	Honshu Isl.
68	2011/10/23	10:41:28.4	38.64	43.40	12.0	19.80	0.95	6.9	6.46	6.60	7.3	7.1	7.48	7.21	Turkey
69	2012/01/10	18:37:13.3	2.59	92.98	23.7	19.88	0.98	6.6	6.52	6.62	7.2	7.2	7.53	7.20	Sumatra
70	2012/02/02	13:34:49.2	−17.69	167.11	20.5	19.64	0.90	6.5	6.41	6.54	7.1	7.0	7.37	7.20	Vanuatu Isl.
71	2012/03/20	18:2:54.9	16.60	−98.39	15.4	20.30	1.12	6.6	6.67	6.76	7.6	7.5	7.81	7.20	Guerrero
72	2012/03/25	22:37:20.9	−35.31	−72.41	33.8	19.78	0.94	6.5	6.46	6.60	7.1	7.1	7.48	7.22	Chile
73	2012/04/11	8:39:31.4	2.35	92.82	45.6	21.96	1.67	7.4	7.26	7.29	8.6	8.6	8.92	7.21	Sumatra
74	2012/04/11	10:43:38.2	0.90	92.31	54.7	21.46	1.51	7.2	7.05	7.14	8.2	8.2	8.57	7.19	Sumatra
75	2012/04/12	7:16:4.6	28.57	−112.76	15.8	19.66	0.91	6.2	6.41	6.54	7.0	7.0	7.38	7.19	Mexico
76	2012/08/27	4:37:38.2	11.91	−89.18	12.0	20.07	1.04	6.5	6.57	6.69	7.3	7.3	7.66	7.21	Salvador
77	2012/08/31	12:47:43.0	11.02	127.00	46.1	20.52	1.20	6.5	6.73	6.82	7.6	7.6	7.94	7.18	Philippine
78	2012/09/05	14:42:23.7	9.87	−85.54	30.8	20.49	1.18	6.8	6.73	6.83	7.6	7.6	7.94	7.21	Costa Rica
79	2012/10/28	3:4:39.2	52.47	132.13	15.0	20.71	1.26	6.2	6.78	6.86	7.7	7.7	8.07	7.19	Charlotte Isl.
80	2012/11/07	16:35:55.2	13.93	−92.47	28.7	20.11	1.06	6.6	6.57	6.69	7.4	7.3	7.68	7.19	Guatemala

$$\log_{10} t_0 = 0.32(\pm0.003)\log_{10} M_0 - 5.43(\pm0.003) \quad (42)$$

From the data that the value M_{Sm} an average of 0.5 more than the M_S, because according to the relation $\log_{10}t_0$ with $\log_{10}M_0$ (from 42) with growth $\log_{10}M_0$ from 16 to 22 on the basis of (19), the value increases from 7.19 $\log\Delta\sigma$ to 7.43 (**Figure 12**), and using equation (38) in the same size ranges of $\log_{10}M_0$ the value of $\log_{10}\Delta\sigma$ increases from 6.5 to 7.10. It is likely that for most crustal earthquakes before 1993 was characterized by the above limits to growth $\log_{10}\Delta\sigma < 7.10$.

Ratio $m_b - M_S$ и $m_{bm} - M_{Sm}$. In **Figure 13** shows the correlation ratio $m_b - M_S$ for crustal earthquakes of the Kuril Islands and Japan for 1993-2011:

$$m_b = 0.52(\pm0.03)M_S + 2.78(\pm0.02) \quad (43)$$

which is in good agreement with the expression:

$$m_{bm} = 0.52 M_{Sm} + 2.74 \quad (44)$$

derived from (42) and (28) for $\theta_t = 0.32$ и $a_t = -5,43$ (**Figure 13**).

Figure 14 shows the correlation ratio $m_b - M_S$ for crustal earthquakes in South America for the years 1993-2012, ($\varphi = -40° - 0°$, $\lambda = -85° - 65°$) by Global CMT Catalogue:

$$m_b = 0.52(\pm0.03)M_S + 2.64(\pm0.02) \quad (45)$$

for this region was obtained ($N = 576$, $r = 0.99$):

$$\log_{10} t_0 = (0.32\pm0.004)\log_{10} M_0 - 5.48(\pm0.003) \quad (46)$$

the substitution of which in (26), $\theta_t = 0.32$ and $a_t = -5.48$ leads to the formula

$$m_{bm} = 0.52 M_{Sm} + 2.77 \quad (47)$$

Equations (43)-(46) are in good agreement with Equations (21) and (22).

Figure 15 shows the correlation $\log_{10}t_0$ of $\log_{10}M_0$ for earthquakes of the Tien Shan ($\varphi = 38.5° - 45°$, $\lambda = 63° - 96°$) for 1960-2012 in interval $13.0 \le \log_{10}M_0 \le 21.5$ ($N = 684$, $r = 0.85$):

$$\log_{10} t_0 = 0.22(\pm0.01)\log_{10} M_0 - 3.45(\pm0.01) \quad (48)$$

which closely coincides with Equations (29), (31) and (36) typical for earthquakes in California (**Figures 1** and **15**).

Therefore, we can expect that the relationship between magnitudes $m_b - M_S$ for earthquakes of the two regions may be similar in this range of seismic moment. Indeed, the data in **Figure 16** confirmed these assumptions and empirical relationship of M_S from m_b for Tien Shan's earthquakes is expressed by the following relation ($N = 1183$, $r = 0.95$, **Figure 16**):

$$M_S = 1.57(\pm0.03)m_b - 3.05(\pm0.02) \quad (49)$$

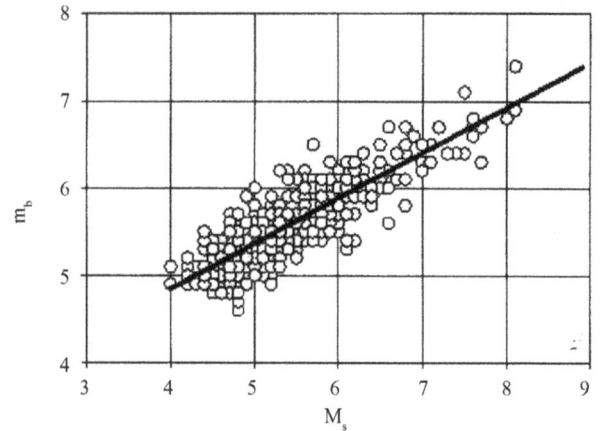

Figure 13. Correlation of magnitudes m_b and M_S for the earthquakes in Japan and Kuril Islands for 1993-2012 years. $m_b = 0.52(\pm0.03)M_S + 2.78(\pm0.02)$, $N = 514$, $r = 0.84$. Calculated dependence $m_{bm} = 0.52M_{Sm} + 2.74$ if $\log_{10}t_0 = 0.32\log_{10}M_0 - 5.43$ (see text).

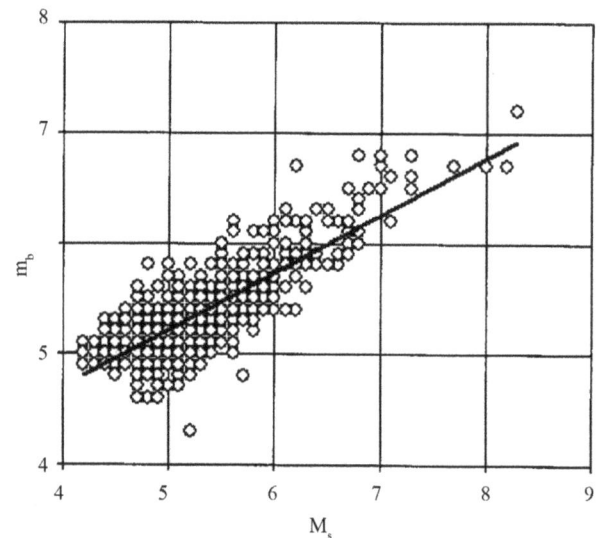

Figure 14. Correlation of magnitudes m_b and M_S for the earthquakes in South America 1993-2012 years. $m_b = 0.52(\pm0.03)M_S - 2.64(\pm0.02)$, $N = 547$, $r = 0.82$. Calculated dependence $m_{bm} = 0.52M_{Sm} - 2.77$ if $\log_{10}t_0 = 0.32\log_{10}M_0 - 5.48$ (see text).

Calculated dependence of M_{Sm} from m_{bm} based on Equations (25), (26) and (47) for the elastic parameters of the standard as follows:

$$M_{Sm} = 1.59 m_{bm} - 3.06 \quad (50)$$

which is in good agreement with Equations (2), (37) and (49).

Therefore, we have adopted model of the relationship of linear relations between M (m_b, M_L, M_S) and $\log_{10}t_0$ with $\log_{10}M_0$ explains many existing empirical formulas.

For a wide range $6 \leq \log_{10} M_0 \leq 23$ changing $\log_{10} t_0$, to a first approximation, can be described by a nonlinear dependence of ($A_0 = \log_{10} M_0$):

$$\log_{10} t_0 = 0.167 A_0 - 2.83$$
$$+ 1/3 \exp\left(2.166 A_0 - 0.045 A_0^2 - 25.09\right) \quad (51)$$

in which the first two terms describes the linear growth $\log_{10} t_0$ in the range $6 \leq A_0 \leq 15$. On the basis of Equations (25)-(27) and (51) in **Figure 17** shows estimates nonlinear dependence m_{bm}, M_{Lm} and M_{Sm} from M_W to (11) for crustal earthquakes. From **Figure 17** shows that in the

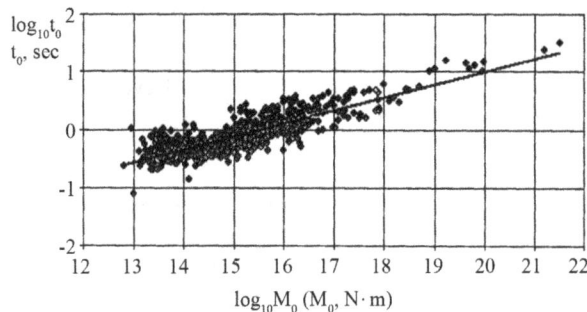

Figure 15. Correlation dependence $\log_{10} t_0$ from $\log_{10} M_0$ for Tien Shan earthquakes (1960-2012 years).
$\log_{10} t_0 = 0.22(\pm 0.01)\log_{10} M_0 - 3.45(\pm 0.01)$, $N = 684$, $r = 0.85$.

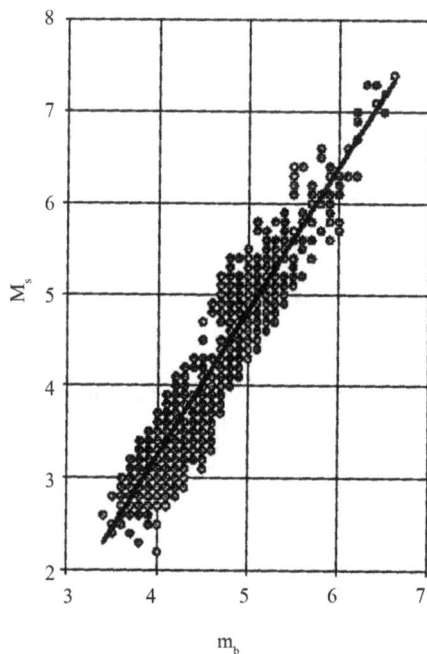

Figure 16. Correlation of magnitudes M_S and m_b for Tien Shan earthquakes (1902-2012 years).
$M_S = 1.57(\pm 0.03) m_b - 3.05(\pm 0.02)$, $N = 1183$, $r = 0.95$.

Calculated dependence $M_{Sm} = 1.59 m_{bm} - 3.06$ if $\log_{10} t_0 = 0.22\log_{10} M_0 - 3.45$ (see the text).

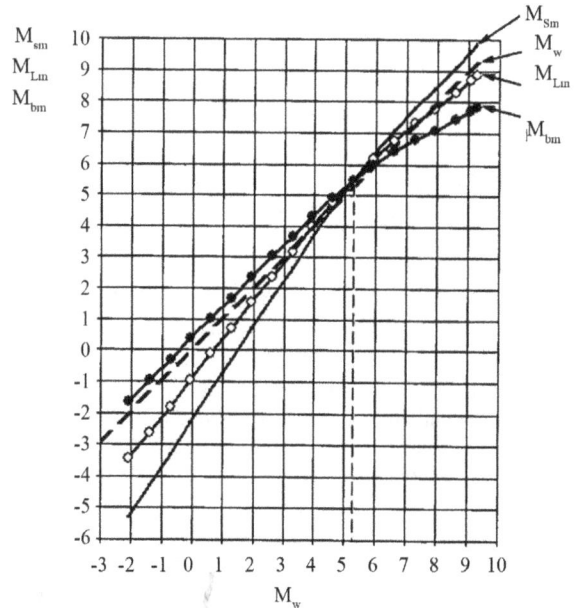

Figure 17. Averaged according M_{Sm} M_{Lm} and m_{bm} from M_W for crustal earthquakes (see text), the dashed line represents the intersection of the curves $M_{Sm} \approx M_{Lm} \approx m_{bm} \approx M_W \approx 5.26 - 5.50$.

interval $4 \leq M_W \leq 6,5$ numerical values of magnitudes $m_{bm} \approx m_b$, $M_{Lm} \approx M_L$, $M_{Sm} \approx M_S$ and M_W within the accuracy of these parameters are close. In accordance with Equations (19) and (51) in the interval $6.0 < A \leq 23.0$ $\log_{10} \Delta\sigma$ value increases from 1.75 to 7.53, and the most intense increase in this parameter is in the range $6.0 \leq A_0 \leq 15.0$.

4. Conclusions

1) A broad range of local Richter magnitude M_L, m_b, and M_S crustal earthquakes in different regions shows a possible functional relationship with the seismic moment magnitude, corner frequency, voltage and depressurized seismic elastic parameters of the geophysical environment. These links justify numerous empirical relationships with magnitudes of seismic moment.

2) It is assumed that an upgraded body-wave magnitude m_{bm} for large earthquakes is proportional to the logarithm of the average displacement along the fault $\log_{10} u$, $\hat{m}_b$, the true magnitude and the maximum amplitude of seismic vibrations A_g; magnitude M_{Sm} is proportional to the logarithm of the square average displacement along the fault ($2\log_{10} u$) and local magnitude proportional $1.5\log_{10} u$.

3) Control parameters of the quantitative relations with seismic moment magnitudes are coefficients depending on the change in corner period of seismic stress drop or discharged from the seismic moment, which provide a self-consistent system of equations between the main source parameters of crustal earthquakes.

REFERENCES

[1] B. Gutenberg, "Amplitudes of P, PP, an S waves and Magnitude of Shallow Earthquakes," *Bulletin of the Seismological Society of America*, Vol. 35, No. 2, 1945, pp. 57-69.

[2] B. Gutenberg and C. F. Richter, "Earthquake Magnitude, Intensity, Energy and Acceleration," *Bulletin of the Seismological Society of America*, Vol. 46, No. 2, 1956, pp. 105-145.

[3] B. Gutenberg and C. F. Richter, "Magnitude and Energy of Earthquakes," *Annali di Geofisica*, Vol. 9, No. 1, 1956, pp. 1-15.

[4] N. V. Kondorskaya, A. I. Zakharova and L. S. Chepkunas, "The Quantitative Characteristics of Earthquake Sources as Determined in the Seismological Practice of the U.S.S.R.," *Tectonophysics*, Vol. 166, No. 1-3, 1989, pp. 45-52.

[5] T. G. Rautian, "Energy of Earthquakes," In: Y. V. Riznichenko, Ed., *Methods for the Detailed Study of Seismicity*, Akademii Nauk SSSR, Moscow, 1960, pp. 75-114.

[6] T. G. Rautian, V. J. Khalturin, K. Fujita, K. G. Mackey, *et al.*, "Origins and Methodology of the Russian Energy K-Class System and Its Relationship to Magnitude Scales," *Seismological Research Letters*, Vol. 78, No. 6, 2007, pp. 579-590.

[7] P. Chen and H. Chen, "Scaling Law and Its Applications to Earthquake Statistical Relations," *Tectonophysics*, Vol. 166, No. 1-3, 1989, pp. 53-72.

[8] L. B. Grant, "Paleoseismology, International Handbook of Earthquake and Engineering Seismology, Part A," Academic Press, Waltham, 2002.

[9] A. A. Gusev and V. N. Melnikova, "Relations between Magnitudes: Global and Kamchatka Data," *Volkanology and Seismology*, Vol. 6, 1990, pp. 55-63.

[10] K. Kasahara, "Earthquake Mechanics," 1985.

[11] P. Mai and G. C. Beroza, "Source Scaling Potpies from Finite-Fault-Rupture Models," *Bulletin of the Seismological Society of America*, Vol. 90, No. 3, 2000, pp. 604-615.

[12] O. W. Nuttli, "Average Seismic Source-Parameters Relation for Mid-Plate Earthquakes," *Bulletin of the Seismological Society of America*, Vol. 73, No. 2, 1983, pp. 519-535.

[13] K. Aki, "Generation and Propagation of G-Waves from the Niigata Earthquake of June 16, 1964. Part 2. Estimation of Earthquake Moment, Released Energy, and Stress-Strain Drop from the G-Wave Spectrum," Bulletin of the Earthquake Research Institute, Tokyo, 1966.

[14] K. Aki and P. A. Richards, "Quantitative Seismology. Theory and Methods, v. t., Moskau: Mir," 1983.

[15] J. N. Brune, "Tectonic Stress and the Spectra of Seismic Shear Waves from Earthquakes," *Journal of Geophysical Research*, Vol. 75, No. 26, 1970, pp. 4997-5009.

[16] V. Keilis-Borok, "On Estimation of the Displacement in an Earthquake Source and Dimension," *Annales Geophysicae*, Vol. 12, No. 1-4, 1959, pp. 205-214.

[17] P. Debye, "Zur Theorie der Spezifischen Wärmen," *Annalen der Physik*, Vol. 344, No. 14, 1912, pp. 789-839.

[18] E. Mamyrov, "Relations among Earthquake Source Parameters Derived from Debye Solid-Body Model," *Journal of Geodynamics*, Vol. 22, No. 1, 1996, pp. 137-143.

[19] H. Kanamori, "The Energy Released in Great Earthquakes," *Journal of Geophysical Research*, Vol. 82, No. 20, 1977, pp. 2981-2987.

[20] G. A. Bollinger, M. C. Chapman and M. S. Sibol, "A Compassion of Earthquake Damage Areas a Function of Magnitude across the United States," *Bulletin of the Seismological Society of America*, Vol. 83, 1993, pp. 1064-1080.

[21] D. M. Boore, "The Richter Scale: Its Development and Use for Determining Earthquake Source Parameters," *Tectonophysics*, Vol. 166, 1989, pp. 1-14.

[22] T. S. Hanks and D. M. Boore, "Moment-Magnitude Relations in Theory and Practice," *Journal of Geophysical Research*, Vol. 89, No. B7, 1984, pp. 6229-6235.

[23] E. M. Scordilis, "Empirical Global Relations Converting M_S and m_b to Moment Magnitude," *Journal of Seismology*, Vol. 10, 2006, pp. 225-236.

[24] T. Utsu, "Relationships between Magnitude Scales, International Handbook of Earthquake and Engineering Seismology, Part A," Academic Press, Waltham, 2002.

[25] C. F. Richter, "An Instrumental Earthquake Magnitude Scale," *Bulletin of the Seismological Society of America*, Vol. 25, No. 1, 1935, pp. 1-32.

[26] N. I. Christensen, "Poisson's Ratio and Crustal Seismology," *Journal of Geophysical Research*, Vol. 101, No. B2, 1996, pp. 3139-3156.

[27] N. I. Christensen and W. D. Mooney, "Seismic Velocity Structure and Composition of the Continental Crust: A Global View," *Journal of Geophysical Research*, Vol. 100, No. B6, 1995, pp. 9761-9788.

[28] E. Mamyrov, "New System of Quantitative Correlations between Seismic Energy, Magnitude and Energy of Seismic Radiation of the Crust Earthquakes in Tien Shan," *The 33rd General Assembly of the European Seismological Commission*, Moscow, 19-24 August 2012, pp. 29-30.

[29] H. Houston and H. Kanamori, "Source Spectra of Great Earthquakes: Teleseismic Constraints on Rupture Process and Strong Motion," *Bulletin of the Seismological Society of America*, Vol. 76, No. 1, 1986, pp. 19-42.

[30] W. Thatcher and C. Hahks, "Source Parameters of Southern California Earthquakes," *Journal of Geophysical Research*, Vol. 78, No. 35, 1973, pp. 8547-8575.

[31] K. K. Zapolskii, J. L. Nersesov, T. G. Rautian and V. I. Halturin, "Physical Basis of Magnitude Classification of Earthquakes," 1974.

[32] F. Scherbaum and D. Stoll, "Source Parameters and Scal-

ing Laws of the 1978 Schwabian Jura (Southwest Germany) Aftershocks," *Bulletin of the Seismological Society of America*, Vol. 73, No. 5, 1983, pp. 1321-1343.

[33] A. Jin, C. A. Moya and M. Ando, "Simultaneous Determination of Site Responses and Source Parameters of Small Earthquakes along the Atotsugawa Fault Zone, Central Japan," *Bulletin of the Seismological Society of America*, Vol. 90, No. 6, 2000, pp. 1430-1445.

[34] H. S. Hasegawa, "Lg-Spectra of Local Earthquake Recorder by the Eastern Canada Telemeters Network and Spectral Scaling," *Bulletin of the Seismological Society of America*, Vol. 73, 1983, pp. 1041-1061.

[35] G. Grunthal and R. Wahlstrom, "An M_W Based Earthquake Catalogue for Central, Northern and North-Western Europe Using a Hierarchy of Magnitude Conversions," *Journal of Seismology*, Vol. 7, No. 4, 2003, pp. 507-531.

[36] J. G. Ristau, G. C. Rogers and F. Cassidy, "Moment Magnitude-Local Magnitude Calibration for Earthquake in Western Canada," *Bulletin of the Seismological Society of America*, Vol. 95, No. 5, 2005, pp. 1994-2000.

[37] J. Ristau, "Comparison of Magnitude Estimates for New Zealand Earthquakes: Moment Magnitude, Local Magnitude, and Teleseismic Body-Wave Magnitude," *Bulletin of the Seismological Society of America*, Vol. 99, No. 3, 2009, pp. 1841-1852.

[38] Y. M. Wu, R. M. Allen and C. W. Wu, "Revised M_L Determination for Crustal Earthquake in Taiwan," *Bulletin of the Seismological Society of America*, Vol. 95, No. 6, 2005, pp. 2517-2524.

[39] G. L. Choy and J. Boatwright, "Global Patterns of Radiated Seismic Energy and Apparent Stress," *Journal of Geophysical Research*, Vol. 100, No. B9, 1995, 18205-18228.

[40] Y. Zhuo and H. Kanamori, "Regional Variation of the Short-Period (1 to 10 Second) Source Spectrum," *Bulletin of the Seismological Society of America*, Vol. 77, No. 2, 1987, pp. 514-529.

[41] O. J. Perez, "Revised World Seismicity Catalog (1950-1997) for Strong ($M_S \geq 6$) Shallow ($h \leq 70$ km) Earthquakes," *Bulletin of the Seismological Society of America*, Vol. 89, No. 2, 1999, 335-341.

8

Strain Energy Release from the 2011 9.0 M$_w$ Tōhoku Earthquake, Japan

Kenneth M. Cruikshank, Curt D. Peterson
Department of Geology, Portland State University, Portland, USA

ABSTRACT

The purpose of this paper is to compare the strain energy released due to elastic rebound of the crust from the tragic 2011 9.0 M$_w$ Tōhoku earthquake in Japan with the observed radiated seismic energy. The strain energy was calculated by analyzing coseismic displacements of 1024 GPS stations of the Japanese GEONET network. The value of energy released from the analysis is 1.75×10^{17} J, which is of the same order of magnitude as the USGS-observed radiated seismic energy of 1.9×10^{17} Nm (J). The strain energy method is independent of seismic methods for determining the energy released during a large earthquake. The analysis shows that although the energy release is concentrated in the epicentral region, about 12% of the total energy was released throughout the Japanese islands at distances greater than 500 km west of the epicenter. Our results also show that outside the epicentral region, the strain-energy was concentrated along known tectonic zones throughout Japan.

Keywords: Japan; Earthquake; Crustal Strain; GPS; Radiated Energy

1. Introduction

Reid's [1] elastic rebound theory indicates that an understanding of the pattern and magnitude of strain in the loading phase of the earthquake cycle is important for evaluating the seismic risk in an area. Some insights into the strain patterns in the loading phase can be gained by examining the pattern of strain in the unloading or earthquake phase. Measurements of tectonic strain release during the 2011 Tōhoku earthquake and tsunami [2] (**Figure 1**) provide important insights into the mechanisms of subduction zone earthquakes. These relations should be of use in other subduction zones where modern strain records might help to constrain predictions of earthquake strain release and energy. To our knowledge, these are the first wide-field comparisons of radiated energy and observed strain energy reported for a subduction zone.

Analysis of strain energy provides a method for estimating the total energy released without assuming a particular earthquake mechanism. For example, estimates of the energy from an earthquake using either the Scalar Moment or Observed Radiated Energy are often different by 5 to 6 orders of magnitude, with the seismic moment being substantially larger than the observed radiated energy [e.g., 3, Figure 12]. (Although, it should be noted that the seismic moment is not a direct measure of the energy released in an earthquake, so it will have a value substantially different from the radiated energy [4]). Direct measure of the strain between points on Earth's surface can be calculated from the relative displacement of Global Positioning System (GPS) stations. GPS measurements of crustal strain provide constraints on the distribution of energy release that are not directly available from seismic stations alone.

In this paper we document the regional distribution of coseismic strain in the upper plate from the 2011 Tōhoku earthquake using length-changes between GPS stations in the Japanese GPS network (GEONET, **Figure 2**). Our estimate of the total strain-energy release, 1.75×10^{17} J, is of the same order of magnitude as the observed radiated seismic energy, 1.9×10^{17} J [2]. Our results also show that a portion of the strain-energy was concentrated along known tectonic zones throughout Japan (compare **Figures 1** and **3**). These tectonic zones apparently served as strain concentrators prior to the 2011 earthquake [5]. The distribution of strain release immediately following the 2011 Tōhoku earthquake is generally consistent with reported patterns of strain accumulation that have been observed over the last 50 years [5,6].

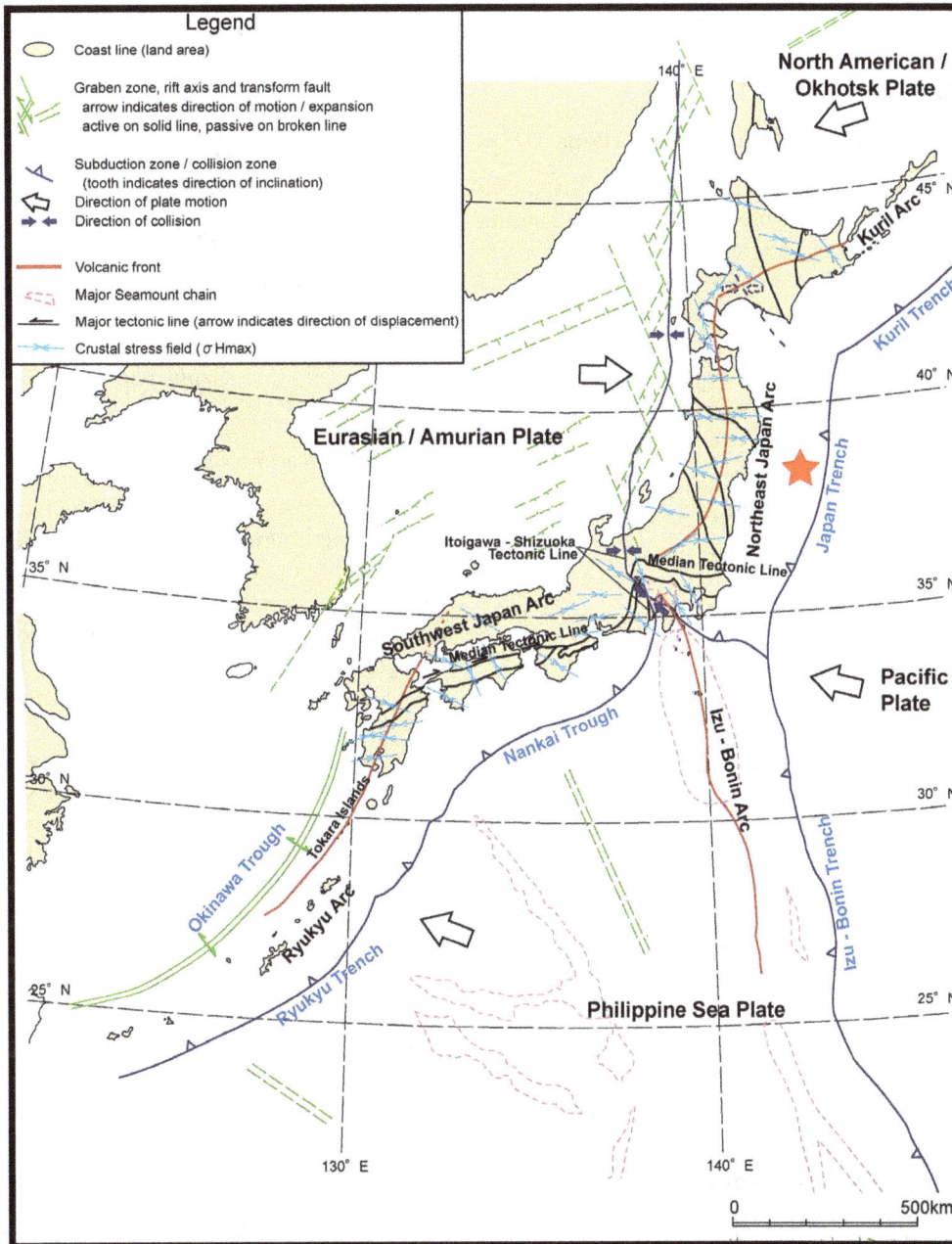

Figure 1. Map showing the current tectonic setting of Japan [Figure 6-2 from 7]. The 2011 epicenter, marked by a red star, is located along the Japan Trench and the Northeast Japan Arc.

2. Background

The energy released in an earthquake is generally believed to come from the release of energy stored by the elastic component of strain in the crust. This is the idea expressed in Reid's elastic rebound theory [1]:

"We know that the displacements which took place near the fault-line occurred suddenly, and it is a matter of much interest to determine what was the origin of the forces which could act in this way. Gravity can not be invoked as the direct cause, for the movements were practically horizontal; the only other forces strong enough to bring about such sudden displacement are elastic forces. These forces could not have been brought into play suddenly and have set up an elastic distortion; but external forces must have produced an elastic strain in the region about the fault-line, and the stresses this induced were the forces which caused the sudden displacements, or elastic rebounds, when the rupture occurred".

A similar mechanism is used to describe the release of energy in phenomenon such as "rock bursts" in mines [8].

Here we consider the general energy balance for the change in shape of an elastic material. We are not concerned with any particular mechanism.

2.1. Energy Stored in Elastic Deformations

An elastic material is one where the original shape (the unstrained state) will be restored after the forces causing a change in shape (strain) are removed. An elastic material can be thought of as a coiled spring, a spring that resists both being shortened and lengthened. The resistance of the spring to a change in length is the spring constant; we take the value of the spring constant to be the same for both shortening and extension, and the relationship force and change in length is described using Hooke's Law,

$$F = k\Delta L,$$

where F is the force, k the "spring" constant, and ΔL the change in length.

An elastic bar can be shortened by compressing it from either end. The force does work in shortening the bar (and so is using energy). Some of the energy may go into heat generation or permanent shape changes but the recoverable energy which we are interested in will be stored in the "springiness" of the elastic bar. Thus, the result of the change in length of the bar is net-energy "storage" in the bar; this is the *elastic strain energy*. A material will show elastic behavior as long as the length change is small compared to the length of the bar, generally less than 10^{-6} of the length of the bar.

In an elastic material the change in length is reversible once the applied forces are removed (unloaded). When the forces are removed from a bar it attempts to return to its original length. The material many not completely revert to its original geometry, since some of the potential energy has been converted to kinetic energy and heat. If removal of the forces is a very slow process the kinetic energy is negligible. If the removal of the forces is rapid in addition to doing work to restore the materials shape a portion of the stored potential energy will be converted to kinetic energy which will be released as seismic waves.

Short term coseismic changes in length are an effect of the elastic behavior of the rock, not the long-term deformation, thus the *elastic strain energy* represents the maximum potential energy that is available to be transferred to kinetic energy and observed as seismic waves.

In summary, the maximum amount of stored potential elastic energy between two points in an elastic material is proportional to change in distance between those two points and the spring constant of the material. Using GPS the change in distance between two points can be determined, which will then be proportional to the change in elastic potential energy between the two points. This change in elastic potential energy is the maximum amount of energy that is available to be released as kinetic energy (*i.e.*, seismic waves).

2.2. Strain and Strain Energy

Strain, ε, is the normalized change in length of a line between two material points

$$Strain = \left(Final\ Length - Initial\ Length\right)\big/Initial\ Length$$

$$Strain = \left(Change\ in\ Length\right)\big/Initial\ Length$$

or

$$Strain = Stretch - 1$$

where

$$Stretch = Final\ Length\big/Initial\ Length$$

Strain, ε, will be negative if the distance between two points gets shorter (shortening), positive if the distance increases (extension), and zero if it is unchanged. The stretch (or stretch ratio), S, will be less than one for shortening, greater than one for extension, and one for no length change.

Over an area, strain (or stretch) provides a description of deformation. This would consist of a series of strains in different directions. Usually the strain is described with two principle components. If we know the principle components and their direction, we can describe the strain in any arbitrary orientation.

Once the principle strains $\left(\varepsilon_1, \varepsilon_2, \varepsilon_3\right)$ are known, the elastic strain energy is given by [8]:

$$e = \frac{1}{2}\left[\lambda\left(\varepsilon_1 + \varepsilon_2 + \varepsilon_3\right)^2 + 2G\left(\varepsilon_1^2 + \varepsilon_2^2 + \varepsilon_3^2\right)\right] \quad (1)$$

where

$$G = \frac{E}{2(1+v)} \quad \text{and} \quad \lambda = \frac{Ev}{(1+v)(1-2v)},$$

where E is Young's Modulus and v is Poisson's Ratio.

In order to determine the principle strains we need the strain in three different directions. The principle component we end up with will represent some average strain over the region enclosed by the three lines needed for the solution.

GEONET stations were grouped into triangles, and changes in position of the vertices were used to determine the magnitude of the principle strains (the solution also gives the orientations, but that information is not needed for this analysis). If we have three points, we can define a triangle.

Consider a plane triangle, the Euclidian distance is:

$$Distance = \sqrt{dE^2 + dN^2 + dZ^2}$$

$$Direction\ (\theta_{base}) = \tan^{-1}\left(\frac{dE}{dN}\right)$$

where $dE = (E_A - E_B), dN = (N_A - N_B), dZ = (Z_A - Z_B)$.

If the direction is needed to be converted to an Azimuth (ϕ)

$$\phi = \begin{cases} if\ 0° < \theta <= 90°; 90° - \theta \\ if\ 90° < \theta < 360°; 450° - \theta. \end{cases}$$

Knowing the coordinates of A, B and C, the distances between points can be determined. The direction of each line element can also be determined.

Given displacements, dX and dY, the change of line length can determined, allowing the strain of each line segment to be determined. The displacement vector will not always be parallel to the line segment, so we need to get the component of the net displacement (imagine point A is now fixed, so we look at the relative movement of point B with respect to point A:

$$\delta N = \delta N_A - \delta N_B$$
$$\delta E = \delta E_A - \delta E_B$$
$$\delta Z = \delta Z_A - \delta Z_B.$$

This describes a net displacement vector with an orientation of

$$(\theta_{disp}) = \tan^{-1}\left(\frac{\delta E}{\delta N}\right),$$

and magnitude $\sqrt{dE^2 + dN^2 + dZ^2}$. The projection of this vector into the direction of the line segment (θ_{base}) is accomplished by:

$$Angle = \theta_{disp} - \theta_{base} + \pi$$

$$\delta L = \sqrt{\delta E^2 + \delta N^2 + \delta Z^2} \cos(Angle)$$

to get the change in the vector direction of the line, or the strain between the two stations

Since we are dealing with Geographic Coordinates, the distance and line direction are determined using a modified Vincenty solution [9].

Evaluation of the strain energy (Equation (1)) requires that the principal strains (ε_i) are known. Displacements of GPS stations that comprise the 3 vertices of triangles in the GEONET array provide sufficient information to calculate the principal strains within the triangle [e.g., 10, 11]. The basic equations solved are (for irrotational strain, [12]):

$$S = 1 + \varepsilon$$

$$= \frac{D}{\left[\frac{\partial y}{\partial Y}\cos\theta - \frac{\partial x}{\partial Y}\sin\theta\right]^2 + \left[\frac{\partial x}{\partial X}\sin\theta - \frac{\partial y}{\partial X}\cos\theta\right]^2} \quad (2a)$$

where

$$D = (\partial x/\partial X)(\partial y/\partial Y) - (\partial x/\partial Y)(\partial y/\partial X). \quad (2b)$$

S is stretch, and θ is the direction cosine for the line,

$\partial x/\partial X$, $\partial y/\partial Y$, $\partial x/\partial Y$, and $\partial y/\partial X$ are components of the deformation gradient tensor [11,13]. Strain is

$$\varepsilon_i = S_i - 1. \quad (2c)$$

Equations (2) are solved using displacements of three GPS stations that make up the triangles shown in **Figure 2**; the evaluation of Equation (1) provides the strain energy density for the triangle. Using an average crustal thickness, the total energy changes in the triangular prism can be calculated.

From the principal strains, the average strain energy density can be computed for a given area, assuming values of Young's modulus (E) and Possion's ratio (v). We use Hanks and Kanamori [14] typical crustal values of 70 GPa and 0.25 respectively. In this paper we use an average crustal thickness of 30 km based on the USGS CRUST 5.1 model [15] for the Japan region. Variations in the value for average thickness do not significantly change the results of estimated energy release (see Discussion).

2.3. Interpreting the "Total Energy Available" Result

The results from Japan show that coseismic displacements produce areas of shortening and areas of extension. Since the elastic response makes no distinction between energy released by the bar restoring to a longer or shorter dimension, we take all of the areas to be generating potential seismic energy.

Another extreme is to consider only areas extending to be releasing stored energy, and area that are shortening to be "absorbing" energy (although there is no *a priori* basis for assuming this). If we do this, then the result is different by 24% from the assumption for all areas releasing energy. The order-of magnitude of the energy released is the same (10^{17}). We will use the all-cells releasing energy figure, since that would represent the upper-limit of available energy, although it is possible that about 12% goes into "loading" other areas of the crust during the earthquake.

3. Data and Methods

The Japanese nation-wide dense GPS network (GEONET, **Figure 2**) has been in operation since 1996 [5]. In this paper we only use displacements recorded in the 9 minutes following the March 2011 Tōhoku earthquake (**Figure 1**). This allows us to exclude any movement from aftershocks. The relative displacement of the GPS stations represents the change in crustal strain. According to Reid's [1] elastic rebound theory the change in strain represents stored elastic energy that is available to be released as seismic waves.

Many of the 1024 stations in GEONET range from 20

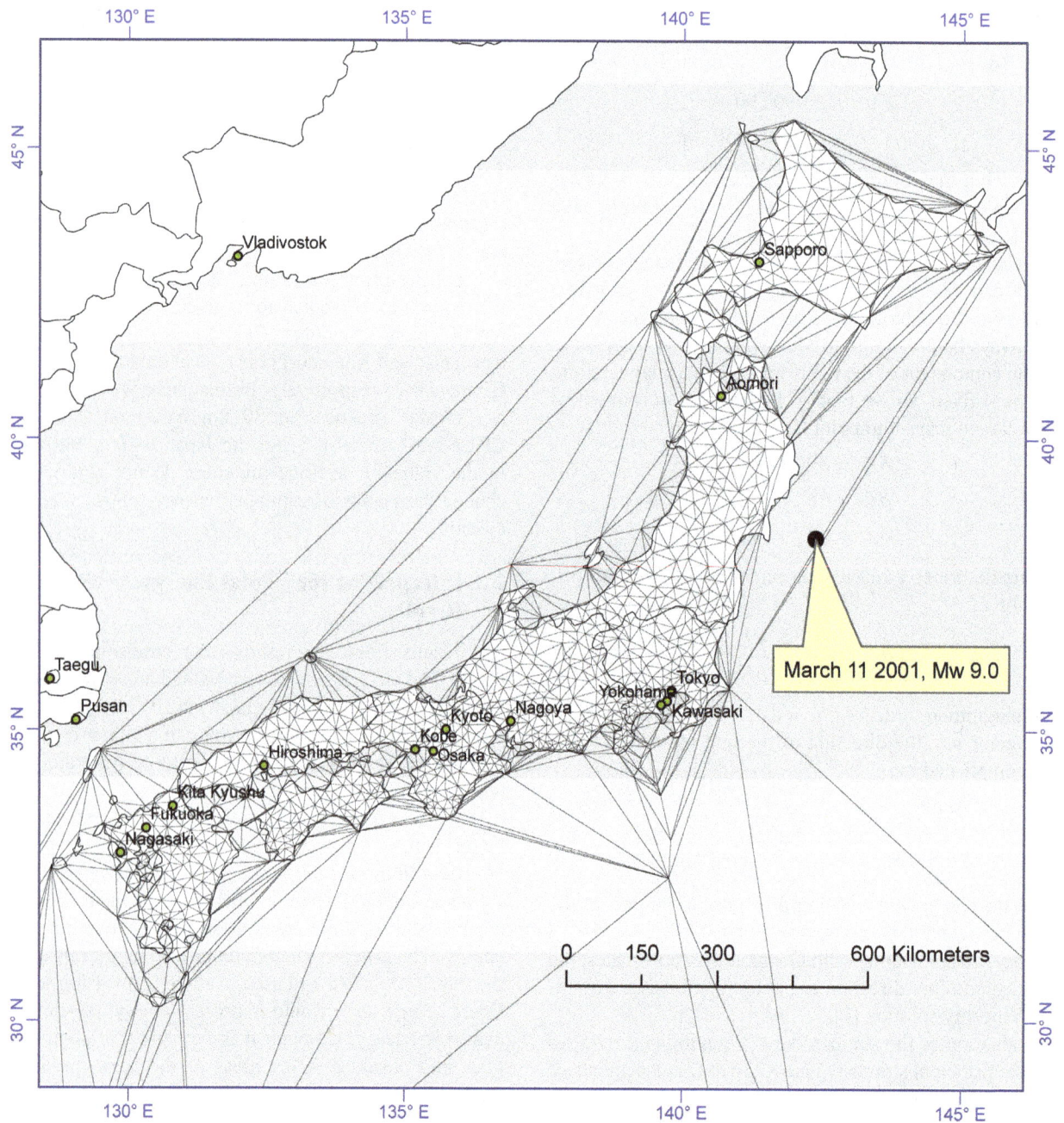

Figure 2. GPS stations that comprise the Japanese GEONET network and the Delauney Triangles that were formed using the 1024 GPS stations. Principal strains for each triangle are calculated, from which the strain energy density for each triangle is computed. The location of the March 2011 M$_w$ 9.0 earthquake is shown of the NE coast of Honshu.

to 40 km apart (**Figure 2**) and provide data coverage up to 1500 km from the 2011 epicenter. A network of 2393 triangles has been used to look at strain accumulation over the last decade [e.g., 16]. Several stations are on islands (**Figure 2**), thus strain in the oceanic crust to the east of Japan can also be calculated, although at lower resolution because of the increased distance between GPS stations. We use the displacement of these stations to calculate the energy released from stored elastic strain as well as the distribution of energy release. This data provides insight where strain energy was released or stored following an earthquake, without the need to assume a particular deformation model, e.g., displacement discontinuities on a fault. GPS gives maps of coseismic displacements that cannot be obtained from traditional re-surveying of control networks that may take years [e.g., 17,18] during which time additional displacements may occur.

Data from GEONET was processed by the Advanced Rapid Imaging and Analysis (ARIA) [19] team at JPL/California Institute of Technology. Version 0.3 of the processed data consisted of coseismic displacements estimated from 5-minute interval kinematic solutions. The coseismic displacements are the difference between the solutions at 5:40 UTC and at 5:55 UTC. The earthquake occurred at 05:46 UTC, starting the 9 minute period of strain measurements. Using this time frame minimizes displacements from aftershocks and other unrelated events. A search of the IRIS database (IRIS, 2011) for the region bounded by 128° to 146°E, 29° to 46°N and the GPS time frame found only the main shock (**Figure 2**). The following steps were taken:

Stations were grouped to form non-overlapping triangles (**Figure 2**) [e.g., 20].

The initial latitude-longitude positions were used to calculate the initial length (l_i) and direction (θ_i) for each side in each triangle.

The displacements at each vertex in a triangle were used to determine the final length of each side of the triangle (L_i) The triangles are an over-determined system. Least-squares were used to solve for the maximum and minimum principal strains (Equation (2)).

Multiplying the area of the triangle by the strain energy density (Equation (1)) gave the strain energy for a 1 m thick slab located within the triangle.

The energy per meter of thickness is multiplied by the thickness of the crust (taken to be 30 km) to give the total strain energy change in the vertical prism. The 30 km crustal thickness was selected based on the USGS CRUST 5.1 model [15].

4. Results

From the analysis of the ARIA data, the sum of the energy per meter thickness of crust is 5.85×10^{12} J·m². Allowing for a crustal thickness of approximately 30 km, the total energy is 1.75×10^{17} J. This value compares well with the total radiated seismic energy (USGS, 2011) of 1.9×10^{17} Nm (or J). Variations in elastic properties and thickness of the crust would change our total strain energy value, but by less than an order of magnitude.

Although the seismically-observed radiated energy is a useful number, the strain-energy analysis shows the distribution of the sources and sinks of the energy (**Figure 3**) without using a specific earthquake model, e.g., a dislocation model. Some areas of the crust under Japan released energy while some areas may have stored energy. This concept of differential loading and unloading has generally been calculated using a specific model of the faulting [e.g., 21]. From the change in strain the maximum available energy can be determined.

The variation in energy distribution can be seen by looking at the "cumulative" energy as a function of distance from the trench axis (**Figure 4**).

Figure 3 shows the distribution of strain energy sources and sinks throughout Japan. The green areas represent triangles that increased in area, where the red triangles represent areas of overall shortening.

5. Discussion

5.1. Surface Observations Related to the Whole Crust

An assumption in GPS strain studies is that surface displacements provide information about the elastic response of the entire crustal material. Over the time-scale of the coseismic energy release, a few minutes in duration, the crust can be regarded as a rigid elastic material. By looking at the crustal response over a few minutes, we do not have to consider the longer-term viscous behavior as observed at the year and decade time scales, of the crustal material. Although, the changes in strain over longer time intervals following an earthquake can provide important insights into longer-term crustal deformation and the partitioning of elastic and viscous behaviors, it is not necessary in this study.

The assumption that the surface displacements are representative of the strains throughout the crustal block do appear to be valid because (1) the total energy calculations agree with the independent radiated energy calculation [2], and (2) the March 2011 Tōhoku earthquake hypocenter is put at 30 km depth, which is near the reported base of the upper-plate crust in the study area.

5.2. Influence of Thickness of Continental Crust

A 30 km crustal thickness was used to calculate the total energy released of 1.75×10^{17} J from the strain energy density of 5.85×10^{12} J per meter thickness of material. The total energy released would range from 0.585×10^{17} J for a 10 km thick crust, to 2.93×10^{17} J for a 50 km thick crust. The value changes by a factor of 5, but retains the same order of magnitude as the observed radiated energy. The value of 30 km was derived from the USGS CRUST 5.1 model [15].

5.3. Pre-Earthquake Strain Measurements

Studies of accumulating strain patterns reflects deformation along several tectonic zones over long time periods (earthquakes occurring during the measurement period), whereas in this paper we only consider the effect of a single earthquake along a single segment of a tectonic zone, so reconciling the two patterns is difficult. However, our results indicate that several tectonic zones, which were areas of increased strain accumulation in previous studies, responded to the March 2011 Tōhoku

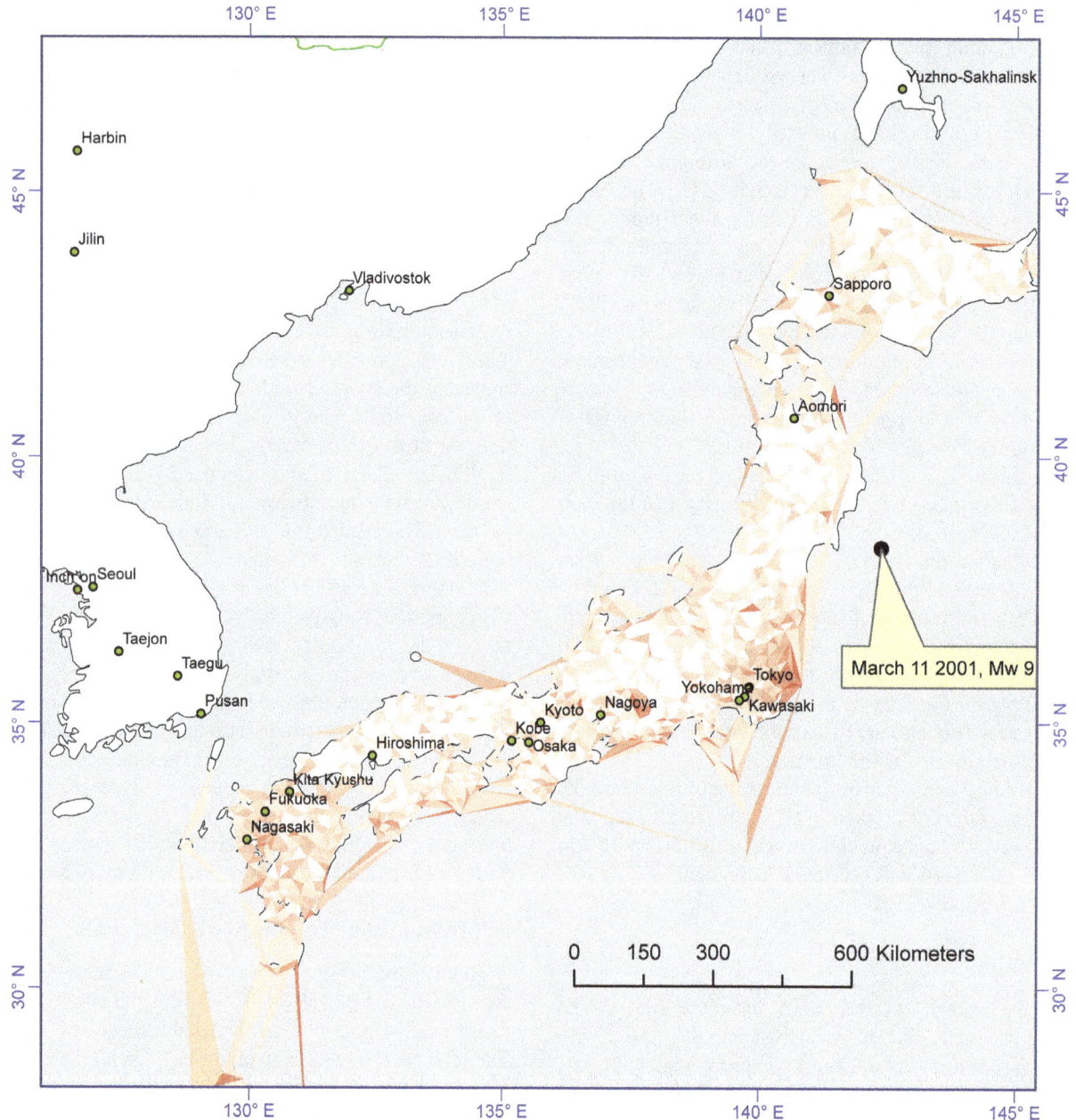

Figure 3. Strain energy density across the Japanese islands. White (land) and grey (oceanic) areas have a very low strain energy density. Darker shading represents greater strain energies. Most of the total energy released is in the large triangles around the epicenter. There is an area of high strain energy density at the southern margin of the rupture zone, just east of Tokyo (Median Tectonic Line, Figure 1). About 5% of the total energy transfer is distributed along the southern coast of Honshu and the West coast of Kyushu (compare with tectonic zones in Figure 1).

earthquake, when and where they became zones of concentrated energy release.

Harada & Simura [6] used three different first-order triangulation surveys conducted over a 94 year range to show the variability in the deformation in areas of Japan. GEONET, which consisted of about 1000 stations at the time, was used by Sagiya and others [16] to look at current crustal deformation in Japan. They concluded that most of the regions of large strain were associated with

tectonic boundaries and volcanoes. GEONET was also used by Hasimoto and others [5] to investigate strain accumulation from a slip-deficit model. All these studies showed that regions of differing amounts of strain coincided with known tectonic zones.

Strain energy release from the 2011 Tōhoku earthquake is shown in **Figure 3**; the denser-colored areas show that the energy density was not uniform across the study area. There is a band of higher strain energy density

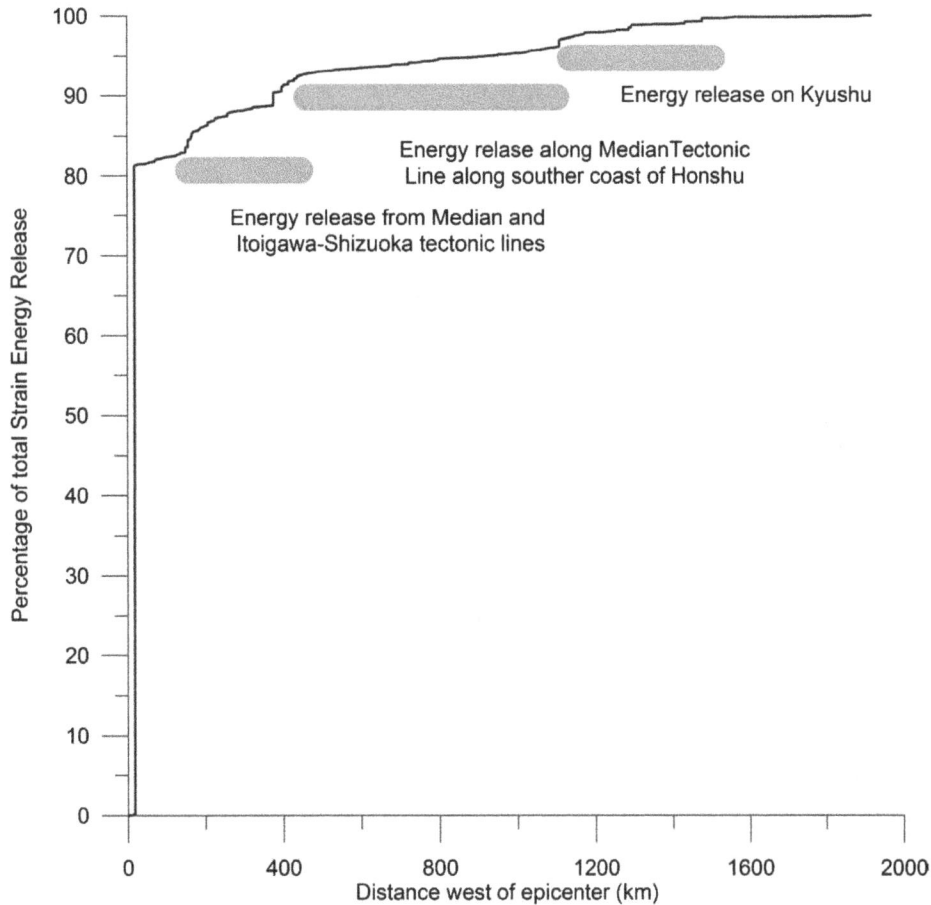

Figure 4. Plot of energy released as a function of distance from the seismic front. The cumulative energy is the total energy released by the entire area of GPS coverage. About 82% of the energy is released within 150 km West of the epicenter. Another 12% is released 150 to 500 km west of the epicenter, and the remaining 6% is released more than 500 km west of the epicenter. The jumps in energy released plot represent the concentrated energy release along tectonic zones.

running WSE-ENE along the southern end of the island of Honshu, which corresponds to the Median Tec tonic Line (**Figure 1**). The island of Kyushu, is another area of increased strain energy density, which is at the intersection of the Ryukyu and SW Japan arcs. At the intersection of the Median and Itoigawa-Shizuoka tectonic lines, located just East of Tokyo, is another area of increased strain energy density. Interestingly, the Niigata-Kobe Tectonic Zone, proposed by Sahiya and others [16], does not appear to be a zone of high strain energy density.

Some of the variability in energy release can also be seen by examining **Figure 4**. Eighty-two percent of the energy is released within 150 km west of the epicenter. Between 150 and 500 km west of the epicenter there was rapid increases in total strain energy (a 6% and 4% rise). These abrupt increases in energy release with distance represent the crossing of the Median tectonic line just east of Tokyo and the energy-dense area near Nagoya. The gradual increase of total strain energy from 500 to 1100 km represents the energy released along the Median Tectonic Line (**Figure 1**). There were smaller increases

of 1% - 2% increases approximately 1200 km west of the epicenter, which represent the energy released in the island of Kyushu.

Although the majority of the energy released in the event was close to the epicenter approximately 12% of the total energy was released 150 to 500 km away from the epicenter (**Figure 4**). This 12% accounts for approximately 2.1×10^{16} J of energy, which is the observed radiated seismic energy for a 7.5 M$_w$ earthquake. This landward release of energy could account for some of the reported local strong shaking at substantial distances from the epicenter.

In summary, pre-earthquake strain studies show non-uniform strain accumulation with increased strain accumulation along known tectonic boundaries. These areas of increased strain accumulation showed high-strain energy densities of released elastic strain during the 2011 earthquake. The strain released in the 2011 9.0 M$_w$ Tohoku earthquake shows the same nature of differential strain accumulation as seen by Harada & Simura [6] and Hasimoto and others [5].

6. Conclusions

The 2011 9.0 M_w Tōhoku earthquake provides a unique dataset of displacements over a large area during a subduction zone earthquake. This study used 1204 GPS stations, with an average station-to-station distance of 20 km. GPS data show coseismic displacements for the earthquake using a 9-minute window of data following the earthquake. Displacements from aftershocks are excluded from the dataset, yielding the spatial variability of co-sesimic displacements. Some conclusions are:

The amount of strain energy released is the same order of magnitude as the observed radiated seismic energy.

About 12% if the total energy released (energy equivalent to a M_w 7.5 earthquake) was released along tectonic zones across the southern margin of Honshu.

Although energy release occurred throughout the Japanese islands, it is concentrated in known tectonic zones.

The pattern of different areas accumulating versus releasing strain is similar to patterns of strain prior to the 2011 earthquake.

This paper has presented a method for calculating the energy released in an earthquake that 1) is independent of seismic energy methods, and 2) matches the seismically observed radiated energy. This method could be applied to areas that are currently accumulating strain to estimate the amount of potential energy, and thus the magnitude of an earthquake, that the area could generate during megathrust fault rupture.

REFERENCES

[1] H. F. Reid, "The California Earthquake of April 18, 1906: The Mechanics of the Earthquake," The Carnegie Institution of Washington, Washington, D.C., 1908.

[2] USGS, US Geological Survey, Vol. 2011, 2011.

[3] H. Kanamori and E. E. Brodsky, "The Physics of Earthquakes," *Reports on Progress in Physics*, Vol. 67, No. 8, 2004, pp. 1429-1496.

[4] S. Stein and M. Wysession, "An Introduction to Seismology, Earthquakes, and Earth Structure," Blackwell, Malden, 2003.

[5] C. Hashimoto, A. Noda, T. Sagiya and M. Matsu'ura, "Interplate Seismogenic Zones along the Kuril-Japan Trench Inferred from GPS Data Inversion," *Nature Geoscience*, Vol. 2, No. 2, 2009, pp. 141-144.

[6] T. Harada and M. Shimura, "Horizontal Deformation of the Crust in Western Japan Revealed from First-Order Triangulation Carried out Three Times," *Tectonophysics*, Vol. 52, No. 1-4, 1979, pp. 469-478.

[7] Nuclear Waste Management Organization of Japan,

"Evaluating Site Suitability for a HLW Repository (Scientific Background and Practical Application of NUMO's Siting Factors)," *Evaluating Site Suitability for a HLW Repository (Scientific Background and Practical Application of NUMO's Siting Factors)*, Vol. NUMO-TR-04-04, 2004, pp. 29-39.

[8] J. C. Jaeger, N. G. W. Cook and R. W. Zimmerman, "Fundamentals of Rock Mechanics," Chapman and Hall, London, 2007.

[9] T. Vincenty, "Direct and Inverse Solutions of Geodesics on the Ellipsoid with Application of Nested Equations," *Survey Review*, Vol. 23, No. 176, 1975, pp. 88-93.

[10] F. C. Frank, "Deduction of Earth Strains from Survey Data," *Bulletin of the Seismological Society of America*, Vol. 56, No. 1, 1966, pp. 35-43.

[11] L. E. Malvern, "Introduction to the Mechanics of Continuous Medium," Prentice-Hall, Inc., Englewood Cliffs, 1969.

[12] K. M. Cruikshank, A. M. Johnson, R. W. Fleming and R. Jones, "Winnetka Deformation Zone. Surface Expression of Coactive Slip on a Blind Fault during the Northridge Earthquake Sequence," Open-File Report No. OF-96-698, 1996, p. 70.

[13] Y. C. Fung, "Foundations of Solid Mechanics," Prentice-Hall, Englewood Cliffs, 1965.

[14] T. Hanks and H. Kanamori, "A Moment Magnitude Scale," *Journal of Geophysical Research*, Vol. 84, No. B5, 1979, pp. 2348-2350.

[15] W. D. Mooney, G. Laske and G. Masters, "CRUST5.1: A Global Crustal Model at 5 × 5 Degrees," *Journal of Geophysical Research*, Vol. 103, 1998, pp. 727-747.

[16] T. Sagiya, S. I. Miyazaki and T. Tada, "Continuous GPS Array and Present-Day Crustal Deformation of Japan," *Pure and Applied Geophysics*, Vol. 157, No. 11-12, 2000, pp. 2303-2322.

[17] G. Plafker, "Tectonics of the March 27, 1964 Alaska Earthquake," Professional Paper No. 543-I, 1969, p. 74.

[18] G. Plafker, "Alaskan Earthquake of 1964 and Chilean Earthquake of 1960: Implications for Arc Tectonics," *Journal of Geophysical Research*, Vol. 77, No. 5, 1972, pp. 901-925.

[19] ARIA, "Preliminary GPS Time Series Provided by the ARIA (Advanced Rapid Imaging and Analysis) Team at JPL and Caltech," 2011. ftp://sideshow.jpl.nasa.gov/pub/usrs/ARIA/

[20] B. Delaunay, "Sur la Sphère Vide," Otdelenie Matematicheskikh i Estestvennykh Nauk, Vol. 7, 1934, pp. 793-800.

[21] G. C. P. King, R. S. Stein and J. Lin, "Static Stress Changes and the Triggering of Earthquakes," *Bulletin of the Seismological Society of America*, Vol. 84, No. 3, 1994, pp. 935-953.

9

The Analysis of the Van-Ercis Earthquake, October 23, 2011 Turkey, for the Transportation Systems in the Region

Hakan Aslan

Department of Civil Engineering, Division of Transportation, Sakarya University, Adapazarı, Turkey

ABSTRACT

This paper investigates the effect of the Van-Ercis, Turkey, (M_w: 7.2) earthquake occurring on 23rd of October, 2011 on the transportation networks in the region. The basic incentive for this research is to conceptualise the reliability and performance of the networks after the earthquake through the operational and topological analysis of the system. The demand and composition of the traffic along with the behaviour of the pedestrians were taken into account to evaluate the performance of the networks. In addition, the general structure of the cities and towns, as far as planning is concerned, is also paid attention and regarded as one of the main elements for the appraisal. The outcomes obtained are thought very important to be guidance for the expected Istanbul earthquake in the near future.

Keywords: Earthquake Hazards; Functionality and Reliability of Transportation Networks; Degraded Road Networks

1. Introduction

The investigation of transportation networks under the possible effect of natural disasters, such as earthquakes, floods, hurricanes etc., became more important after the Kobe earthquake in Japan in 1995. The basic motivation of these researches is to investigate the reliability and durability of the road networks from the perspective of infrastructure planning and traffic management when some of the components, such as links, of the system are partly closed to traffic with reduced capacity or completely out of use resulting in degraded networks.

Nicholson and Du (1997) proposed an integrated equilibrium model for the networks with long term capacity degradation to allow the traffic flow to adjust itself and move towards a new equilibrium situation [1]. The proposed model is a multi-modal network in the sense that the disaster will not necessarily affect all the transportation modes equally. The model mainly assesses the socioeconomic impacts of road networks under degradation and produces a unique equilibrium solution when a standard concave programming algorithm is used.

Asakura (1998) developed an approximation algorithm for nondeterministic network states as the probability of the failure of a link in the system cannot be precisely determined due to the nature of the earthquake or any other type of disasters [2]. The algorithm investigated the performance of the network through possible network states in terms of the stochastic travel time of an OD pair given in the network. The comparison of generated travel time in a degraded road network considering every combination of the link failures to normal network state is based on the reliability measure defined as the probability of whether the ratio of the travel times obtained from both degraded and normal network states remain within a predetermined acceptable level. The proposed algorithm converges to the exact expected value through upper and lower bounds of reliability measure by employing the most probable state vectors reflecting optimistic and pessimistic expectation of the operated/failure function, respectively.

Kiremidjian et al. (2007) proposed a method for the performance evaluation of a highway transportation system based on the expected loss from damage to bridges and resulting travel times [3]. The method analysed the network by assuming both the travel demand is fixed and variable after earthquake. Risk assessment of transportation system was based on the consideration of both direct cost of the damage and the costs resulting from time delays in the degraded system. The evaluation of the travel times for all O-D pairs in the system was carried out by using the commercial software EMME/2. The study confirmed the fact that the total vehicle hours increased for the post-earthquake state even the demand in the system is supposed to be the same with the normal network. The variable travel demand model on the other hand, produced decreased vehicle travel hours as the model assigned fewer trips to the system.

Samadzadegan and Zarrinpanjeh (2008) suggested a pre-event digital vector map and post-event high resolution satellite imagery usage to design and assess the damages to the road networks [4].

Shoji and Toyota (2012) analysed the functionality of road networks over which the restoration process of lifeline systems is carried out [5].

As having huge potential of major earthquakes, Turkey faces high probability of devastating earthquakes in the near future. The one being expected to hit the Marmara region where Istanbul is located seems to be the most important with the possible magnitude around M_w: 7.0 and is thought to have a significant effect on the daily life of the people causing possibly many of them to be killed and the general economy of the country. This is simply due to the fact that the population density and industrial activity in the region represents the highest level for the country.

This paper investigates the technical points of the Van-Ercis earthquake with regard to transportation problems, assesses the network topology and proposes some likely countermeasures to be taken to relieve the adverse effects of the earthquakes by bearing in mind the planning characteristics of the cities in Turkey.

2. The Transportation Network Performance of City of Van

The earthquake which hit the southern part of Turkey on 23rd of October, 2011 and affected Van, Igdır, Agrı, Bitlis, Kars, Batman, Siirt, Hakkari, Erzurum, Sanlıurfa, Mardin Mus, Diyarbakır and Northern Iraq had a magnitude of 7.2 Riechter Scale (*United States Geological Survey, Kandilli Observatory and Earthquake Research Institute*). This earthquake is one of the top three earthquakes along with Kocaeli, M_w 7.6 in 1999 and Duzce, M_w 7.1 in 1999, and among the first ten strongest for the last 110 years in terms of moment magnitude. According to official figures, 604 and 2000 people lost their lives and wounded, respectively. The number of people rescued from the rubbles is given as 222.

Although there was not any problem to arrive Van, the most affected city in the region, by plane, some one-hour delays were reported due to the restriction on parking facilities at the airport during the first couple of days after the disaster. In general, however, it should be stated that the role of Van Ferit Melen Airport was vital for the transportation of the rescue teams and emergency materials from both Turkey and all over the world to the city.

As the destructive effect of the earthquake in the Van city centre was limited and the number of buildings collapsed was just a few, the city transportation network system performed its function without significant decrease in its capacity. This provided an efficient transportation environment to reach the places where people were in urgent need. The transportation infrastructure interms of both roads and traffic management applications, such as traffic signals, was functional and traffic flows were quite running. **Figure 1** illustrates the functionality of the transportation network at the heart of the city of Van.

The relative high number of the people left the city due to the psychological reasons after the earthquake caused the demand to decrease significantly by having positive effect on the quality of the city transportation system. The main form of the public transportation, paratransit systems, performed its function and there was not any degradation on service quality.

The state highways connecting Van-Muradiye, Muradiye-Caldiran, Caldiran-Ercis, Ercis-Adilcevaz, Adilcevaz-Ahlat, Ahlat-Tatvan, Tatvan-Edremit, Edremit-Van were not directly affected by the earthquake, and there was not any decrease in terms of the level of service apart from those caused by routine maintenance and construction works. Although some minor cracks caused by the faulting mechanism were determined on the Ercis-Van state highway, located very close to the earthquake epicentre, the required repairs were done quite swiftly and effectively so that no major negative impact experienced by the traffic. **Figure 2** clarifies the effect of fault on the land and state highway.

As for village roads; Van-Alakoy road was determined as the one physically affected by the fault. Nevertheless, as the cracks were repaired quite rapidly, the traffic flow and connection were not affected. The **Figure 3** gives an idea about the size of the cracks on the road and their repairs providing smooth traffic flow.

Figure 1. Signalised intersection between the airport and the Van city centre after earthquake.

Figure 2. The effect of fault mechanism on Van-Ercis state road and its repair.

Figure 3. The cracks and repaired road sections on Van-Alakoy village road.

The effect of the earthquake being local on a few points in the region along with quick and efficient responses should not be deceptive and not to cause the underestimation of the requirement of the enough human sources, mechanical equipment stock as well as taking the measures of the maintenance work before the earthquake for the transportation systems prone to high scale

earthquake impact.

Having this size of earthquake in major cities like Istanbul may result in significant cracks causing the loss of continuity of roadway and highway surface. As a result, there might be major connectivity problems along the network preventing the people reaching and leaving the highly affected areas. The connectivity plays very important role to rescue people from the rubbles and take them to the hospitals in the vicinity at the earliest time possible. The infrastructure must be kept as strong and robust as possible to minimise this adverse outcomes stemming from the cracks to be occurred as a result of strong earthquakes. The major intersection points may be regarded as the key locations with this regard and utmost effort should be paid to get the minimal connectivity problems in case of a strong earthquake.

3. The Analysis of the Effects of the Earthquake on Ercis Network System

The district of Ercis where the highest destructive effect of disaster was observed became the mostly affected residential area with regard to transportation network performance. Although the ring road, **Figure 4**, passing the outside of Ercis and connecting the cities of Van and Agrı was not affected, central transportation system of the town was badly deteriorated.

The traffic on the main arteries of the town; Zeylan, Inonu, Cinarli, Emniyet and Ataturk Avenues, were disrupted due to the partly closures caused by the collapsed buildings on both sides of the roads. **Figure 5** demonstrates the main arteries of Ercis town centre.

The average travel speed is measured about 15 km/h (around 10 mph) at the town centre. The reason behind this low speed is attributed toon the one hand the capacity reduction of the roads, on the other hand the insufficiency for the applied traffic management and arrangement techniquesalong with the disobedience of the drivers and pedestrians to the traffic rules. The effective and efficient arrangements of horizontal and vertical traffic signs within a short period of time would result in a better traffic movement in terms of the physical capacity usage of the road network.

Figure 4. The Ercis ring-road after the earthquake.

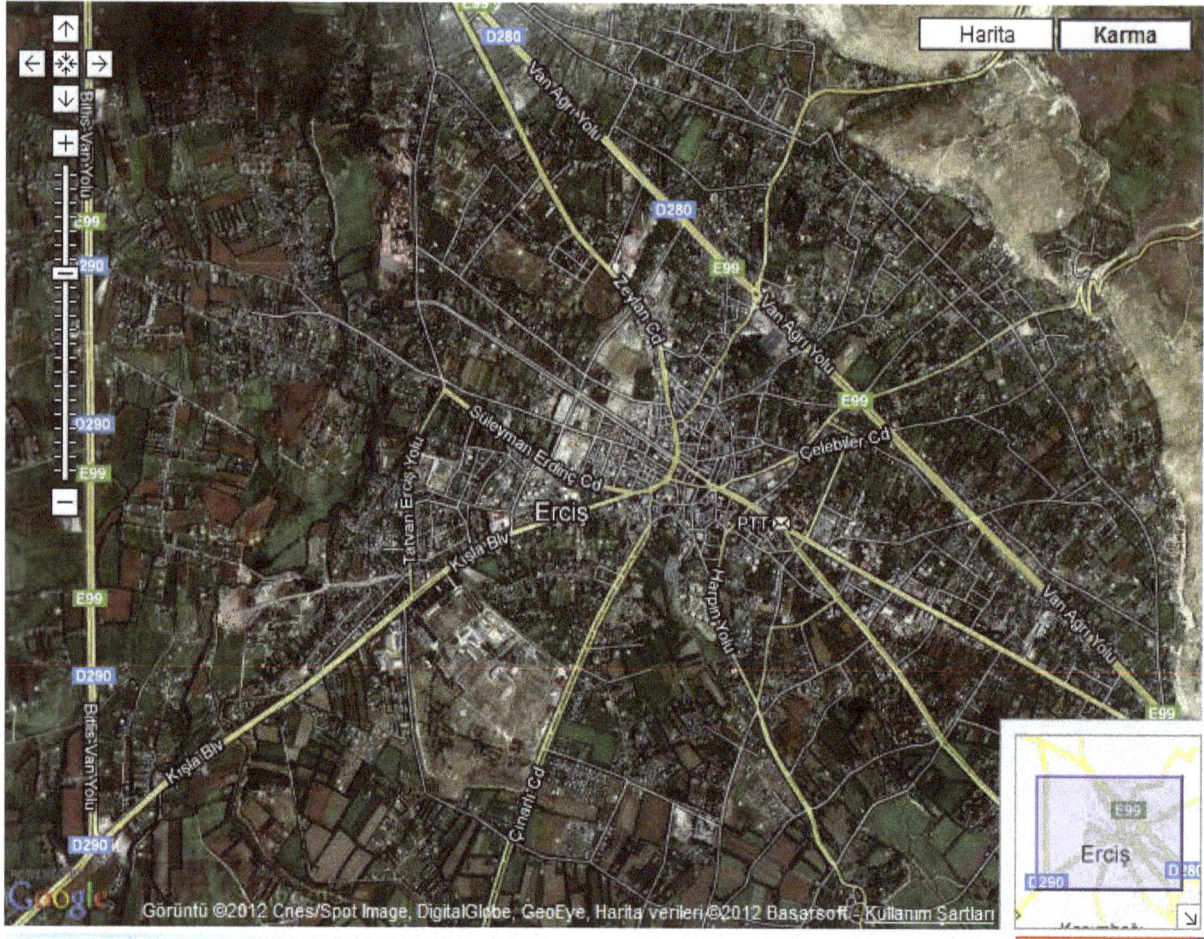

Figure 5. Ercis town centre and transportation network.

Figure 6 next page illustrates the chaotic nature of the traffic in Ercis town centre explaining the slow travel speeds.

Since the daily needs of the people, such as cooked food, water, blankets etc., are provided on the roads of the town by the specially designed lorries, there were long queues at the town centre blocking the through traffic to a degree causing slower traffic on some parts of the network.

As the earth-movers used to remove the rubbles had to use the network unlike from the usual daily traffic conditions, the concomitant effect of this usage on the average travel speed was unavoidable. This effect comes from both the slow speed and the size of these types of vehicles. It has been observed that the capacity of the roads is heavily occupied by these earth-movers on every time of the day.

Since the alternative road sections were not available for the through traffic, there occurred long queues right behind of these vehicles with even more reduced average travel speed of 15 km/h along Zeylan street. It has been advised to the local authorities that the development plan

for the topology of the transportation network in the Ercis town centre must consider this point and alternate connections must be available for the town centre. The radial network system seemed to be quite effective and feasible with this regard for the future planning of the town centre in Ercis providing alternate connections between Zeylan street and the ring road.

Figure 7 clearly shows the effect of non-daily traffic vehicles and movements on the traffic after earthquake.

The present bridges, viaducts and tunnels on the pre-mentioned alignments of highways in the region performed their functions and none of them has been reported being totally or partly collapsed causing significant accessibility problem. **Figure 8** illustrates the bridge on the road between Van and Alakoy, one of the villages mostly affected by earthquake.

As being the key elements of whole transportation network in terms of providing essential connectivity, bridges and viaducts are extremely important road sections. Thus they have to be constructed carefully and strongly by paying attention on their role and inevitable importance in case of an earthquake.

Figure 6. The chaotic structure of the central traffic flow at the town of Ercis.

Figure 7. The effect of earth-movers on Ercis town centre traffic.

Figure 8. A functional bridge just before Alakoy.

4. The Effect of City Planning and Collapsed Buildings on the Traffic Flow

It was fortunate that there were only a few buildings collapsed blocking the main arteries to a certain extent by not causing in total connectivity failure. In addition, al-

most every main artery was accessible through the available side roads providing a passage for the emergency and health service vehicles to take the required action on time. If the destructive effect of the earthquake had been more severe with much more building falling down, the topological characteristics of the transportation network would not have lessened the negative aspects of the outcomes to be experienced. The town centre, where the governmental offices are located to be used for managing the earthquake, was reachable only via Zeylan street. The total closure of this single artery in the future earthquakes means that the management of the earthquake would be much less efficient in terms of both providing the basic life saving equipments to the people and the coordination of the organisations being in the region to help the residents.

Another important point to be taken into account is related with the development plan of the cities and towns especially with the high-density population. The required distance between the edge of the roads and the buildings with regard to the height of those buildings should be determined in a way that the collapse of the buildings would not cause the entire even partly closures, reducing the capacity significantly, of the roads to be used for effective response in case of earthquake. The minimisation of this negative effect for the main arteries plays a vital role for the performance of the transportation network for the future and stronger earthquakes in the region. On some main roads in Ercis, it was observed that residential buildings started just the end side of the main roads without even pavement sections for the pedestrian usage. The collapse of such buildings had direct impact on the partly closure of the vital roads.

Figure 9 illustrates the effect of the collapsed buildings on the main arteries in Ercis.

Another striking point observed was the lack of parking spaces for the private cars both day time and night time. It has been detected that the number of cars squashed under the rubbles of the buildings is quite numerous. This fact must urge the local authorities to determine and provide safe parking areas and facilities for the people living in Ercis. This would surely minimise the cost of the car owners and insurance companies along with providing excellent opportunity to use the parking areas as the accommodation of the people in the tent cities. Day-time usage of the parking areas as part of the park-and-ride traffic would result in the demand reduction revealing the traffic congestion in the town centre. According to the traffic survey conducted by us the required parking area for the day time is for 5000 vehicles in terms of car equivalent around the town centre.

Figure 10 might give an idea about the collapsed buildings and trapped vehicles under the rubbles.

In order to maximise the mobility of those vehicles,

Figure 9. The effect of collapsed buildings on the town centre traffic in Ercis.

Figure 10. Vehicles trapped under the collapsed buildings.

especially the heavy ones, carrying aid materials, the traffic assignment and parking locations of these vehicle must be predetermined, which was not the case in Ercis. This would certainly provide the fastest arrival of these vehicles to the points where they are needed by minimising their possible adverse effects to the daily traffic on the network. The fact that around 150 articulated lorries demanded to use the network daily in the first three weeks to provide the urgent needs for the people of Ercis with the population of 90,000 caused the network to be much busier and slower than the pre-earthquake state. The size and number of these vehicles were beyond the capacity and physical structure of the town network for a smooth and efficient traffic flow. It is quite clear that this becomes much more important for a possible earthquake affecting major cities with a population more than couple of millions.

The present law forces the compulsory construction of the parking areas under the apartments through the basement. This provides both parking facilities for the vehicles and stronger buildings against earthquakes. However, the investigation of this point in Ercis town

centre revealed the fact that even major buildings did not have such facilities. The importance of this emphasised during the technical meetings with the local authorities and they were asked to show zero tolerance for the non-construction of these basements as the construction would increase the cost of the building to some extent.

5. Pedestrian Movements

Another important traffic element that should be mentioned is the movement of the pedestrians. The disorganized nature of the traffic right after the earthquake influenced the pedestrians as well. The observed pattern of the pedestrians in Ercis town centre was quite disorderly without obeying the basic traffic rules, including disobedience to traffic lights, walking on the pavements, sudden crossings the roads, etc. Although this can be tolerable to some extent considering the slow moving nature of the traffic and the psychological situation of the people right after the disaster, all measures should be taken to prevent this traffic pattern to be the daily habitual after the earthquake's adverse effects disappear in the long term.

6. Discussions and Conclusions

The problems caused by earthquake should be dealt with through holistic approach as it is related with many aspects of the daily life. Security, health services, geotechnical engineering, structural engineering, Emergency Aid and Rescuing, infrastructure of life-lines, transportation etc. are some of the parts of all these multi-categorised nature of the earthquake. All these sub-elements point out the effective coordination between the official and private organisations for the risk and hazard analysis of the earthquake.

As far as the transportation system and risk analysis are concerned, the connectivity, travel time and capacity reliability evaluations with regard to different scenarios must be done well before the earthquake hits and the strategies should be developed in terms of the worst case scenario. This analysis would not only relieve the transportation system and traffic flow but will also provide an efficient rescuing activity along with the fact that the aid materials reach their required destinations efficiently after the earthquake.

These studies have not been done in the region thoroughly and the cities and towns were not ready for the devastating effects of the earthquakes. Luckily, the effect of the earthquake was not proportional to the magnitude of it. If the effect had been bigger, the result that has been faced with would not have been as moderate as it happened.

The optimal locations of the rescue centres are another important problem in the field of transportation to be solved. This is vital as far as the prompt actions to be taken to save the lives.

The realistic evaluation of the transportation system as a whole in the region revealed the following points to be considered to lessen the current and possible future effect of the earthquakes in the region and as a whole.

- The network topology must be analysed to make sure that maximum accessibility is available in case of earthquake for the future planning of the cities and towns. The vital official organisations are to be located to minimise the adverse effects of the present topology's accessibility limitations. The determination of these locations is an optimality problem to be solved.
- The constructional arrangements of the buildings especially on the major arteries should be done to minimise the possible road closures in case of collapse.
- The parking facilities should be analysed and required capacity must be provided at the feasible locations of the cities and towns by considering the needs of the aid convoys, too, as an element in emergency transportation planning in long-term planning.
- Horizontal and vertical traffic signs must be repaired and kept operational right after the earthquake to provide a safe and efficient traffic environment for both vehicles and pedestrians.
- The availability of the public transportation system plays an extremely important role after the earthquake for the mobility of the population in an effort to reduce the earthquake effect on the mobility. Hence, it is vital that the infrastructures of the public transporttation systems are functioning. The provision of this results in orderly and efficient traffic system on the network by encouraging people using public transportation systems rather than being on the roads without obeying the traffic rules with their private cars.
- In order to give the impression to the people that the traffic system is still under control and operated by professionals even after the earthquake, these officials must be ready for the post-earthquake network management through proper training sessions. The contents of these sessions are to be determined by taking into account of the plants and paths of the public transportation, general behaviour of the public, physical structure of the transportation network and possible magnitude of the earthquake itself.

As the major transportation corridor to Iran is in this region, the possible future earthquakes might affect this highway corridor inextricably tied to transportation network causing significant trade loss. The required maintenance works must be properly scheduled and the plants must be kept ready all the time to minimise the adverse effects of the earthquake in this sense.

REFERENCES

[1]   Z. P. Du and A. Nicholson, "Degradable Transportation Systems: Sensitivity and Reliability Analysis," *Transportation Research Part B*, Vol. 31, No. 3, 1997, pp. 225-237.

[2]   Y. Asakura, "Reliability Measures of an Origin and Destination Pair in a Deteriorated Road Network with Variable Flows," *Selected Proceedings of the 4th EURO Transportation Meeting*, 1999, pp. 273-287.

[3]   A. Kiremidjan, J. Moore, Y. Y. Fan, O. Yazlalı, N. Basoz and M. Williams, "Seismic Risk Assessment of Transportation Network Systems," *Journal of Earthquake Engineering*, Vol. 11, 2007, pp. 371-382.

[4]   F. Samadzadegan and N. Zarrinpanjeh, "Earthquake Destruction Assessment of Urban Roads Network Using Satellite Imagery and Fuzzy Inference Systems," *The International Archives of the Photogrammetry, Remote Sensing and Spatial Information Sciences*, Vol. 37, Part B8, Beijing, 2008, pp. 409-414.

[5]   G. Shoji and A. Toyota, "Function of Emergency Road Networks during the Post-Earthquake Process of Lifeline Systems Restoration," *Journal of Disaster Research*, Vol. 7, No. 2, 2012, pp. 173-182.

10

Analysis of the Low-Energy Seismic Activity in the Southern Apulia (Italy)

Pierpaolo Pierri[1], Salvatore de Lorenzo[1], Gildo Calcagnile[1,2]

[1]Dipartimento di Scienze della Terra e Geoambientali, Università degli Studi di Bari "Aldo Moro", Bari, Italy
[2]Osservatorio Sismologico, Università degli Studi di Bari "Aldo Moro", Bari, Italy

ABSTRACT

In this paper we analysed the historical and instrumental seismicity of the seismic district "Penisola Salentina" (Salento peninsula) in the southern part of the Apulia region, making use of the most recent seismological database. Relocation of available dataset points out that the events are spatially distributed over a belt of deformation that approximately corresponds to Soglia Messapica (Taranto-Brindisi depression). Besides, computed source characteristics indicate dextral strike-slip solutions with an approximately E-W orientation that seismologically confirm previous geodynamic studies indicating a NE-SW extension in the Taranto-Brindisi depression. In particular, the tensional stress associated to the present seismic activity could be the consequence of the relaxation of the buckling process following the extensional re-arrangement of the Apenninic belt masses. Moreover attenuation Q_P values obtained in this study are much greater than those inferred in other parts of Italian peninsula; this result agrees with previous macroseismic investigations and indicates a greater efficiency of the studied area in the transmission of body waves.

Keywords: Salento Peninsula (Southern Italy); Seismicity; Focal Mechanisms; Source Parameters

1. Introduction

The Apulian region stretches for about 350 km in the southern part of Italy, between the Adriatic and Ionian Sea. It is constituted by an emerged sector of the Apulian plate [1] or Adriatic subplate [2] or Adria [3], characterized by a relatively thick lithosphere [4] and by a weakly deformed sedimentary cover [5].

This plate, which shows a marked elongation from NW to SE, represents the Plio-Pleistocene foreland (Apulian foreland) of the Southern Apennine orogenic system to west and of the Dinaric and Hellenic chains to east, generated along its boundaries by the interaction with more deformable lithospheric structures. These regions are affected by diffuse seismicity whose characteristics are correlated to a general counter-clockwise motion of Adria [6,7] (**Figure 1**).

The internal part of Adria shows a minor, but not negligible, seismic activity [8,9]; in particular this region is near to different areas in which seismicity is frequent and intense; for example the Salento peninsula is less than 100 km far from Albanian and Greek coasts where many energetic earthquakes occurred. Besides the propagation

characteristics of the lithosphere of the foreland permit to the energy irradiated by hypocenters distant a few hundreds of kilometres of arriving into the Salento peninsula only weakly attenuated, as demonstrated from felts of Albanian and Greek earthquakes [10,11] and from macroseismic field of recent earthquakes (e.g. 8 January 2006, 3 February 2007, 25 March 2007).

The southern part of Apulia has been generally considered practically aseismic [12]. The seismic history of Salento peninsula shows that only one event of magnitude higher than 6.0 is reported in the historical catalogues, killing hundreds of people, in particular in the town of Nardò and Francavilla Fontana: this earthquake, which occurred on 20 February 1743, caused damage maximum effects of IX-X degree on the Mercalli-Cancani-Sieberg scale (MCS) and also a tsunami [13], with boulder accumulations along the Otranto-Leuca coast [14]; it has been felt in an abnormally wide area, from Greece, Albanian, Malta to northern Italy; its magnitude was in the range 7.13 ± 0.19 [15]. It is affected by remarkable uncertainty in the location, but most likely occurred south of the Salento Peninsula (Apulian Ridge), in the

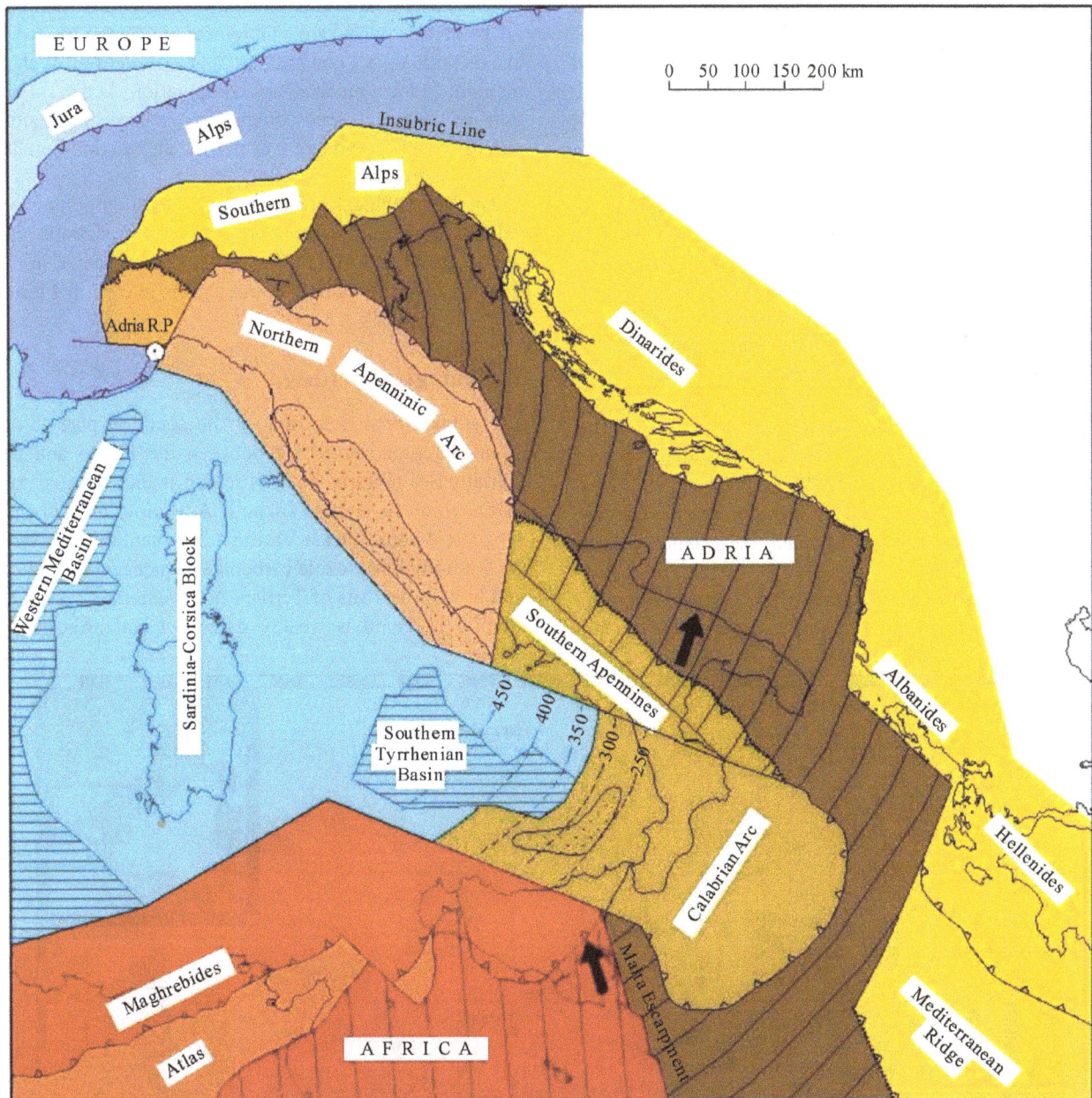

Figure 1. Structural sketch of Italy and surrounding areas [6]. Black arrows indicate the slip vectors of Africa vs. Europe and of Adria vs. Europe obtained from geodetic data. Adria RP is the Adria rotation pole.

same area where there have been several other recent earthquakes (e.g. 20 October 1974, 23 November 1974). According to Argnani et al. [16] the main triggering factor is the local stress accumulation due to the small radius of curvature of the Adriatic-Apulian plate under the double load of the Hellenides and Apennines-Calabrian arc. This event has been studied in detail by Galli and Naso [11]: it seems that the highest intensities in Salento peninsula have been controlled by local amplification (double resonance) that occurred in all localities characterized by thin Pleistocene basins filled with soft sedi-

ments, such as Nardò, Francavilla Fontana and Leverano. The depth of the seismogenic source (30 - 40 km), its directivity effects (toward Salento, in NW-SE direction) and the strong site amplification were considered the reasons of both the large areal distribution of effects and the locally high gravity of damages causing the devastating shaking [11].

The occurrence of strong earthquakes in this area is also suggested by Pieri et al. [17], considering the sparse occurrence of seismites in the Tyrrhenian deposits along the Adriatic-Apulian coast; besides additional data on the possible presence of a strong seismic activity that would affect the Salento peninsula have been put forward on the basis of geomorphological (e.g. through the study of speleothems and tsunami traces) and archaeological evidence (recent excavations have revealed the possible origin of collapse of walls).

The seismic activity inside Salento peninsula is almost absent or of low-energy and therefore it is rarely recorded, due to the lack of seismic stations (until about 15 years ago the "Istituto Nazionale di Geofisica e Vulcanologia" (INGV—Rome) managed only 2 stations in an area more than 5000 km^2). In the last years, four seismic stations were implemented in the southern-central part of Apulia region and added to the network managed by the "Osservatorio Sismologico dell'Università di Bari" (OSUB). The recordings of this network, in conjunction with those of the Italian National Seismic Network (RSNC), operated by the INGV, allowed detecting several low-energy events, such as those occurred on 5 May, 2012 near Ostuni and felt by many inhabitants, of local magnitude M$_L$ 2.8.

In this paper we present an analysis carried out on this earthquake and on other events located in the seismic district named "Penisola Salentina" (**Figure 2**). The main goals of this work are: 1) relocation of the recorded events with Vp/Vs computation using a modified Wadati method; 2) focal mechanisms computation; 3) determination of source (corner frequency, source dimension, seismic moment and stress drop) and attenuation (quality factor) parameters.

The geographical position of the most important localities of the studied area and the seismic stations (belonging to OSUB and INGV network), considered in the revision of instrumental seismicity, are shown in **Figure 2**.

2. Geological Setting

The studied area is located in the southern part of the Apulian region, stretching between the Ionian and the Adriatic Sea (**Figure 2**). This region is the emerged part of the foreland domain of both Apenninic and Dinaric orogens; it constitutes a Variscan basement covered by a 3 - 5 km thick Mesozoic carbonate sequence and is overlain by thin deposits of Tertiary and Quaternary age. The Apulian foreland is weakly deformed and affected by

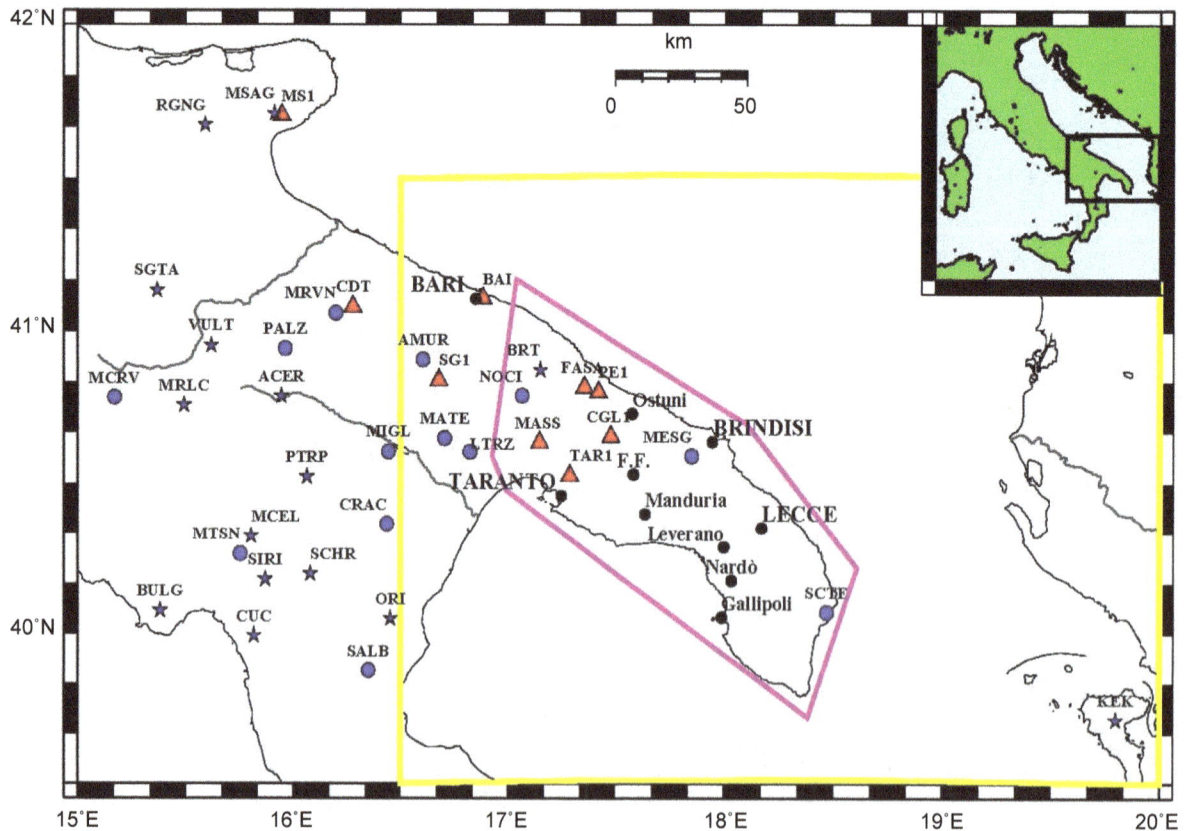

Figure 2. Geographic location of the study area with the seismic stations used in the relocation. Red triangles represent the stations of OSUB network, while blue circles indicate the stations of INGV network. The yellow rectangle is the area selected for the extraction of seismic events shown in Figure 4; the fuchsia line encompasses the "Penisola Salentina" seismic district. Geographical position of some localities mentioned in the paper is also shown (F.F. indicates Francavilla Fontana).

Apenninic and anti-Apenninic trending faults which subdivide it into five main structural blocks with different uplift rates (from NW to SE Gargano, Tavoliere, Murge, Taranto-Brindisi plain and Salento peninsula).

The Salento peninsula and the Murge block are separated by the so-called "Soglia Messapica" [17], also named Taranto-Brindisi depression [5], which is a high scarp, mainly oriented E-W; the relationship between Salento and Murge is complicated by their different rotations [18] and by strike-slip movements along their boundary (**Figure 3(a)**, North and South Salento Fault Zone by Gambini and Tozzi [19]).

Normal faults, with trend NW-SE, are present in both the Salento peninsula and the Salento plateau, represented by the "Apulian Ridge or Swell" [16,20] (**Figure 3(b)**); these faults, active in Plio-Pleistocene, almost transversal to the strike-slip fault in which occurred energy transfer, dissect the large antiform structure due to the different regional uplift and the Pleistocene deposits, witnessing a tectonic activity in Quaternary. The Apulian Ridge is a morphological element that separates the deep Ionian basin from the shallower Southern Adriatic basin, extending from Salento peninsula to the island of Kefallinia.

3. Data Selection

All the events located in the area having a latitude between 39.5° and 41.5°N and a longitude between 16.5° and 20.0°E in the period 2003-2012 were extracted from the Bulletin of the Instrumental Seismicity (http://bollettinosismico.rm.ingv.it/) and from the Italian Seismic Instrumental and parametric Data-basE (ISIDE) (http://iside.rm.ingv.it) of INGV.

These events are distributed over the southern Apulia and surrounding regions. In **Figure 4** the 743 extracted events, subdivided in 9 "seismic districts" (or epicentral regions), are shown; the districts where the highest number of events occurred are "Murge" (273) and "Penisola Salentina" (163).

However this map may be severely biased by the recording of quarry blasts. It has been in fact shown [21] that many of these events were caused by artificial explosions. This observation is supported by both the distribution of the events with respect to the day of occurrence and to the hour of occurrence.

In this paper we considered the seismic activity occurred inside the district named "Penisola Salentina", where 163 earthquakes were located with a maximum magnitude $M_L = 2.8$.

Owing to the poor coverage of INGV stations in this area, some low-energy events were not detected by the INGV network. As an example, the events occurred on 6 August, 2012 (at 23:14 GMT) and on 6 September, 2012 (at 04:48) were detected only by OSUB stations.

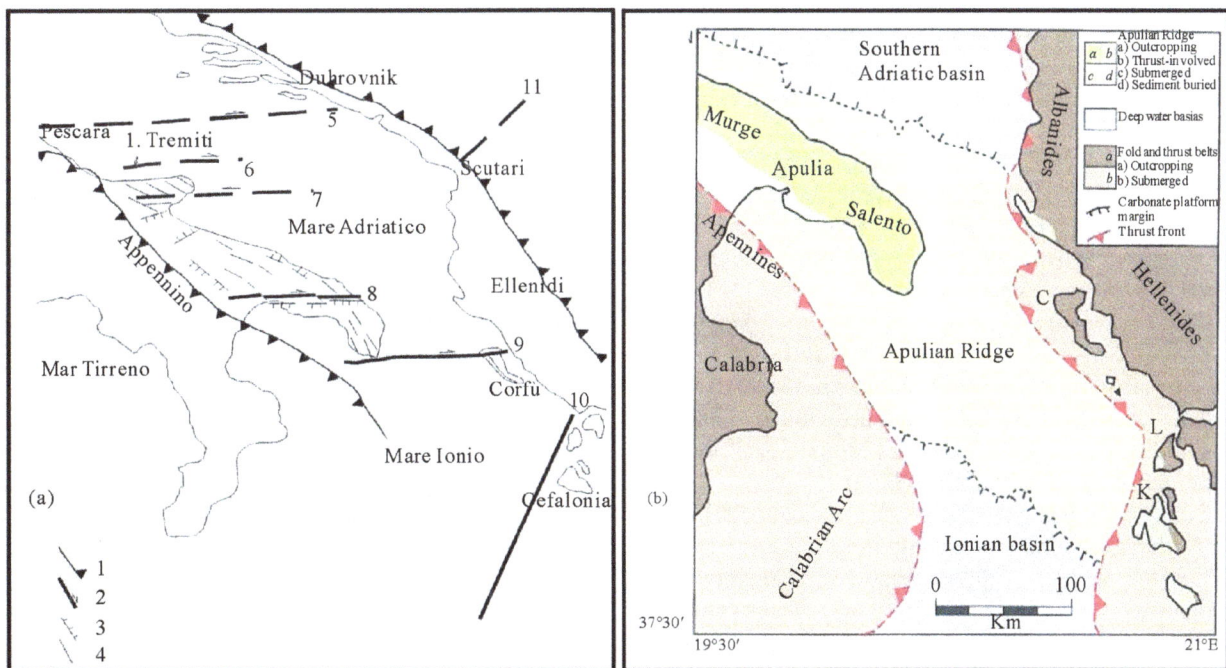

Figure 3. (a) Structural map of the Apulian foreland [19, simplified]: 1) Front of the external Calabrian Arc; 2) main strike-slip faults (the arrows indicate the movement versus); 3) main extensional faults; 4) other fault alignments; 5) Pescara-Dubrovnik fault; 6) Tremiti fault; 7) Mattinata fault; 8) North-Salento fault; 9) South-Salento fault; 10) Kefallinia fault; 11) Scutari fault. (b) Geological setting of the study area [16]: the Apulian Ridge (or Apulian Swell) represents the foreland of the Apennines and Hellenides fold and thrust belts. With C, L e K are indicated Corfu, Lefkas and Kefallinia Island.

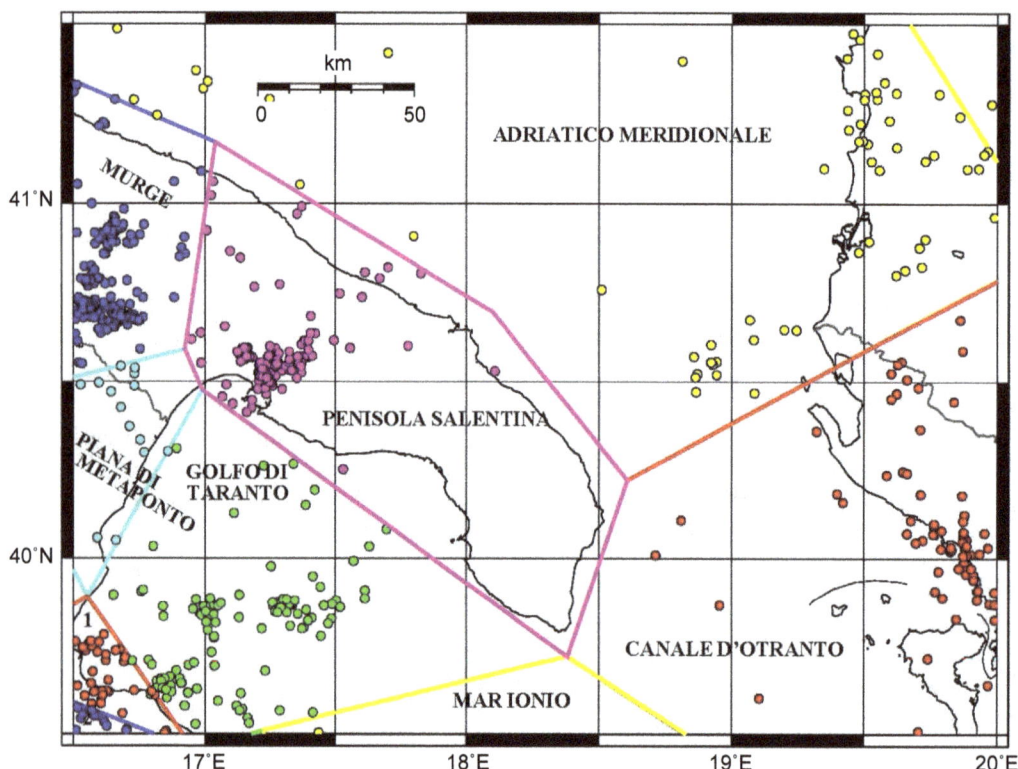

Figure 4. Map of instrumentally detected earthquakes of the Southern Apulia and surrounding regions extracted from the INGV on-line bulletins between 2003-2012. The 743 epicentres are localized in 9 seismic districts (1 and 2 represent the seismic districts "Piana di Sibari" and "Sila" in the SW corner).

The southern and central part of Apulia is historically characterised by low seismicity [8]. **Figure 5** shows the spatial distribution of historical earthquakes extracted from the last version of the Parametric Catalogue of Italian Earthquakes (CPTI11) [15], which covers the period 1000-2006; **Table 1** lists a few events located in "Penisola Salentina" seismic district or very close, including the instrumental (M_L = 5.3) earthquake occurred in 1983 near Gallipoli (see **Figure 2**).

In this catalogue the strong event occurred in 1743 is located (on macroseismic base obviously) offshore the southern-eastern Salento coast (at about 50 km), on the Salento plateau, in the ZS931 (named Otranto Channel) of the ZS9 seismogenic zonation [22], near well located epicentres of other earthquakes (**Figure 5**).

Besides, analyzing the CFTI4Med catalogue [23] other 25 earthquakes located in the examined area occurred in XX century with M ≥ 3.0; it is important to stress that the strong earthquake occurred in 1743 is relocated in Salento peninsula, about 20 km north of Gallipoli (grey star in **Figure 5**).

The seismic activity reported in the catalogue PFG [24] is also shown in **Figure 5**: this catalogue (which covers the period 1000 - 1980) contains 170 earthquakes occurred in the extraction area, of which 47 earthquakes

with I ≥ VI MCS; several events are not present in the other 2 catalogues, for example the earthquakes occurred on 10 October, 1858 near Brindisi (I = VI MCS, M = 4.1) and 26 April, 1970 in the Ionian Sea (I = VII MCS, M = 4.6).

On the whole, the existing pre-instrumental seismicity indicates the occurrence of many events of low magnitude (M ≈ 4.0) in Southern Apulia.

The spatial distribution of instrumentally located events, extracted from the CSI catalogue [25] in the period 1981-2002, has been also analysed; the seismicity is present everywhere in all the districts, although, considering only the events with M ≥ 3.0, cluster of events can be noted, for example in the area near Lecce. Since data relative to these years and to "Penisola Salentina" seismic district are rather poor and with a remarkable uncertainty on the phase readings, we re-analysed the instrumental seismicity only in the period 2003-2012.

4. Discrimination between Tectonic Earthquakes and Quarry Blasts

The concentration of events near Taranto (**Figure 4**) is mainly due to an anthropic activity (explosions). We considered the approach of Wiemer and Baer [26] to analyse

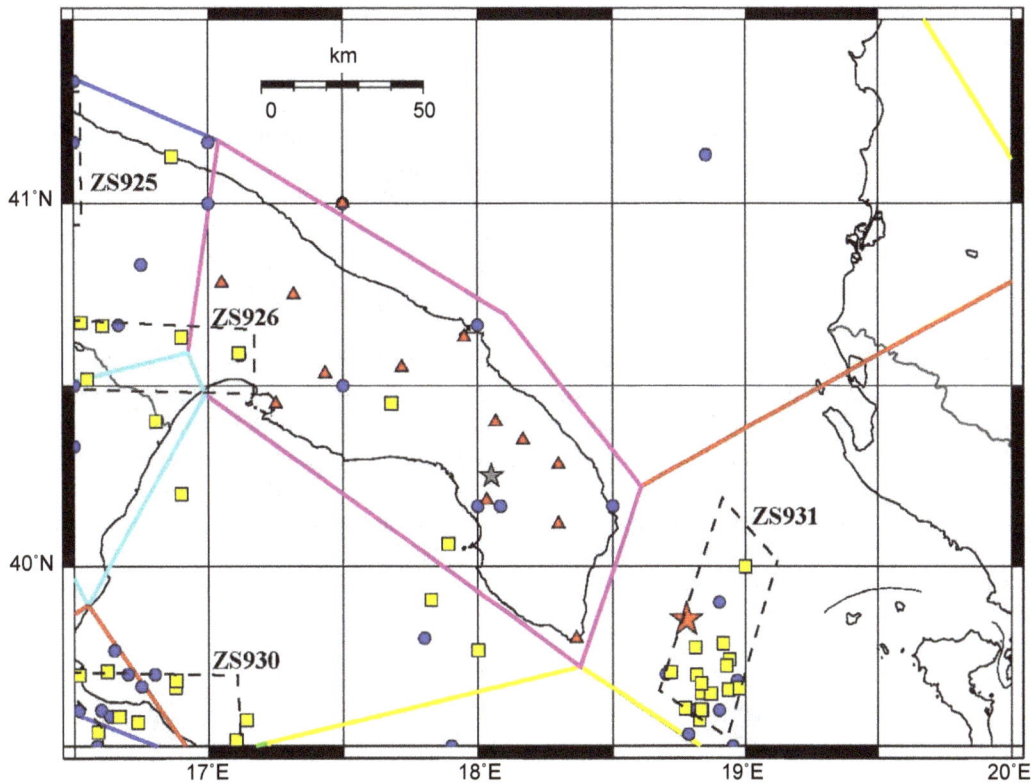

Figure 5. Map of historical detected (red circles and blue triangles for events with M ≥ 3.0) earthquakes of the Southern Apulia and surrounding regions extracted from the CPTI11 [15] (yellow squares), PFG catalogue with I_{min} ≥ VI MCS [24] (blue circles) and CFTI4Med catalogues [23] (red triangles) respectively. Red and grey stars indicate the epicentre of the 1743 earthquake located from CPTI11 and CFTI4Med respectively. The delimitation and the number of the seismogenic zones according to the ZS9 [22] (dashed lines) is reported. The boundaries of 9 seismic districts are shown.

Table 1. List of historical earthquakes extracted from the Parametric Catalogue of Italian Earthquakes [15] occurred in the "Penisola Salentina" seismic district or very close to it. See Figures 2 and 9 for locations.

Date	Or. Time	Lat. N	Long. E	Mw	I_{max}	Location
1087/09/10	-	41.128	16.864	4.93	VI-VII	Bari
1713/01/03	-	40.589	17.113	4.51	VI-VII	Massafra
1826/10/26	18	40.451	17.678	5.36	VI-VII	Manduria
1932/03/30	09:56	40.633	16.900	4.80	VI	Castellaneta
1983/05/07	22:09	40.062	17.890	4.96	-	Gallipoli
1983/11/08	20:11	39.907	17.825	4.56	-	Pen. Salent.
2001/09/23	21:16	39.767	18.001	4.96	-	Pen. Salent.

the frequency distribution of the events. It is known that dividing the number of events occurred in working hours (generally between 08:00 a.m.-04:00 p.m. for mining activity) or days (from Monday to Friday) by the number of events recorded in not-working hours (00:00 a.m.-08:00 a.m. and 04:00 p.m.-00:00 a.m.) or days (Saturday, Sunday or holiday) the indicator Rq can be computed

[26]:

$$Rq = \left(N_d / N_n \right) / \left(L_d / L_n \right) \qquad (1)$$

N_d is the number of events recorded in diurnal hours, N_n is the number of events recorded in night hours and the term L_d / L_n is a normalization factor equal to the ratio between the number of diurnal and night hours (for hypothesis 8 and 16 respectively). It has been shown that **Rq** is equal to unit in areas where are recorded only tectonic events, whose occurrence is clearly casual at any time or day, while anomalous values (>> 1) are obtained if also quarry explosions are present (effectuated usually between 08:00 a.m.-04:00 p.m.).

In our case we inferred N_d equal to 151, N_n equal to 12, and therefore Rq ≈ 25; this result indicates that the major part of recorded events is represented by quarry blasts. This outcome is supported by the histograms representing the distribution of the number of the events versus daytime and weekday (**Figure 6**), where the anomalous distribution, having a minimum on Saturday and Sunday and in intervals of not-working hours, is inferred.

It is, however, highly probable that not all 151 diurnal events are due to quarry explosions, but the attention has been paid only to events surely of tectonic origin, very often easily distinguishable by the simple inspection of waveforms and by spectral analysis. It's important to

remember that exist a guideline for the discrimination of earthquakes from mine explosions on the web (http://earthquake.usgs.gov/earthquakes/eqarchives/mine blast/evidence.php); a typical feature of explosion seismograms is the envelope with a "fish-like" shape (**Figure 7(a)**), P-onset emergent with only compressive polarity, lack of a clear S wave signature (the secondary phase is due to converted phases), magnitude less than 2.0, and waveforms very different from those of tectonic events (**Figure 7(b)**).

Among the 151 diurnal events, an earthquake surely of tectonic origin, as suggested from different location and magnitude respect to classical "anthropic" events and as confirmed from waveform analysis (with some tensional polarities), is the event occurred on 13 May, 2011 in working time (06:21 GMT-08:21 local time) and day (Friday); it was one of the most energetic events, felt by the population.

Adding other 9 events occurred in not working-days

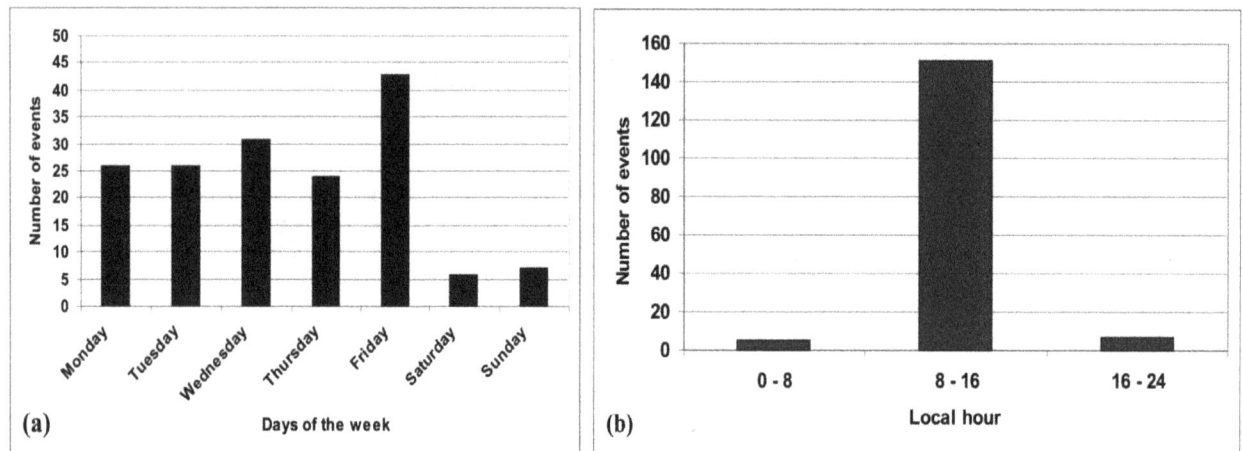

Figure 6. Distribution of the 163 "Penisola Salentina" events in the period 2003-2012 (a) versus days of the week and (b) three local hour intervals.

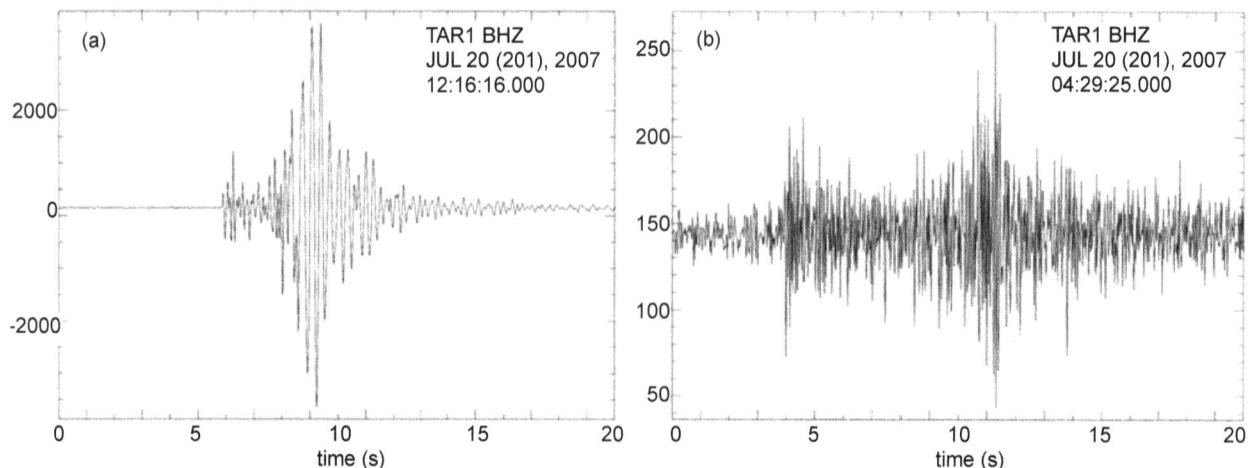

Figure 7. Comparison between the waveform of (a) an anthropic event (quarry blast) and (b) a tectonic earthquake recorded at TAR1 station on 20 July, 2007 at 12:16:22 and at 04:29:29 (ID 6 in Table 2) respectively (GMT hour).

(but in working hours), the resultant dataset includes 22 events.

5. Event Relocation

First of all, we estimated the average Vp/Vs ratio, required by the program HYPOELLIPSE [27]. In the determination of this ratio we required that the following three criteria were jointly satisfied by each earthquake: at least 5 stations, 8 phases and 2 OSUB phases. The last criteria is fundamental to better constrain the solution using the nearest stations. Based on these rules, about 700 phases of 14 earthquakes have been included and 8 earthquakes have been excluded (ID number 1, 2, 8, 10, 11, 12, 13 and 14 in **Table 2**), despite an accurate re-picking of their arrival times for both the INGV and the OSUB stations. In any case we relocated all 22 selected earthquakes, with maximum local magnitude equal to 2.8, using only data of stations having epicentral distance up to 200 km, giving the maximum weight to stations placed within 100 km.

The Vp/Vs ratio was estimated by using a modified Wadati method [28]. In this method the average **Vp/Vs** ratio is equal to **DTs/DTp** ratio, where **DTs** and **DTp** are the time difference between phases S_i and S_j and P_i and P_j in two stations i and j; the slope of the least square fit of DTs versus DTp for all available pairs of stations (**Figure 8**) gives the value of the slope Vp/Vs (1.78 with a linear correlation coefficient of 98 %).

For the choice of a local velocity model, trials were made adopting different models: the best results, in terms of minimum RMS and error ellipsoid, were obtained (considering whether 14 or 22 events) with the model OSUB (reported in [8]) which was definitely assumed in all the following processing.

Compared to the locations provided by the INGV (214 arrival times) without the OSUB stations, relocations have an average RMS and azimuthal gap smaller (from

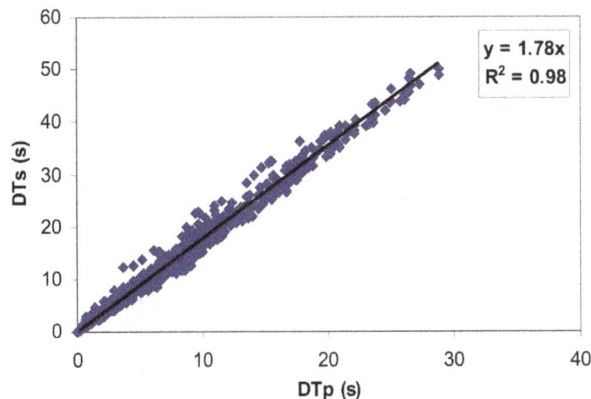

Figure 8. Wadati diagram with average Vp/Vs ratio obtained (1.78) using linear least squares method. Linear correlation coefficient is equal to 98%.

0.43 s to 0.29 s and, respectively, from 245° to 218°); in our study we exploited 324 phases.

The accuracy in the location of events is about 2 km for the epicentral coordinates and about 5 km for the focal depth; it becomes better for the most recent events, owing to the implementation of 3 seismic stations (CGL1, MASS, FASA) in the period October, 2010 December, 2010. As an example, for the event occurred on 11 May, 2008 the minimum source to receiver distance is almost 60 km (see D_{min} in **Table 2**), while the minimum distance becomes about 15 km for the events of 2011-2012. As expected, almost all discarded events have rather high errors.

The depths of hypocentres, about 15 km on the average, indicate that this activity is located within the crystalline basement and in the deepest part of the upper crust.

The distribution of the relocated epicentres is rather sparse, but roughly follows an E-W striking trend that corresponds to Soglia Messapica (**Figure 9**).

6. Focal Mechanisms

Since the earthquakes recorded in the "Penisola Salentina" seismic district have a low magnitude ($M_L \leq 2.8$), it is difficult to collect enough data for the determination of focal mechanisms; therefore fault plane solutions were computed only for three events (ID 9, 19, 21) having at least eight clear observations. We used FPFIT [29] to infer focal mechanisms. The velocity model used for the computation of the azimuths and take-off angles is the same as that used for the location of events.

In all cases both the low number of seismic stations (the average number of polarities per event is 11) and their spatial distribution does not allow to obtain well constrained fault plane solutions. For all the events, the best provided solution (**Figure 10** and **Table 3**) shows for the pressure axis P a trend of about 300° and a plunge of about 30°, whereas the tension T axis has a trend of about 45° (NE-SW direction) and a plunge of 20°. All events reveal strike-slip faulting mechanisms along E-W striking planes and in particular the best constrained mechanism of the 5 May, 2012 earthquake is well representative of this kind of solution.

Due to the limited number of polarities and the eastern network gap (Adriatic Sea area) the reliability of the fault plane solution is not fully assessed. However the score of correct polarities is rather high (equal to 100% for the third event) and the uncertainty is very low for strike and dip of the two nodal planes (<15°), whilst the uncertainty is rather high for the rake (in two cases > 20°) even if it is worth noting that the fault plane solutions carried out remain approximately the same.

For other events occurred in the "Penisola Salentina"

Table 2. List of instrumentally earthquakes located from INGV in the "Penisola Salentina" seismic district and relocated in this study. ID = identification number. Time is the earthquake origin time (UTC). In the column "Days" H represents holiday, while days are numerated from 1 (Monday) to 7 (Sunday). Depth is the earthquake depth (fixed when marked as *). Nd1 and Nd2 represent the number of used phases of the OSUB and INGV stations. Ns is the number of used stations. Dmin is the minimum distance (in km). Md and Ml are the duration magnitude and the local magnitude (taken from INGV Seismic Bulletin). The 8 events highlighted in grey have been excluded in the estimate of the Vp/Vs ratio.

ID	Date	Time	Days	Lat. N	Lon. E	Depth	Nd1	Nd2	Ns	Dmin	Gap	Md	Ml	RMS	AZ	SEH1	SEH2	SEZ
1	2005/08/29	15:05:54.2	1	40.833	17.166	5.0*	0	6	3	4.7	258	1.7	1.4	0.14	−66	1.7	0.9	2.1
2	2006/01/06	10:21:25.9	H	40.916	17.013	10.9	0	5	3	12.3	204	1.9	1.1	0.13	−43	2.8	1.8	6.6
3	2006/01/21	09:14:10.9	6	41.059	17.044	10.1	7	6	7	14.9	201	1.9	1.6	0.25	33	1.1	0.6	2.6
4	2006/04/23	07:55:07.7	7	40.679	17.397	7.8	6	6	6	13.7	203	1.9	1.3	0.38	−87	1.7	1.0	5.2
5	2007/04/10	15:05:31.8	2	40.542	17.260	10.5	5	6	6	2.4	144	2.2	1.6	0.41	1	2.2	1.7	2.9
6	2007/07/20	04:29:21.2	5	40.858	17.552	15.2	5	8	7	12.9	290	2.2	1.9	0.30	21	2.5	1.4	1.2
7	2007/12/23	11:53:35.8	7	40.400	18.008	25.4	6	14	11	53.4	164	2.8	2.5	0.40	33	3.6	0.8	1.7
8	2008/04/30	14:22:58.5	3	40.577	17.177	11.8	0	6	3	25.3	298	2.0	1.6	0.07	−48	0.9	0.5	1.9
9	2008/05/11	23:03:13.2	1	40.843	17.830	8.0	5	19	17	58.1	303	/	2.6	0.39	30	2.5	2.0	7.4
10	2008/10/16	15:19:21.5	4	40.528	16.990	5.0*	0	6	3	16.7	287	1.6	1.1	0.16	−31	3.3	1.0	14.5
11	2009/08/23	10:15:14.6	7	40.801	17.841	4.3*	0	6	3	23.6	237	2.2	2.0	0.51	30	18.2	2.0	99.0
12	2009/11/15	05:07:50.4	7	40.619	16.937	2.5	0	5	3	10.1	231	1.5	0.7	0.21	−37	9.2	1.4	65.9
13	2010/04/30	19:32:46.1	5	40.604	17.715	2.1	0	8	4	58.7	340	2.1	1.6	0.35	11	4.6	2.8	99.0
14	2010/05/14	03:53:16.6	5	40.625	16.998	3.7	0	6	3	15.3	244	1.4	1.1	0.36	−37	6.4	1.4	34.6
15	2010/09/11	02:47:25.1	6	40.789	17.275	8.0	7	6	7	12.1	126	2.0	1.6	0.23	40	0.8	0.6	2.9
16	2010/10/31	07:16:19.8	7	41.120	17.082	3.5	8	14	13	17.1	225	2.2	2.1	0.19	32	0.8	0.5	8.8
17	2011/05/07	13:19:53.0	6	40.481	17.023	22.6	4	6	7	19.7	228	1.8	1.1	0.22	4	1.7	1.0	3.1
18	2011/05/07	13:40:21.8	6	40.521	17.073	12.7	4	8	8	13.9	165	1.7	1.0	0.40	−16	1.9	0.9	5.7
19	2011/05/13	06:21:29.8	5	40.772	17.529	18.7	15	19	18	13.8	178	2.5	2.3	0.38	20	1.2	0.8	0.8
20	2011/09/25	01:54:57.5	7	40.946	17.088	9.5	12	8	10	17.6	167	2.2	1.4	0.34	38	1.0	0.5	3.4
21	2012/05/05	12:44:03.9	6	40.533	17.542	7.3	14	38	29	14.7	99	-	2.8	0.43	15	1.1	0.5	3.3
22	2012/12/22	19:31:28.2	6	41.004	17.365	23.6	12	8	10	20.9	212	-	2.1	0.27	26	1.0	0.6	1.8

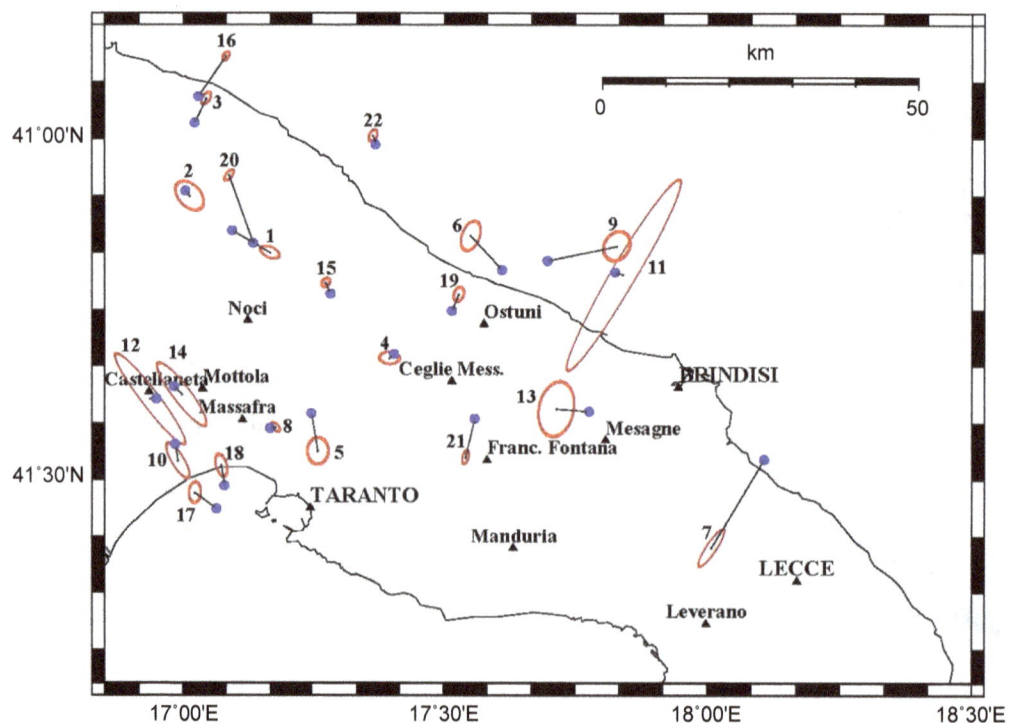

Figure 9. Events re-located in the "Penisola Salentina" seismic district with representation of the error ellipse estimated for each location (with ID number). In blue the INGV location is shown. Geographical position of main localities mentioned in the paper is also shown.

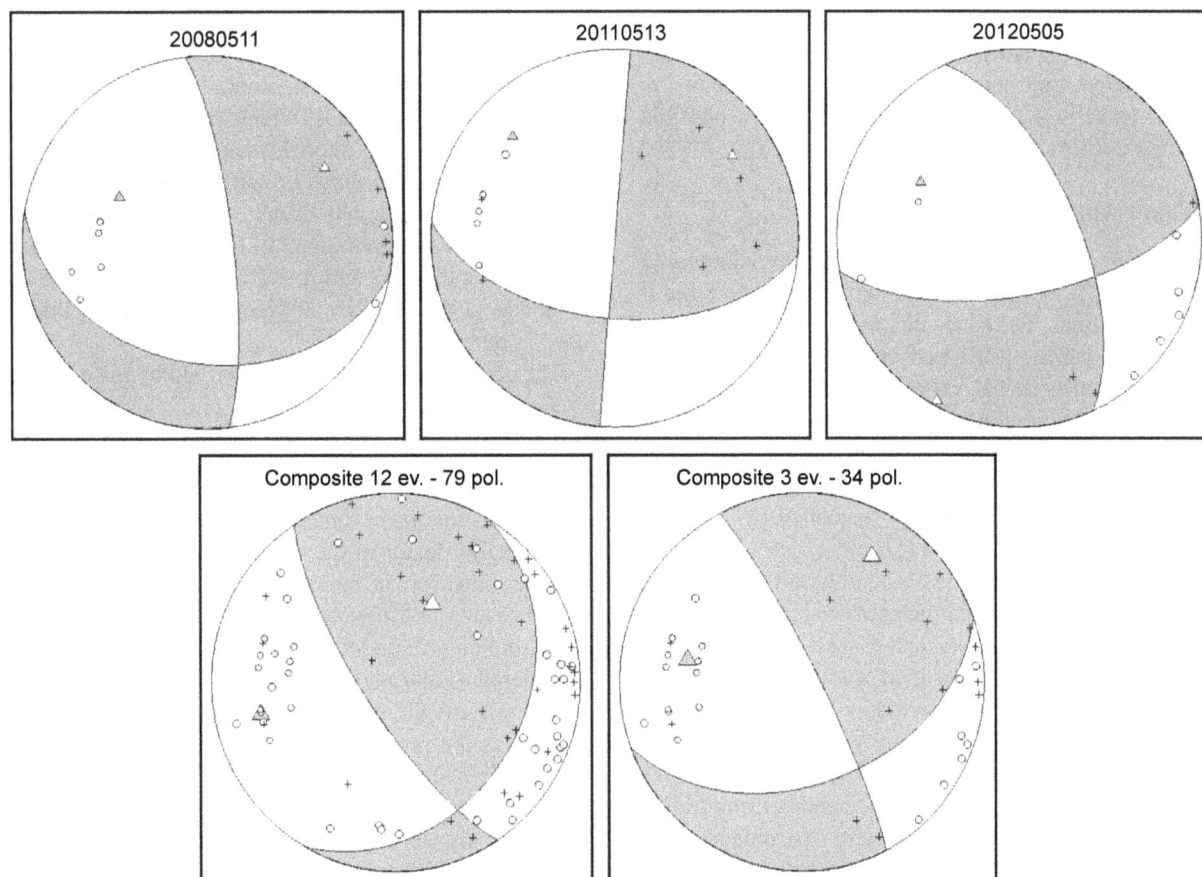

Figure 10. Focal mechanisms calculated with the code FPFIT [29] for three events (11 May 2008, 13 May 2011, 5 May 2012) and for the composite solution (two different tests): white and grey triangles mark the T and P axes, white circles and small crosses represent dilatational and compressive first arrivals.

Table 3. "Penisola Salentina" fault plane solutions carried out for the 3 major events and 2 composite solutions (A and B): strike (S), dip (D) and rake (R) of a nodal plane, score polarities (ratio between correct and total polarities), trend and plunge of P and T axes.

ID	Date	S	D	R	Score polarities	P-axis tr.	P-axis pl.	T-axis tr.	T-axis pl.
9	2008/05/11	100	35	−160	10/11 = 0.91	297	46	58	26
19	2011/05/13	95	55	−180	11/13 = 0.85	314	24	56	24
21	2012/05/05	80	65	−150	10/10 = 1.00	299	38	208	1
A	13 events	30	30	150	56/79 = 0.71	257	26	26	52
B	3 events	70	45	−170	28/34 = 0.82	281	36	30	24

seismic district (very probably originated from the same fault system) it has not been possible to infer stable estimates of the focal mechanism owing to the small number of available polarities. To overcome this limitation a composite fault plane solution has been obtained by combining the P-onset polarities of these events. Some tests with different number of polarities (including or excluding groups of polarities belonging to events with the highest errors of relocation) have been carried out: in **Figure 10** and **Table 3** the two most opposite tests are

shown, the first using all 79 polarities of 13 events (score 0.71) and the second joining only the polarities of the 3 above mentioned events (with 6 polarities in disagreement out of a total of 34). In both cases the composite solution fits well the solutions obtained for the single major events.

7. Source Parameters

In this section we discuss the results of an analysis aimed

at inferring source parameters (corner frequency, source dimension, seismic moment and stress drop) and attenuation parameters from the inversion of P wave spectra. As we will detail in a next section, this has been possible only for 3 events (ID 9, 19, 21).

7.1. Data Analysis

The seismic spectra were computed by considering a time window $T_L = 2.56$ s which starts 0.1 s before the P wave arrival time. To reduce distortions due to the finite length of the signals [30], a cosine taper window with a 15% fraction of tapering was applied to the P wave time window before computing its amplitude spectrum. Traces were deconvolved for the instrumental response.

In a preliminary analysis we noted that seismic noise dominates over the signal at low frequencies (f < 2 Hz) and at high frequencies (f > 40 Hz).

For this reason, we applied a band-pass filter 2 < f < 40 Hz to data and removed the mean value before computing the spectra. The same data processing was applied for the calculation of seismic spectra of noise. The analysis was performed on a time window having a length of 2.56 s that is adjacent to the time window used for the calculation of P wave spectra. Finally, an average moving window with a full width of five neighbouring points [31] was used to smooth the spectra. To obtain sufficiently stable estimates of model parameters, we selected only those signals having an average signal to noise ratio greater than 3.

7.2. Method

The velocity spectrum $U_{i,j}(f)$ of the i-th event observed at j-th station can be expressed as [e.g. 32]:

$$U_{i,j}(f) = S_{i,j}(f) B_{i,j}(f) R_j(f) \qquad (2)$$

where **S** is the source spectrum, **B** is the attenuation spectrum and **R** is the site spectrum. In the far-field approximation, the source spectrum can be written as:

$$S_{i,j}(f) = \frac{\Omega_{0,i,j}}{\left[1 + \left(\frac{f}{f_{c,i}} \right)^{\gamma n} \right]^{1/n}} \qquad (3)$$

In the previous equation, $f_{c,i}$ is the corner frequency, $\Omega_{0,i,j}$ is the low-level spectral amplitude, γ is the high-frequency spectral fall-off and **n** is a constant. The low-level spectral amplitude is related to the seismic moment by the equation:

$$M_0 = \frac{4\pi\rho r c^3 \Omega_0}{R_{\theta,\varphi}} \qquad (4)$$

where **r** is the source to receiver distance, ρ is the density

of rocks, $R_{\theta,\Phi}$ is the radiation pattern and **c** is the velocity of the considered wave.

The source spectrum is generally described by the Brune source model [33], which corresponds to the so called "omega square" model (*i.e.* n = 1 and γ = 2). As, in some cases, the recordings of earthquakes show a corner sharper than the original Brune model (for a review see [34]), a different approximation to the source model that corresponds to γ = 2 and n = 2 is used. This choice corresponds to the Boatwright source model [35], that we used in our analysis.

The attenuation spectrum is described by the equation:

$$B_{i,j}(f) = \exp\left(-\frac{\pi f t_{i,j}}{Q} \right) \qquad (5)$$

where $t_{i,j}$ is the travel time of the considered wave and **Q** is the quality factor of the waves. **Q** may depend on frequency; however, in many cases [e.g. 36] a constant **Q** model results in a best compromise between data fitting and simplicity of model. For this reason, in this study, we considered a constant **Q** model.

As concerns the site response $R_j(f)$, this is usually expressed as the product of the near-site attenuation function $K_j(f)$ and the local site amplification $A_j(f)$. The near-site attenuation is usually described above a limiting frequency known as f_{max} [37] in terms of the k_j attenuation factor, which was first introduced by Anderson and Hough [38]:

$$K_j(f) = \exp\left(-\pi f k_j \right) \qquad (6)$$

The local site amplification $A_j(f)$ is not described by any particular mathematical relationship, and it depends on the elastic and geometrical properties of the rocks near the recording site (e.g. [39]). It is generally computed by averaging residuals between theoretical and observed spectra, when many recordings of several earthquakes at a given station are available (e.g. [40,41]). In our case, we do not computed $R_j(f)$ owing to the small number of available recordings.

7.3. Inversion Technique

We used the 2-step inversion strategy developed by de Lorenzo *et al.* [36]. At the first step the whole physically admissible model parameter space is explored using a coarse grid. The range of model parameter values in the coarse grid is selected on the basis of the a priori analysis of the whole data set. At each point of this grid, a misfit function between the observed and the theoretical spectrum is computed. We compared the results obtained using several misfit function, as proposed by Edwards *et al.* [40] and inferred that the best results are obtained using the following **L1** misfit function of decimal logarithms of amplitude:

$$\sigma\left(f_c, t^*, \Omega_0\right) = \frac{1}{N} \sum_{i=1}^{N} \left| \log_{10}\left(U_{obs}\right) - \log_{10}\left(U_{teo}\right) \right| \quad (7)$$

This approach allows obtaining an initial estimate of model parameters m_{best_ini} without being subjected to the choice of an initial corner frequency.

At the second step, a refined grid is built around m_{best_ini} and the misfit function is computed at each point of this grid, allowing us to estimate the best fit model parameters m_{best} for each station.

Figure 11 shows the fit of model to data for the event occurred on 11 May, 2008.

After the inversion the event corner frequency is obtained as the average of the station corner frequency, whereas average Q_P and the seismic moment are obtained by the logarithmic averages of station values.

The estimated source parameters are summarized in **Table 4**.

8. Discussion and Concluding Remarks

The main goal of this study was to deduce the charac-teristics of the low-energy seismic activity that affects the "Penisola Salentina" seismic district, a generally con-sidered substantially aseismic area. In the historical cata-logues is reported only one event of magnitude greater than 6.0, occurred in 1743, and several other earthquakes of lower magnitude.

As regards the instrumental seismicity until 2002 the spatial coverage of the seismic stations was too poor in this area to obtain reliable relocations. Notwithstanding, the study of the set of seismic parameters, albeit in an area with a low seismic activity, might provide useful information on its seismological characteristics. There-fore we analysed the seismic activity from 2003 to 2012: the larger number of stations available in the last years, through the improvement of the OSUB network and the implementation of new INGV stations, allowed us to obtain more constrained relocation of low magnitude events compared to official locations (provided in Bulle-tin of the Instrumental Seismicity of INGV).

The "Penisola Salentina" seismic district is subjected to very frequent quarry blasts that make more difficult

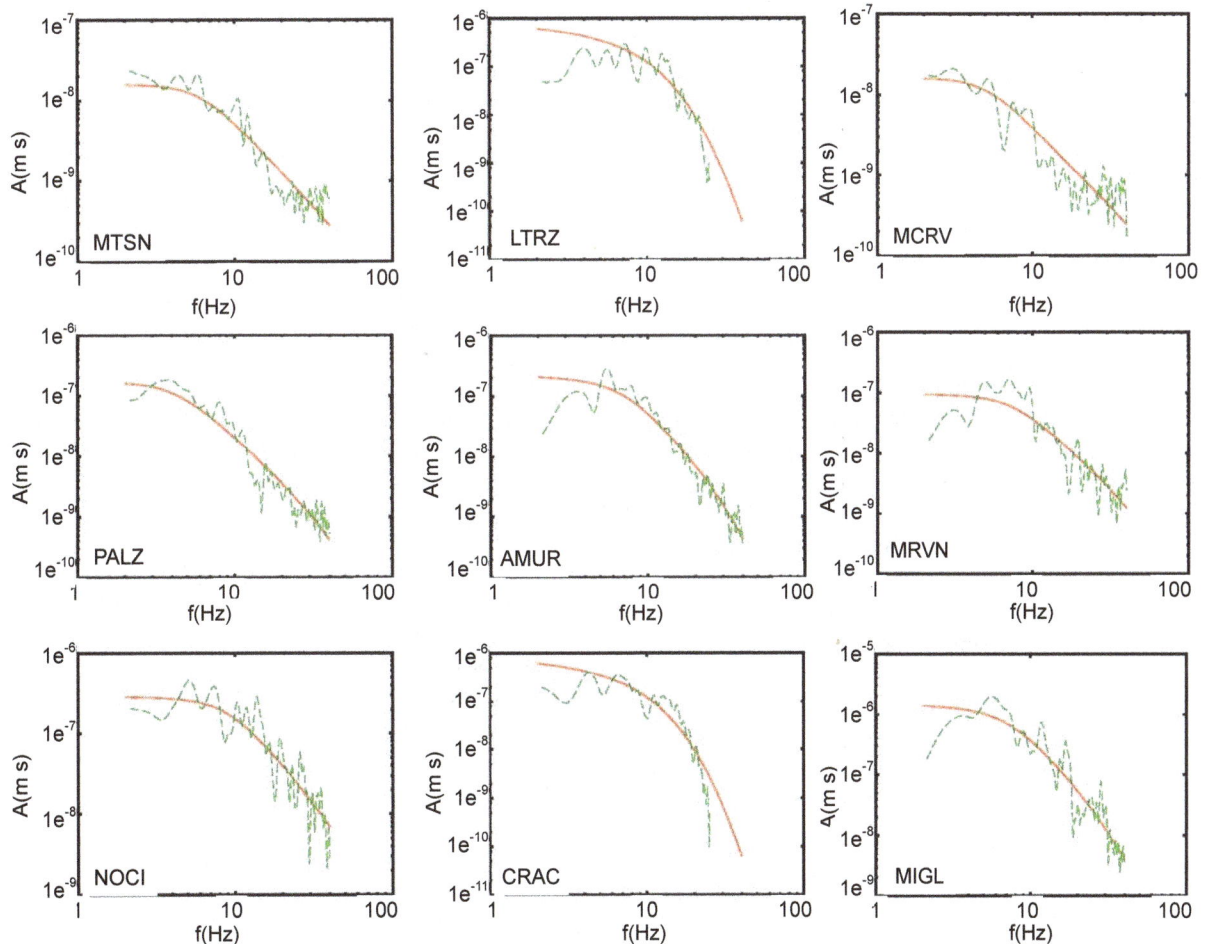

Figure 11. Fit of the P wave displacement spectra at different stations after the inversion for the event occurred on 11 May, 2008.

Table 4. Source and attenuation parameters for the studied events determined using Boatwright source model [35]. M_0, f_c, L, $\Delta\sigma$ and $<Q_P>$ are the seismic moment, the corner frequency, the source dimension, the stress drop and the average attenuation of the P-wave respectively.

ID	Date	M_0 (Nm)	f_c (Hz)	L (m)	$\Delta\sigma$ (Mpa)	$<Q_P>$
9	2008/05/11	$5.8E + 14 \pm 1.5E + 14$	8.2 ± 3.4	137 ± 57	18	2188
19	2011/05/13	$3.1E + 12 \pm 1.3E + 10$	10.4 ± 4.4	107 ± 45	0.2	826
21	2012/05/05	$4.2E + 13 \pm 1.3E + 12$	11.3 ± 3.5	99 ± 31	3	3982

the systematic collection of tectonic events; quarry blasts (probably 141 out of 163 events) and tectonic earthquakes (at least 22) have been discriminated. These 22 events have been relocated to obtain more reliable estimations on their hypocenters.

After several tests, a local velocity model, suitable for this area, has been adopted, with the Vp/Vs ratio equal to 1.78 estimated using a modified Wadati diagram. This value is slightly higher than the 1.73 Poissonian value and may indicate that the crust is partially fluid-permeated (e.g. [42]). Despite the high azimuthal gap and the quality of instruments (in some cases mono-component) the relocations are more precise. In fact a RMS reduction of 33% is obtained and an accuracy of about 2 km for the epicentral coordinates and about 5 km for the focal depth is inferred. The higher degree of accuracy in the computed locations, with respect to those available in the INGV database, is due to the use of data from both the INGV network and the OSUB network. The majority of these events (including the three most energetic) are located in a weakness zone, the Taranto-Brindisi depression, near the North Salento Fault Zone (see **Figure 3(a)**).

Information on the stress regime controlling this low-energy seismic activity is rather poor, because only for three events it has been possible to determine the focal mechanisms. The re-picking of the recordings allows us to detect 79 clear polarities on the whole, of which 34 for three major events.

Our best constrained solutions of the 5 May, 2012 earthquake reveal strike-slip faulting mechanisms along E-W striking planes. All five focal mechanisms inferred in this study have a common characteristic with regard to the trend of the T axes (of about 45°), which delineates the existence of an NE-SW extension direction. This pattern is similar to that found by some authors in Murgian area [8,43], despite the tectonic differences between the two adjacent areas. The presence of tensional stress in the Taranto-Brindisi depression is consistent with the hypothesis that the margin of the Adriatic plate has undergone a buckling process [44,45] following the extensional rearrangement of the Apenninic belt masses. The obtained focal mechanisms suggest that a tensional regime could be still active. The NE-SW active extension

in this area was previously inferred also by Di Bucci *et al*. [46] on the basis of mesostructural analysis; they proposed other different geodynamic models to justify it, for example a consequence of the convergence between Africa and Europe.

We also computed source (corner frequency, source dimension, seismic moment and stress drop) and attenuation (quality factor) parameters of the three major events. We inferred that the quality factor is generally at least a magnitude order higher than the average value ($Q \approx 300$) determined in other parts of Italian peninsula (Central Apennines, Campi Flegrei, Southeastern Sicily) [36,47, 48]. Therefore, even if the energy radiated by these earthquakes is generally smaller than that of tectonic or volcanic earthquakes, it is transferred with a greater efficiency, as was previously hypothesized on the basis of macroseismic fields of Albanian and Greek events. This result probably reflects the differences between the geodynamic context of the areas where the studied earthquakes occurred. In fact, whereas earthquakes occurring in the Apennine mark the limits among plates and therefore occur in strongly deformed areas, the events occurred in the southern Apulia are probably located inside a plate. The variability of stress drop estimates (roughly between 10 and 100 bar) may indicate that stress drop is not a selective indicator of the geodynamic context, as early proposed by Kanamori and Anderson [49].

9. Acknowledgements

We warmly thank Franco Mele and Alberto Basili of the "Istituto Nazionale di Geofisica e Vulcanologia" who promptly satisfied our data requests.

Most of figures were obtained by employing the GMT freeware package by Wessel and Smith [50].

This work was supported by MIUR (Italian Ministry of Education, University and Research) and by FCRP (Fondazione Cassa di Risparmio di Puglia, Bari).

REFERENCES

[1] J. M. Lort, "The Tectonics of the Eastern Mediterranean: A Geophysical Review," *Reviews of Geophysics*, Vol. 9, No. 2, 1971, pp. 189-216.

[2] J. E. T. Channel and F. Horváth, "The African/Adriatic Promontory as a Palaeogeographical Premise for Alpine Orogeny and Plate Movements in the Carpatho-Balkan Region," *Tectonophysics*, Vol. 35, No. 1-3, 1976, pp. 71-101.

[3] J. E. T. Channel, B. D'Argenio and F. Horváth, "Adria, the African Promontory, in Mesozoic Mediterranean Palaeogeography," *Earth-Science Reviews*, Vol. 15, No. 3, 1979, pp. 213-292.

[4] G. Calcagnile and G. F. Panza, "The Main Characteristics of the Lithosphere-Asthenosphere System in Italy and Surroundings Regions," *Pure and Applied Geophysics*, Vol. 119, No. 4, 1980/81, pp. 865-879.

[5] G. Ricchetti, N. Ciaranfi, E. Luperto Sinni, F. Mongelli and P. Pieri, "Geodinamica ed Evoluzione Sedimentaria e Tettonica dell'Avampaese Apulo," *Memorie della Società Geologica Italiana*, Vol. 41, 1988, pp. 57-82.

[6] D. Slejko, R. Camassi, I. Cecić, D. Herak, M. Herak, S. Kociu, V. Kouskouna, J. Lapajne, K. Makropoulos, C. Meletti, B. Muço, C. Papaioannou, L. Peruzza, A. Rebez, P. Scandone, E. Sulstorova, N. Voulgaris, M. Živčić and P. Zupančić, "Seismic Hazard Assessment for Adria," *Annals of Geophysics*, Vol. 42, No. 6, 1999, pp. 1085-1107.

[7] C. Meletti, E. Patacca and P. Scandone, "Construction of a Seismotectonic Model: The Case of Italy," *Pure and Applied Geophysics*, Vol. 157, No. 1-2, 2000, pp. 11-35.

[8] V. Del Gaudio, P. Pierri, G. Calcagnile and N. Venisti, "Characteristics of the Low Energy Seismicity of Central Apulia (Southern Italy) and Hazard Implications," *Journal of Seismology*, Vol. 9, No. 1, 2005, pp. 39-59.

[9] V. Del Gaudio, P. Pierri, A. Frepoli, G. Calcagnile, N. Venisti and G. Cimini, "A critical revision of the seismicity of Northern Apulia (Adriatic Microplate—Southern Italy) and Implications for the Identification of Seismogenic Structures," *Tectonophysics*, Vol. 436, No. 1-4, 2007, pp. 9-35.

[10] S. Castenetto, E. Di Loreto, L. Liperi and C. Margottini, "Studio Macrosismico e Risentimento in Italia dei Terremoti del Mediterraneo Centro-Orientale del 26 Giugno 1926 e del 17 Gennaio 1983," Atti 4° Conv. GNGTS, Rome, 1986, pp. 439-456.

[11] P. Galli and G. Naso, "The 'Taranta' Effect of the 1743 Earthquake in Salento (Apulia, Southern Italy)," *Bollettino di Geofisica Teorica ed Applicata*, Vol. 49, No. 2, 2008, pp. 177-204.

[12] N. Ciaranfi, A. Cinque, S. Lambiase, P. Pieri, L. Rapisardi, G. Ricchetti, I. Sgrosso and L. Tortorici, "Proposta di Zonazione Sismotettonica dell'Italia Meridionale," *Rendiconti della Società Geologica Italiana*, Vol. 4, 1981, pp. 493-496.

[13] S. Tinti, A. Maramai and L. Graziani, "The New Catalogue of Italian Tsunamis," *Natural Hazards*, Vol. 33, No. 3, 2004, pp. 439-465.

[14] G. Mastronuzzi, C. Pignatelli, P. Sansò and G. Selleri, "Boulder Accumulations Produced by the 20th of February, 1743 Tsunami along the Coast of South-Eastern Salento (Apulia Region, Italy)," *Marine Geology*, Vol. 242, No. 1-3, 2007, pp. 191-205.

[15] A. Rovida, R. Camassi, P. Gasperini and M. Stucchi, "CPTI11, the 2011 Version of the Parametric Catalogue of Italian Earthquakes," Milano, Bologna, 2011.

[16] A. Argnani, F. Frugoni, R. Cosi, M. Ligi and P. Favali, "Tectonics and Seismicity of the Apulian Ridge South of Salento Peninsula (Southern Italy)," *Annals of Geophysics*, Vol. 44, No. 3, 2001, pp. 527-540.

[17] P. Pieri, V. Festa, M. Moretti and M. Tropeano, "Quarternary Tectonic Activity of the Murge area (Apulian Foreland—Southern Italy)," *Annals of Geophysics*, Vol. 40, No. 5, 1997, pp. 1395-1404.

[18] M. Tozzi, "Assetto Tettonico dell'Avampaese Apulo Meridionale (Murge Meridionali—Salento) Sulla Base dei dati Strutturali," *Geologica Romana*, Vol. 29, 1993, pp. 95-111.

[19] R. Gambini and M. Tozzi, "Tertiary Geodynamic Evolution of Southern Adria Microplate," *Terra Nova*, Vol. 8, No. 6, 1996, pp. 593-602.

[20] S. Merlini, G. Cantarella and C. Doglioni, "On the Seismic Profile CROP M5 in the Ionian Sea," *Bollettino della Società Geologica Italiana*, Vol. 119, 2000, pp. 227-236.

[21] F. Mele, L. Arcoraci, P. Battelli, M. Berardi, C. Castellano, G. Lozzi, A. Marchetti, A. Nardi, M. Pirro and A. Rossi, "Bollettino Sismico Italiano 2008," *Quaderni di Geofisica*, Vol. 85, 2010, 45 p.

[22] Gruppo di Lavoro MPS, "Redazione Della Mappa di Pericolosità Sismica Prevista dall'Ordinanza PCM 3274 del 20 Marzo 2003," Rapporto Conclusivo per il Dipartimento della Protezione Civile, INGV, Milano-Roma, 2004, 65 p.

[23] E. Guidoboni, G. Ferrari, D. Mariotti, A. Comastri, G. Tarabusi and G. Valensise, "CFTI4Med, Catalogue of Strong Earthquakes in Italy (461 B.C.-1997) and Mediterranean Area (760 B.C.-1500)," INGV-SGA, 2007. http://storing.ingv.it/cfti4med/

[24] D. Postpischl, "Catalogo dei Terremoti Italiani dall'Anno 1000 al 1980," C.N.R.—Progetto Finalizzato Geodinamica—Quaderni della Ricerca Scientifica, 114 2B, Bologna, 1995, 239 p.

[25] B. Castello, G. Selvaggi, C. Chiarabba and A. Amato, "CSI—Catalogo Della Sismicità Italiana 1981-2002," Versione 1.1, INGV-CNT, Roma, 2006. http://csi.rm.ingv.it/

[26] S. Wiemer and M. Baer, "Mapping and Removing Quarry Blast Events from Seismicity Catalogs," *Bulletin of the Seismological Society of America*, Vol. 90, No. 2, 2000, pp. 525-530.

[27] J. C. Lahr, "HYPOELLIPSE—Version 2.0: A Computer Program for Determining Local Earthquakes Hypocentral Parameters, Magnitude, and First-Motion Pattern," US Geol. Surv. Open-File Rep. 89-116, 1989, 92 p.

[28] J. L. Chatelain, "Etude Fine de la Sismicité en Zone de Collision Continentale à l'Aide d'un Réseau de Stations Portables: La Région Hindu-Kush-Pamir," Ph. D. Thesis, Univ. Paul Sabatier, Toulouse, 1978.

[29] P. Reasenberg and D. Oppenheimer, "FPFIT, FPPLOT and FPPAGE: FORTRAN Computer Programs for Calculating and Displaying Earthquake Fault-Plane Solutions," US Geol. Surv. Open-File Rep. 85-739, 1985.

[30] S. Stein and M. Wysession, "An Introduction to Seismology, Earthquakes and Earth Structure," Blackwell, Malden, 2003.

[31] W. H. Press, B. P. Flannery, S. A. Teukolsky and W. T. Vetterling, "Numerical Recipes: The Art of Scientific Computing (Fortran Version)," Cambridge Univ. Press, Cambridge, 1989.

[32] F. Scherbaum, "Combined Inversion for the Three-Dimensional Q Structure and Source Parameters Using Microearthquake Spectra," *Journal of Geophysical Research*, Vol. 95, No. B8, 1990, pp. 12423-12438.

[33] J. N. Brune, "Tectonic Stress and the Spectra of Seismic Shear Waves from Earthquakes," *Journal of Geophysical Research*, Vol. 75, No. 26, 1970, pp. 4997-5009.

[34] R. E. Abercrombie, "Earthquake Source Scaling Relationships from −1 to 5 M_L Using Seismograms Recorded at 2.5 km Depth," *Journal of Geophysical Research*, Vol. 100, No. B12, 1995, pp. 24015-24036.

[35] J. Boatwright, "A Spectral Theory for Circular Seismic Sources: Simple Estimates of Source Dimension, Dynamic Stress Drop and Radiated Seismic Energy," *Bulletin of the Seismological Society of America*, Vol. 70, No. 1, 1980, pp. 1-27.

[36] S. de Lorenzo, A. Zollo and G. Zito, "Source, Attenuation, and Site Parameters of the 1997 Umbria-Marche Seismic Sequence from the Inversion of P Wave Spectra: A Comparison between Constant QP and Frequency Dependent QP Models," *Journal of Geophysical Research*, Vol. 115, No. B09, 2010.

[37] T. C. Hanks, "fmax," *Bulletin of the Seismological Society of America*, Vol. 72, 1982, pp. 1867-1879.

[38] J. G. Anderson and S. E. Hough, "A Model for the Shape of the Fourier Amplitude Spectrum at High Frequencies," *Bulletin of the Seismological Society of America*, Vol. 74, 1984, pp. 1969-1993.

[39] N. Tsumura, A. Hasegawa and S. Horiuchi, "Simultaneous Estimation of Attenuation Structure, Source Parameters and Site Response Spectra—Application to the Northeastern Part of Honshu, Japan," *Physics of the Earth and Planetary Interiors*, Vol. 93, No. 1-2, 1996, pp. 105-121.

[40] B. Edwards, A. Rietbrock, J. J. Bommer and B. Baptie, "The Acquisition of Source, Path, and Site Effects from Microearthquake Recordings Using Q Tomography: Application to the United Kingdom," *Bulletin of the Seismological Society of America*, Vol. 98, No. 4, 2008, pp. 1915-1935.

[41] B. Edwards and A. Rietbrock, "A Comparative Study on Attenuation and Source-Scaling Relations in the Kantõ, Tokai, and Chubu Regions of Japan, Using Data from Hi-Net and Kik-Net," *Bulletin of the Seismological Society of America*, Vol. 99, No. 4, 2009, pp. 2435-2460.

[42] C. Chiarabba and A. Amato, "Vp and Vp/Vs Images in the Mw 6.0 Colfiorito Fault Region (Central Italy): A Contribution to the Understanding of Seismotectonic and Seismogenic Processes," *Journal of Geophysical Research*, Vol. 108, No. B5, 2003, p. 2248.

[43] C. Maggi, A. Frepoli, G. B. Cimini, R. Console and M. Chiappini, "Recent Seismicity and Crustal Stress Field in the Lucanian Apennines and Surrounding Areas (Southern Italy): Seismotectonic Implications," *Tectonophysics*, Vol. 463, No. 1-4, 2009, pp. 130-144.

[44] G. Ricchetti and F. Mongelli, "Flessione e Campo Gravimetrico Della Micropiastra Apula," *Bollettino di Geofisica Teorica ed Applicata*, Vol. 36, 1980, pp. 381-398.

[45] C. Doglioni, F. Mongelli and P. Pieri, "The Puglia (SE Italy) Uplift: An Anomaly in the Foreland of the Apenninic Subduction Due to Buckling of a Thick Continental Lithosphere," *Tectonics*, Vol. 13, No. 5, 1994, pp. 1309-1321.

[46] D. Di Bucci, R. Caputo, G. Mastronuzzi, U. Fracassi, G. Selleri and P. Sansò, "Quantitative Analysis of Extensional Joints in the Southern Adriatic Foreland (Italy), and the Active Tectonics of the Apulia Region," *Journal of Geodynamics*, Vol. 51, No. 1-2, 2011, pp. 141-155.

[47] S. de Lorenzo, A. Zollo and F. Mongelli, "Source Parameters and Three-Dimensional Attenuation Structure from the Inversion of Microearthquake Pulse Width Data: Qp Imaging and Inferences on the Thermal State of the Campi Flegrei Caldera (Southern Italy)," *Journal of Geophysical Research*, Vol. 106, No. B8, 2001, pp. 16265-16286.

[48] S. de Lorenzo, G. Di Grazia, E. Giampiccolo, S. Gresta, H. Langer, G. Tusa and A. Ursino, "Source and Qp Parameters from Pulse Width Inversion of Microearthquake Data in Southeastern Sicily, Italy," *Journal of Geophysical Research*, Vol. 109, No. B07, 2004.

[49] H. Kanamori and D. L. Anderson, "Theoretical Basis of Some Empirical Relations in Seismology," *Bulletin of the Seismological Society of America*, Vol. 65, No. 5, 1975, pp. 1073-1095.

[50] P. Wessel and W. H. F. Smith, "New, Improved Version of Generic Mapping Tools Released," *Eos, Transactions American Geophysical Union*, Vol. 79, No. 47, 1998, 579 p.

11

A Wavelet Transform Method to Detect P and S-Phases in Three Component Seismic Data

Salam Al-Hashmi[1*], Adrian Rawlins[2], Frank Vernon[3]
[1]Earthquake Monitoring Center, Sultan Qaboos University, Muscat, Oman
[2]University of Newcastle upon Tyne, Newcastle upon Tyne, UK
[3]Institute of Geophysics and Planetary Physics, University of California, San Diego, USA

ABSTRACT

The discrete time wavelet transform has been used to develop software that detects seismic P and S-phases. The detection algorithm is based on the enhanced amplitude and polarization information provided by the wavelet transform coefficients of the raw seismic data. The algorithm detects phases, determines arrival times and indicates the seismic event direction from three component seismic data that represents the ground displacement in three orthogonal directions. The essential concept is that strong features of the seismic signal are present in the wavelet coefficients across several scales of time and direction. The P-phase is detected by generating a function using polarization information while S-phase is detected by generating a function based on the transverse to radial amplitude ratio. These functions are shown to be very effective metrics in detecting P and S-phases and for determining their arrival times for low signal-to-noise arrivals. Results are compared with arrival times obtained by a human analyst as well as with a standard STA/LTA algorithm from local and regional earthquakes and found to be consistent.

Keywords: Discrete Time Wavelet Transform; P and S-phases; Automatic Detection; Rectilinearity Function

1. Introduction

Seismic events such as earthquakes cause a release of energy represented by seismic waves that can be recorded by seismic monitoring stations. Detection of seismic waves and estimation of their arrival times provides information about earthquake location and magnitude. Analysis and detection of seismic waves may be done by visual inspection by a trained analyst or automatic analysis software.

Rapid and accurate detection of seismic waves is of great importance in locating earthquakes [1-10]. Automatic detection techniques are of interest because they can be processed in near real-time; they apply a consistent set of metrics and are repeatable. However, it is often very difficult to determine consistent estimates of P and S waves if they have low signal-to-noise characteristics, particularly if arriving at seismic stations at regional distances.

The objectives of this research are:

1) To design an algorithm for the detection and analysis of the two main types of seismic body waves (P and S-phases).

2) To test the algorithm using seismic events recorded by various stations at local and regional distances, and then compares the results with the results obtained by other methods.

Several methods have been tried recently these that have involved digital signal processing in both time and frequency domains [11-13]. The method adopted here is wavelet transform using an approach initially developed by K. Anant and F. Dowla [11].

The algorithms have been developed in Matlab 6.5[TM] (http://www.mathworks.com) including signal processing and WaveLab[TM] (http://www-stat.stanford.edu/~wavelab) toolboxes. WaveLab is a library of Matlab routines for wavelet analysis, wavelet-packet analysis, cosine-packet analysis and matching practice. The algorithms can be run on a laptop with 256 MB RAM or on a mainframe computer.

Seismic Data Used in Testing

Seismic data from Earthquake Monitoring Center in Oman has been used for testing the algorithms. The ANZA Broadband Seismic Network (http://eqinfo.ucsd.edu) also provided analyst reviewed data and STA/LTA automatic detected data used for comparison with the wavelet detector.

*Corresponding author.

The paper structured as follows: An introduction about wavelet transform is given in Section 2. Section 2 also includes description of the discrete time wavelet transforms, its application on seismic signals to match specific features and how the selection of wavelet type has been done in this study. The multiresolution analysis technique and the perfect reconstruction filter banks are explained too. The developed software and methodology are explained in Section 3. The flow charts of the algorithms are illustrated in Appendix B and a detailed annotation is given in Section 3 too. The results are discussed in Section 4. Comments about the testing results and the comparison results with other methods are presented in Section 5.

2. Wavelet Transform Analysis

2.1. Introduction

The wavelet transform is a very useful tool for analysing noisy and transient signals and has the ability to represent the signal in both its time and frequency domains. It is a very useful tool in the analysis of non-stationary signals such as seismic signals due to its ability to resolve features at various scales [14].

The wavelet theory arises in 1909 when Haar constructs the first orthonormal system of compactly supported functions called the Haar basis [15].

The term wavelet has been coined in the field of seismology in 1940 by Ricker, N. [15]. The wavelet theory has been applied on seismic signals by Grossmann and Morlet in 1984 [11]. In 1988, Daubechies developed an orthonormal, compactly supported wavelet basis that is smoother than Haar basis [14-16]. Application of wavelet transform on seismic signals has been done by Anant and Dowla [11], Oonincx, P. J. [12] as well as by Zhang et al. 2003 [13].

The wavelets form a family. The basic form is called the mother wavelet. All the daughter wavelets are derived from this wavelet according the following equation:

$$\Psi_{s,\tau}(t) = \frac{1}{\sqrt{s}} \Psi\left(\frac{t-\tau}{s}\right) \qquad (2.1)$$

where, s and τ are the scale and translation of the daughter wavelet. The term $s^{-1/2}$ normalizes the energy for different scales, whereas the other terms define the width and translation of the wavelet.

Digital signal analysis using wavelet transforms begins with the construction of a single parent wavelet. The signal is then decomposed into a series of basis functions of finite length consisting of dilated (stretched) and translated (shifted) versions of this parent wavelet function, i.e., wavelets of different scales and positions in time or space. This process is similar to Fourier analysis, where the parent wavelet is analogous to the sine wave,

and the basis functions in Fourier decomposition are sine waves of various amplitude, phase, and frequency variations of the parent sine wave [14,17].

Scaling a wavelet simply means stretching (or compressing) it. The smaller the scale factor, the more "compressed" the wavelet. The more stretched the wavelet, the longer the portion of the signal with which it is being compared, and thus the coarser the signal features being measured by the wavelet coefficients. Thus, there is a correspondence between wavelet scales and frequency as revealed by wavelet analysis:

- Low scale s $\Rightarrow$ compressed wavelet $\Rightarrow$ rapidly changing details $\Rightarrow$ High frequency ω.
- High scale s $\Rightarrow$ Stretched wavelet $\Rightarrow$ Slowly changing, coarse features $\Rightarrow$ Low frequency ω.

The decomposition advantage is very useful in dealing with signals contain features with various frequency characteristics. Another advantage of the wavelet transform is that its analysis can be chosen based on the application [11].

Figure 1 shows seismic signal at top and wavelet coefficients of six scales where "Daubechies 8" wavelet has been used. The smallest scale is the one that contains the highest frequency while the largest scale is the one that contain the lowest frequency.

2.2. The Discrete Time Wavelet Transform

A discrete type of wavelet transform exists, termed the discrete time wavelet transform (DTWT), where scales and shifts take on discrete values. It allows fast computation of the transform for the digitized signals and gives perfect signal reconstruction. It is a form of multiresolution analysis and is related to perfect reconstruction filter banks [14,18,19].

The wavelet transform can be applied to seismic signals in terms of the type of decomposition as well as in terms of pattern matching [11,12].

In wavelet analysis, we often speak of approximations and details. The approximations are the high-scale, low-frequency components of the signal while the details are the low-scale, high-frequency components. **Figure 2** shows that the input signal is filtered by a low-pass and high-pass filters in one stage wavelet transform. Then, the signal is down-sampled by a factor of 2 to produce DWT coefficients.

2.2.1. Subband Coding

The DTWT is implemented using the sub-band coding scheme represented in **Figure 3**. The signal is decomposed into 2 frequency bands: low (approximations) and high (details). For the approximation and detail coefficients to represent a discrete-time wavelet decomposition H and G filters must belong to a perfect reconstruction filter bank and required to be regular [14,18,20,21].

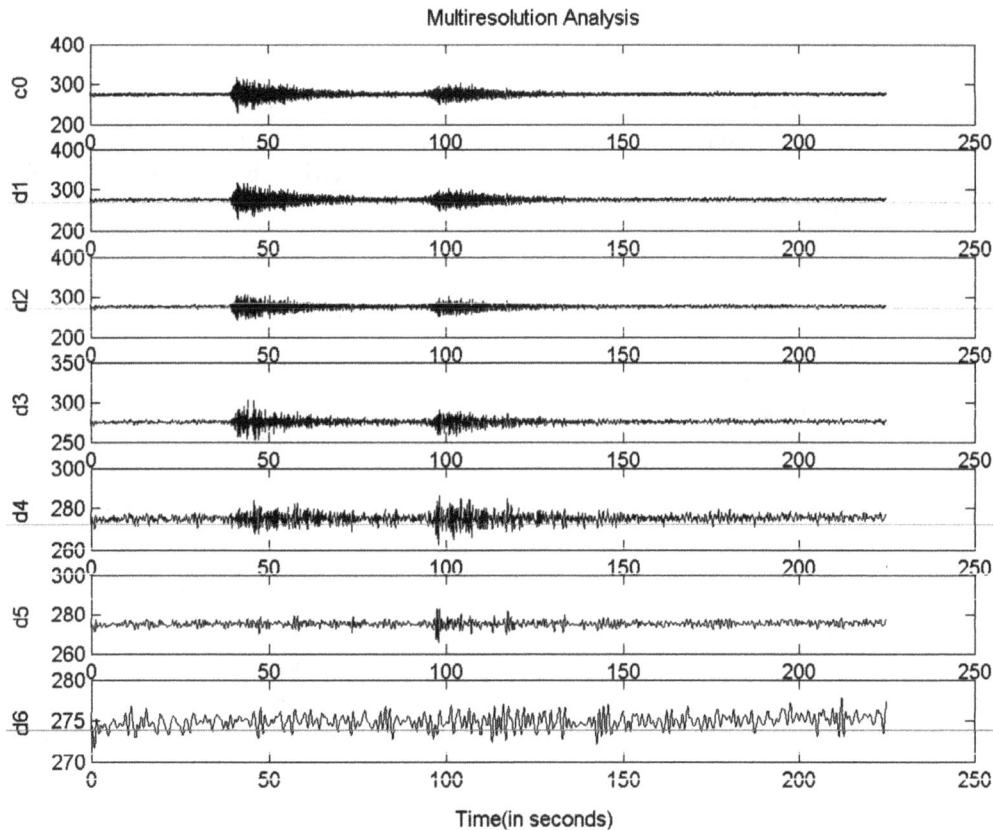

Figure 1. Seismic signal decomposed into several scales.

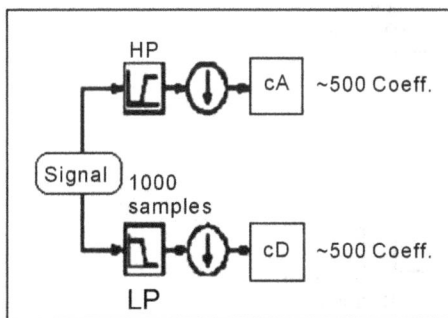

Figure 2. One stage wavelet transforms.

The wavelet analysis involves filtering and down-sampling while the wavelet reconstruction process consists of up-sampling and filtering. Up-sampling is increasing the number of samples by inserting zeros between samples.

The following filters then perform an interpolation by filling in the zero points with the appropriate signal information.

The filtering part of the reconstruction process bears some discussion, because it is the choice of filters that is crucial in achieving perfect reconstruction of the original signal.

Figure 4 is a typical example of such filters magnitude responses.

The down-sampling of the signal components performed during the decomposition phase introduces a distortion called aliasing. It turns out that by carefully choosing filters for the decomposition and reconstruction phases that are closely related, we can "cancel out" the effects of aliasing.

In order for the low- and high-pass decomposition filters (H and G), together with their associated reconstruction filters (L* and G*) shown in **Figure 3** to perform such decomposition and reconstruction a system of what is called Quadrature Mirror Filters is used [14,17,19].

2.2.2. Quadrature Mirror Filters

QMF banks are a set of finite impulse response (FIR) filters that enable a signal to be decomposed into sub-bands and allow reconstruction of the signal from the sub-bands without distortion. It is an example of decimation and interpolation and allows designing two-channel perfect reconstruction filter banks and therefore to generate wavelet bases.

The analysis filters have typically low pass and high pass frequency responses with a cutoff at $\pi/2$ (the "quadrature frequency") as illustrated in **Figure 4**. The philosophy of QMFs is to allow aliasing to be introduced by using overlapping filters for the analysis bank and design the synthesis filters in such a way that any aliasing is exactly cancelled out in the reconstruction process. Filters are also designed so that overall amplitude and phase distortion are minimized or eliminated.

2.3. Wavelet Type Selection

The selection of wavelet type is crucial in the processing. It can depend on seismic arrival shapes. Therefore, the wavelets used in the processing are chosen based on matching the seismic phase's arrival shapes. Thus, the selection will rely on the event as well as on the station that recorded this event. This is because the arrival shapes will vary from event to event and between the different stations as the seismic waves travel through different media and distances for each event and for each monitoring station.

Figure 5 shows some of the wavelets used in the wavelet transform decomposition.

In some cases where P-arrival shape clearly identified, it can be noticed that the wavelet that produce a composite function CT (the function that is used to locate S-arrival time) with the highest magnitude is closest in shape to P-arrival.

Figures 6 and **7** illustrate examples of P-arrival shapes of two different events. In the first example the P-arrival

shape seems similar to Daubechies 8 while in the second example the P-arrival shape seems similar to Daubechies 6.

In our analysis the wavelet type selection was verified to be more important with respect to S-phase arrival, while the P-arrival is not affected significantly by the wavelet that is applied. This was also proved by Anant and Dowla [11].

3. Software Development

3.1. Introduction

The software is developed in Matlab 6.5™ including signal processing and Wavelab™ toolboxes. It consists of two algorithms: the P-phase detection algorithm and S-phase detection algorithm described in the sections below. The P-phase detection algorithm can be categorized into three main modules:

1) DTWT Processing that includes the following:
➤ Multiresolution Analysis
  • Signal Decomposition (For east, north and vertical components).
  • Coefficients Reconstruction (For east, north and vertical components).

2) The Composite Rectilinearity Function Construction

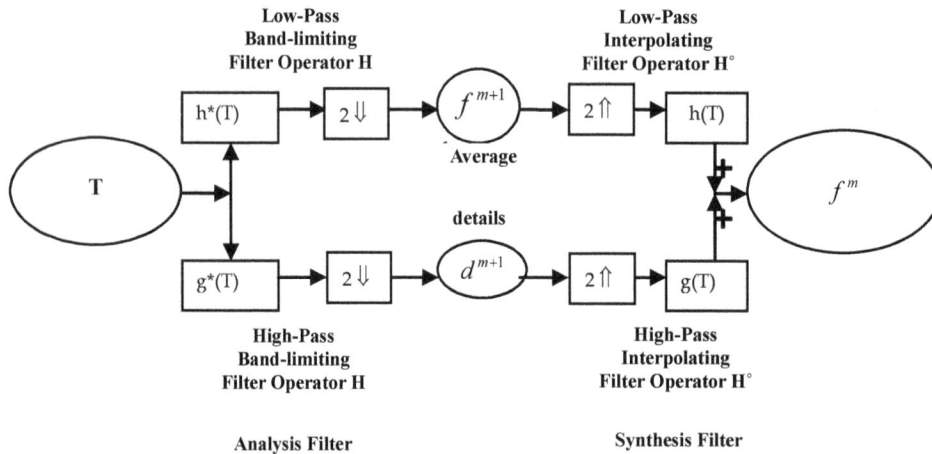

Figure 3. Sub-band coding scheme.

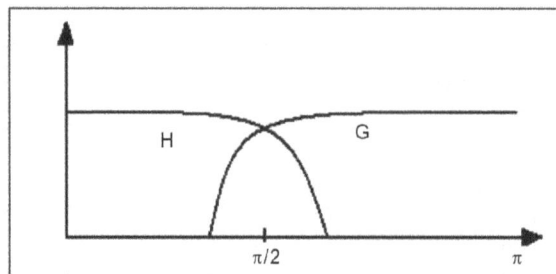

Figure 4. Magnitude responses of analysis filters.

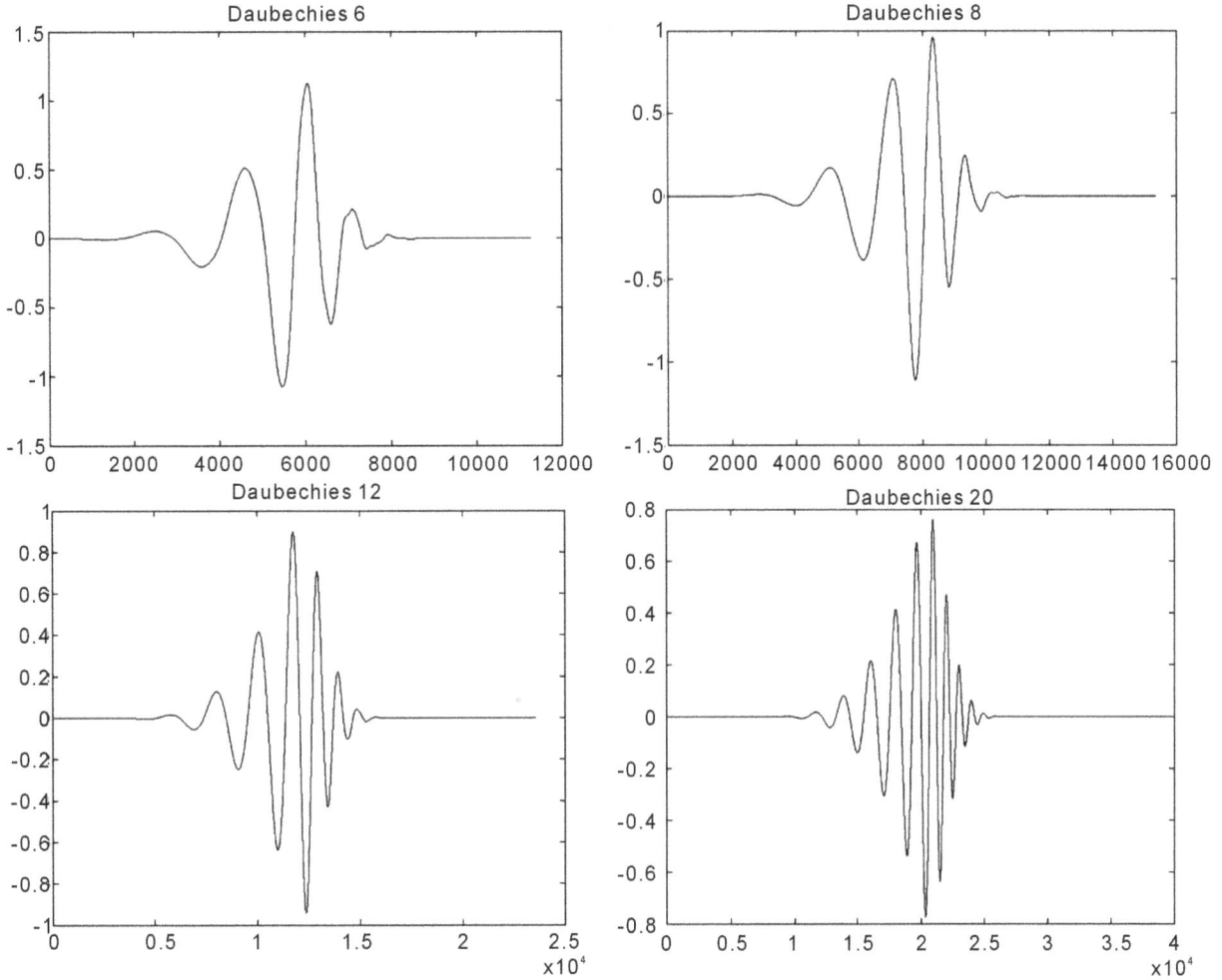

Figure 5. Some of the wavelets used in the research.

3) Backazimuth calculation

The S-phase detection algorithm can also be categorized into three main modules:

1) Components Rotation
2) DTWT Processing that includes the following:
   ➤ Multiresolution Analysis
      • Signal Decomposition (For radial and transverse components).
      • Coefficients Reconstruction (For radial and transverse components).
3) Constructing the Composite Function of Transverse over Radial Amplitude Ratio.

3.2. P-Phase Detection Algorithm

It is known that the P-phase is linearly polarized with respect to the direction of propagation. Therefore, a metric that measures the degree of linear polarization is helpful to detect the P-phase arrival. Such a metric known as the rectilinearity function is defined by Kanasewich [22].

Its equation is:

$$\mathbf{F} = \mathbf{1} - \left(\lambda_2 / \lambda_1 \right) \tag{3.1}$$

where λ_1 and λ_2 are the largest and the second largest eigenvalues of the covariance matrix respectively.

If the covariance matrix (Equation 3.2) is diagonalized, an estimate of the rectilinearity of particle motion trajectory over the specified time window can be obtained from the ratio of the principal axis of this matrix [22] i.e. the rectilinearity estimation can be obtained from the ratio of the largest and the second largest eigenvalues of the covariance matrix. The covariance matrix [23] is defined as:

$$\mathbf{M} = \begin{bmatrix} Cov(X,X) & Cov(X,Y) & Cov(X,Z) \\ Cov(Y,X) & Cov(Y,Y) & Cov(Y,Z) \\ Cov(Z,X) & Cov(Z,Y) & Cov(Z,Z) \end{bmatrix} \tag{3.2}$$

where,

X: east component wavelet coefficients at scale j;
Y: north component wavelet coefficients at scale j;
Z: vertical component wavelet coefficients at scale j.
While the covariance of X and Y is defined as:

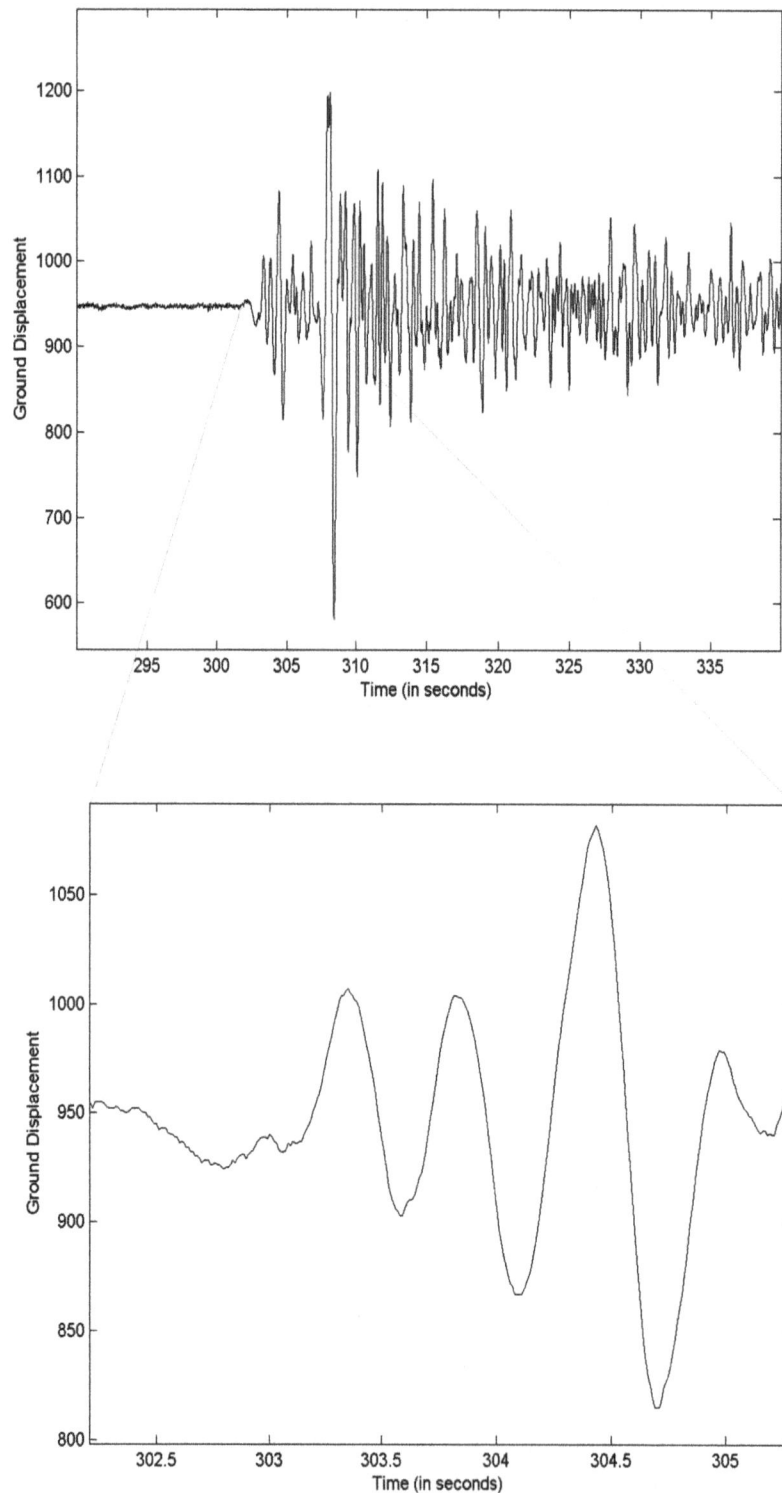

Figure 6. P-phase arrival of an earthquake record and the lower plot is a zoomed view of the arrival.

$$Cov(X,Y) = \frac{1}{N} \sum_{i=1}^{N} (X(i) - \mu_X)(Y(i) - \mu_Y) \qquad (3.3)$$

where μ_X and μ_Y are the mean values of X and Y respectively.

The direction of polarization may be measured by considering the eigenvector of the largest principal axis [22]. If λ_1 is the largest eigenvalue and λ_2 is the next largest eigenvalue of the covariance matrix, then a function of the form as in (Equation 3.1) would be close to

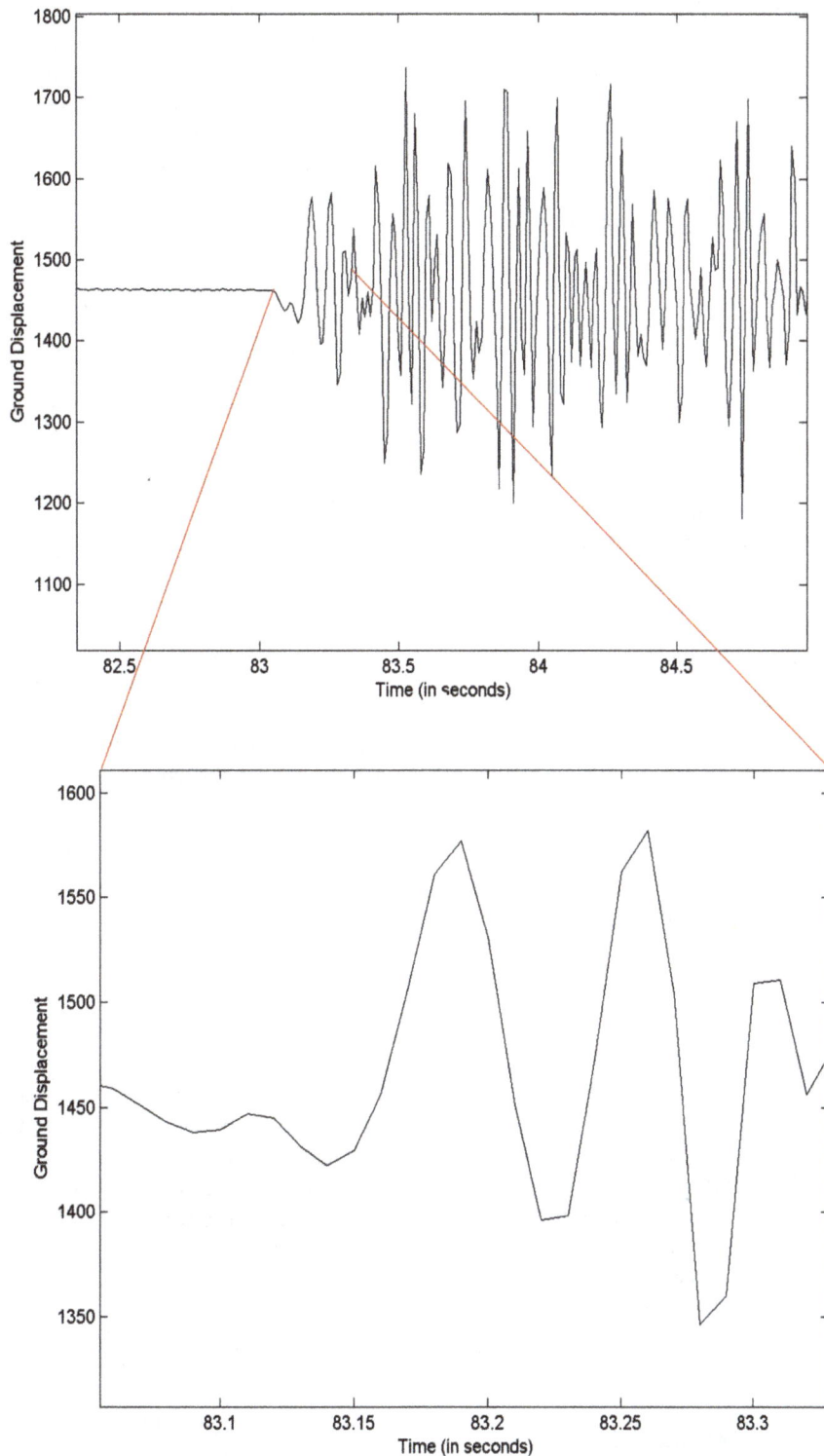

Figure 7. P-phase arrival of an earthquake record and a zoomed view of the arrival.

unity when rectilinearity is high ($\lambda_1 \square \lambda_2$) and close to zero when two principal axes approach one another in magnitude (low rectilinearity).

The direction of polarization can be determined by considering the components of the eigenvector associated with the largest eigenvalue with respect to the coordinate directions X, Y and Z.

The P-phase algorithm is developed as can be described in the following steps:

1) Processing the three component seismic signals

using Discrete Time Wavelet Transform (DTWT)

A three component seismic signals represented by east (*X*), north (*Y*) and vertical (*Z*) are processed by the discrete time wavelet transform (DTWT) to calculate the wavelet coefficients wc1, wc2 and wc3 respectively.

"Daubechies" wavelets are used in this process because of their compact support basis functions and their shape that coincide with seismic wave arrivals shapes.

Multiresolution Analysis

Each component is decomposed into several scales. So, the results are the wavelet coefficients x_1^j, x_2^j and x_3^j for east, north and vertical components respectively (*j* is the number of scales).

2) Constructing the rectilinearity function

At each scale, a pointwise moving window is constructed over the three components as shown in **Figure 8**.

The window length is determined using a measure called the varimax norm that is described in Appendix A. At each window, a 3-by-3 covariance matrix is constructed by using Equation (3.2) with x_1^j substituted for *X*, x_2^j substituted for *Y* and x_3^j substituted for *Z*. Then, eigenvalues and their corresponding eigenvectors are calculated.

At each window the rectilinearity function is constructed so that the result is a rectilinearity function (*Fj*) for each scale. Then, a composite rectilinearity function (*Cf*) is constructed so that the rectilinearity function of each scale contributes in this function. This composite function is constructed by (Equation 3.4).

$$Cf = \prod_j Fj \qquad (3.4)$$

where *j* is the scale number.

The location where this function gets its maximum value is taken as the P-arrival time.

3) Calculating the back azimuth angle

Back-azimuth is the angle measured from north to the direction from which the energy arrives at a given station [24,25]. It is used to determine the longitudinal and transverse directions for an incoming ray at a prescribed station. For an incident P wave the backazimuth is calculated by constructing the rectilinearity function and finding the eigenvector associated with the maximum eigenvalue for the detected P-wave. This eigenvector repre-

sents the direction of linear polarization that specifies the back azimuth angle. This calculation can be conducted at each wavelet scale as well as on the original signals. But, since the original three component signals generally contain noise, the calculation is carried out at the third and higher scales while the first two scales that contain high frequency noise are excluded. It is found that the polarization is conserved along different scales *i.e.* the same backazimuth angle is obtained in the different scales. So, one value is considered in the processing.

3.3. S-Phase Detection Algorithm

The S-phase is a shear wave with particle motion in the transverse direction [22]. Accordingly, it has higher amplitude in the transverse component relative to its amplitude in the radial or vertical component. So, in order to locate the S-phase arrival, the amplitude ratio of the transverse to radial components of the earthquake record is analyzed.

The S-phase algorithm is developed and can be described as follows:

1) Rotating the east and north components

The backazimuth angle (θ) calculated in the previous part is used to rotate the east and north components to radial and transverse components respectively using the following equation:

$$\begin{bmatrix} dr \\ dt \end{bmatrix} = \begin{bmatrix} \sin\theta & \cos\theta \\ -\cos\theta & \sin\theta \end{bmatrix} \begin{bmatrix} X \\ Y \end{bmatrix} \qquad (3.5)$$

where *dr* and *dt* are the radial and transverse component respectively.

2) Processing the radial and transverse components by the discrete time wavelet transform (DTWT)

The radial and transverse components processed by the DTWT. So, the result is several scales for each component. Referring to the algorithm the result is x^j and y^j that represents the different scales for radial and transverse components respectively (*j* is the number of scales).

3) Constructing the amplitude ratio

At each scale, a transverse over radial amplitude ratio is calculated as follows:

$$Atr^j = \frac{envt^j}{\left(envt^j + envr^j\right)} \qquad (3.6)$$

where $envt^j$ and $envr^j$ are the envelope functions of transverse and radial components respectively.

The envelope function is used to avoid divide-by-zero problems. It is defined as

$$env(x) = \sqrt{x^2 + h^2} \qquad (3.7)$$

where h is the Hilbert transform of *x*.

All scales are combined to construct a second composite function (CT) to be used to locate the S-arrival.

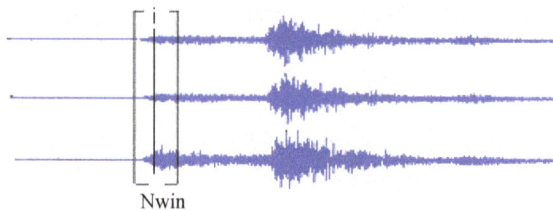

Figure 8. Illustration of windowing used to calculate covariance matrices.

$$\mathbf{CT} = \prod_j Atr^j \qquad (3.8)$$

The point after the P arrival time that has a value that is at least one half the maximum of CT is chosen for the S arrival time. The peak of CT itself is not used because it represents the time when S-wave attains its highest amplitude which is few seconds after its arrival time [11].

Several wavelets are used in the wavelet transform decomposition because the choice of wavelet that used in the processing is very important in how well CT locates the S arrival time.

4. Results

4.1. Introduction

The algorithm has been tested by a real seismic data represented by several events recorded on various three components stations. The tested events apart from event 17 are distributed as shown in **Figure 9**. The Figure illustrates the tested events in orange circles as they located according to the stations in dark blue and black triangles for short period and broadband stations respectively. The purpose of this testing is to investigate if the project objectives listed in section 1 are achieved. To measure the performance of the algorithm, the algorithm results are compared with manual inspection results as well as with another automatic algorithm results (STA/LTA method).

The performance of the system is measured by two ways: the first way is testing the ability of the software to detect P and S waves. However, the second way is comparing the results with another system results to see how accurate the system is.

4.2. P-Phase Detection Algorithm Results

The rectilinearity function proved to be an effective metric in locating the P-arrival time. **Figure 10** shows the rectilinearity functions of six different scales obtained by testing the algorithm with event 21. It can be observed that the rectilinearity function is approximately equal to 1 when the waveform is linearly polarized and equal to zero when there is no linear polarization. This can be seen clearly in the first four scales.

Figure 11 shows the testing result of an earthquake (event 24). It illustrates the composite rectilinearity function Cf and how it is used to locate the P-arrival time.

Figure 12 shows the tested result of another earthquake (event 1) and the composite rectilinearity function Cf. It shows that the peak of this function is used to determine the P-arrival time. It can be observed from the figure that the position at this function is a maximum is where P-phase arrives.

Figure 13 illustrates the direction of Arabian Sea earthquake (event 4) as the result of testing the algorithm by three component data recorded by **ABT** station (one of the southern stations). Referring to **Figure 9** it can be observed that ABT station is located in Southern Oman, which means that the algorithm indicates the event direction accurately as the event occurred in Arabian Sea.

In order to test the performance of the algorithm, it is

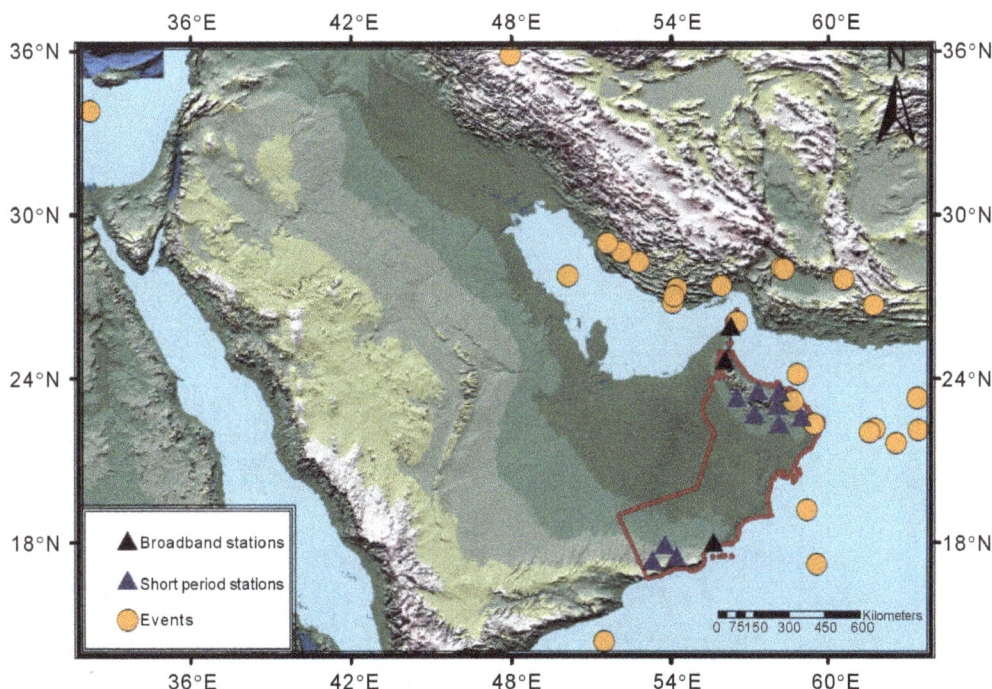

Figure 9. Distribution of tested events according to seismic stations.

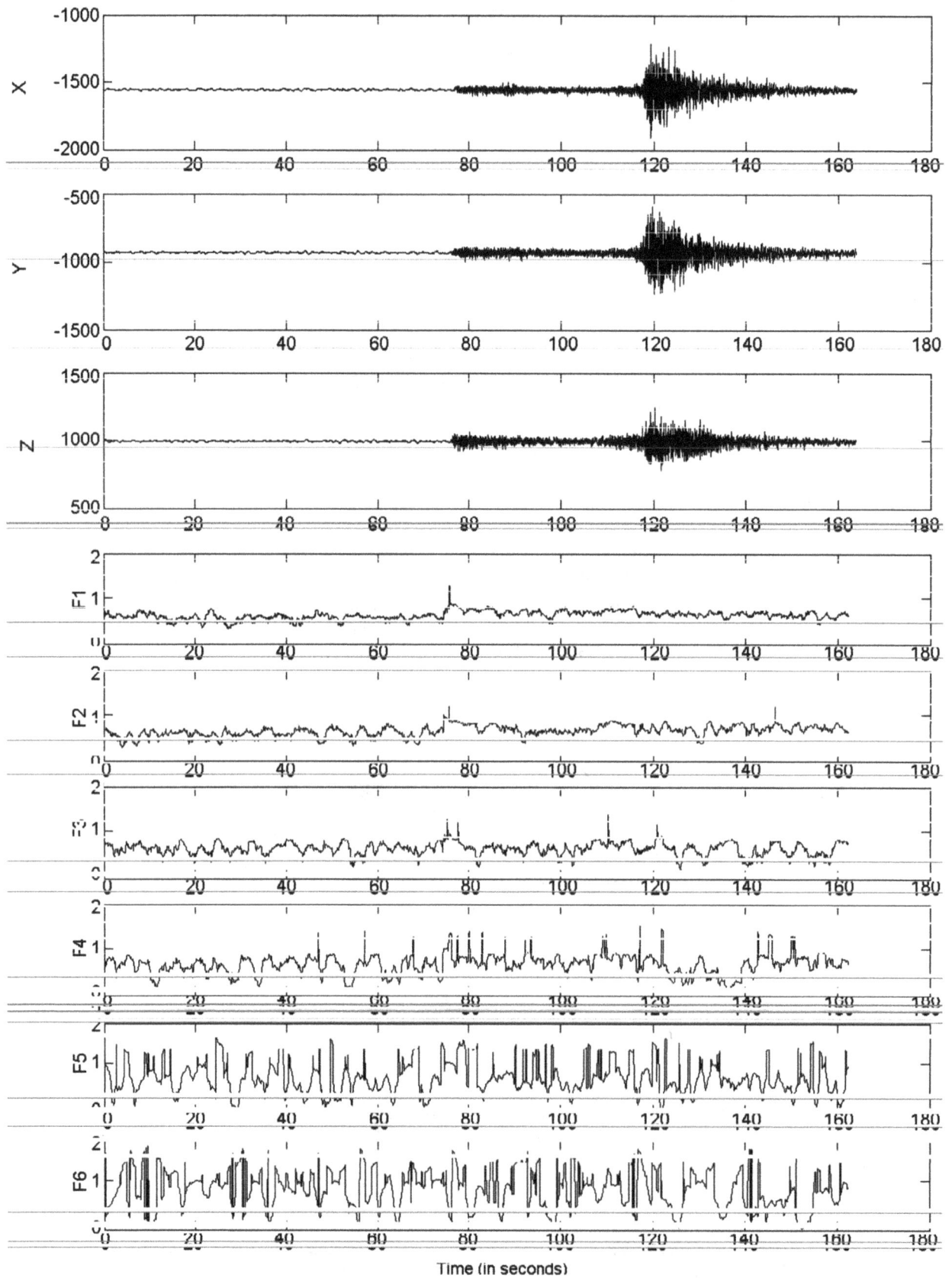

Figure 10. Three component waveforms of event 21 and its related rectilinearity functions of six different scales.

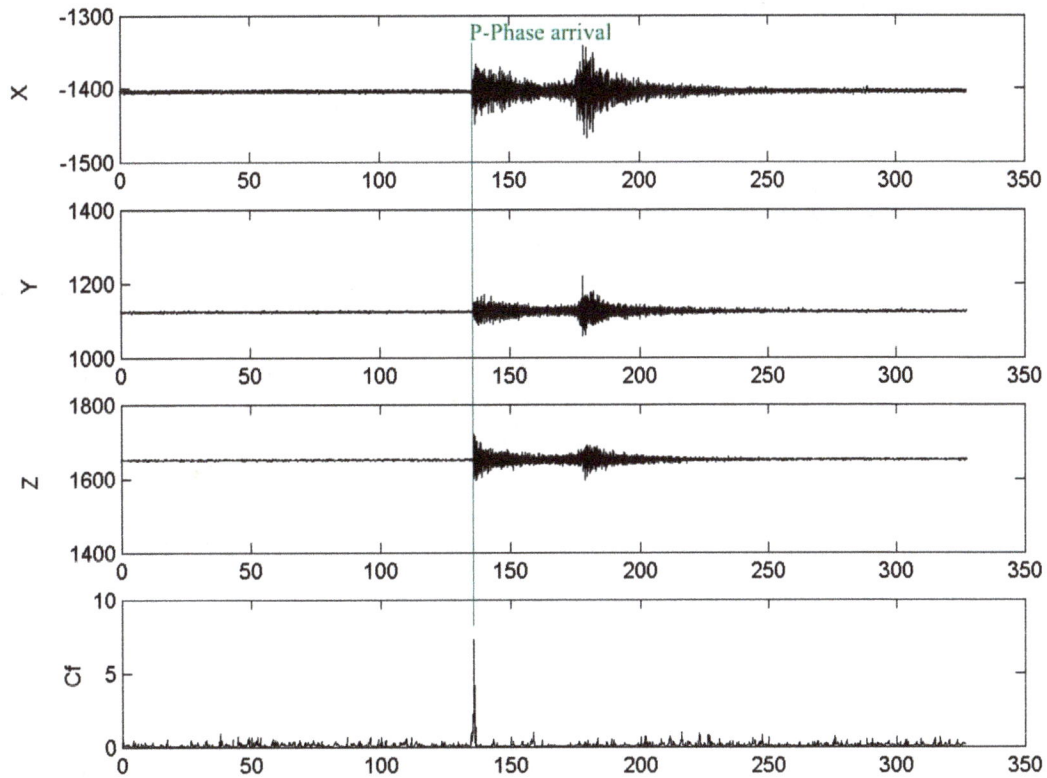

Figure 11. Three component data (east, north and vertical) of event 24 recorded by WBK station (one of the northern stations) with epicentral distance of 386.4 km and backazimuth of 106.9 degrees and the composite rectilinearity function *Cf*.

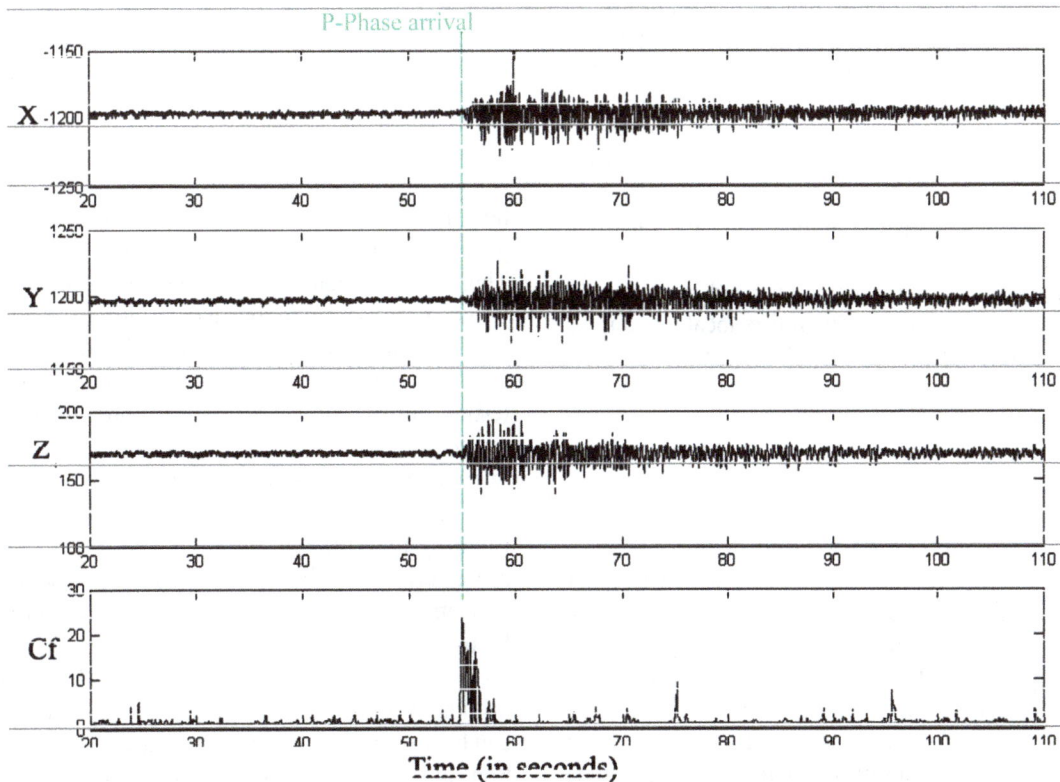

Figure 12. Three component data (east, north and vertical components) of event 1 recorded by JMD station with epicentral distance of 545.898 km and back azimuth of 96.03 degrees and the composite function *Cf*.

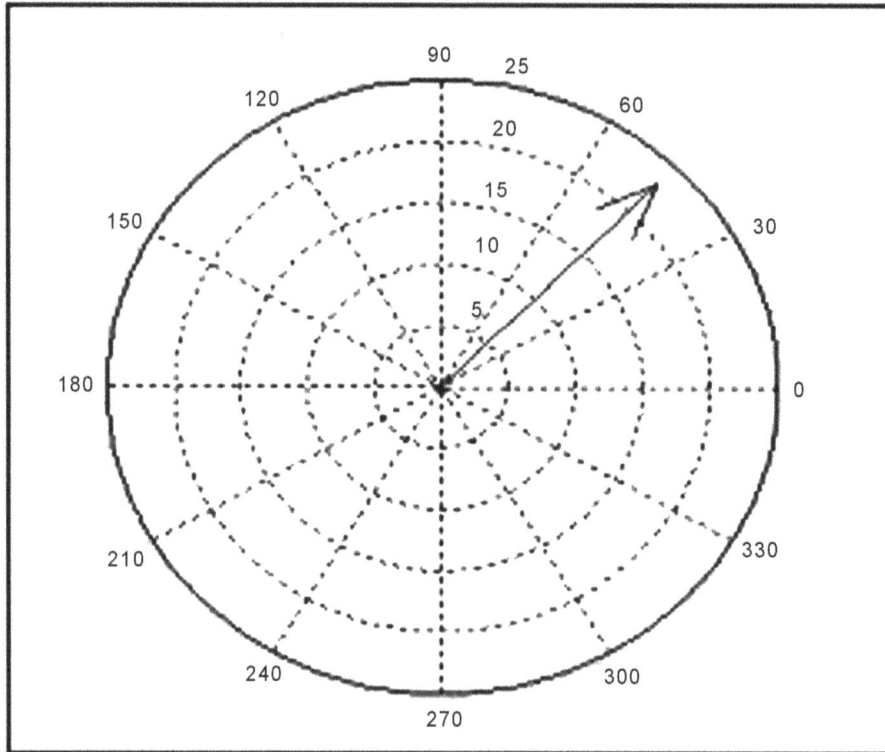

Figure 13. The direction of event 4 from ABT station.

tested by an event of magnitude 1.35 that occurs in California-Mexico border region (event 17). The purpose of this testing is to compare the algorithm results with the results obtained by another automatic detector and manual inspection that has been done by an analyst from the Broadband Array data collection center at the University of California, San Diego. The comparison between the analyst result and the algorithm result is illustrated in **Figure 14**. It can be noticed that the results are consistent with each other.

The three components signals illustrated in **Figure 14** are recorded by a station with noisy local conditions (BVDA2 of ANZA network). That's why; the STA/LTA (short term average/long term average) method couldn't give a result for P-arrival time for the data recorded by this station. However, the P-phase detection algorithm gives a good result though of this noise condition.

The algorithm result is compared with the analyst result as well as with STA/LTA results. The comparison shows that the three results are consistent with each other as illustrated in **Figure 15**.

To enhance the algorithm results, they compared with the results of STA/LTA and analyst results of all the events listed in **Table 1**. The time difference in the arrival pick of the algorithm and that of the analyst is shown in **Figure 16**. It can be observed that the mean magnitude of algorithm error is low (0.1952 seconds) as compared with STA/LTA mean magnitude error (0.3982 seconds).

This proves that the performance of the algorithm is efficient.

4.3. S-Phase Detection Algorithm Results

As stated earlier in Section 3, various wavelets are used in the wavelet transform decomposition. The wavelet is chosen to match the pattern of S-phase to provide the best match in the wavelet scales and thus give the composite function CT with the greatest dynamic range. Therefore, the composite function with the highest amplitude peak is used to locate the S-arrival.

The S-phase detection algorithm is tested by the same events used to test the P-phase detection algorithm. Its testing results of event 17 are compared with the analyst results as shown in **Figure 17** while STA/LTA method failed to give any result for S-arrival.

Figures 18 and **19** show the results of testing the algorithm by two different events (event 4 and event 18 respectively). They illustrate the radial and transverse components and the composite function that constructed to be used to detect the S-arrival time. They show that the composite function of transverse to radial amplitude ratio accurately succeeded to determine the S-phase arrival as can be noticed from the figures that the peak of this function indicates the position where S-phase attains its highest magnitude while the onset time of S-phase is few seconds before this peak.

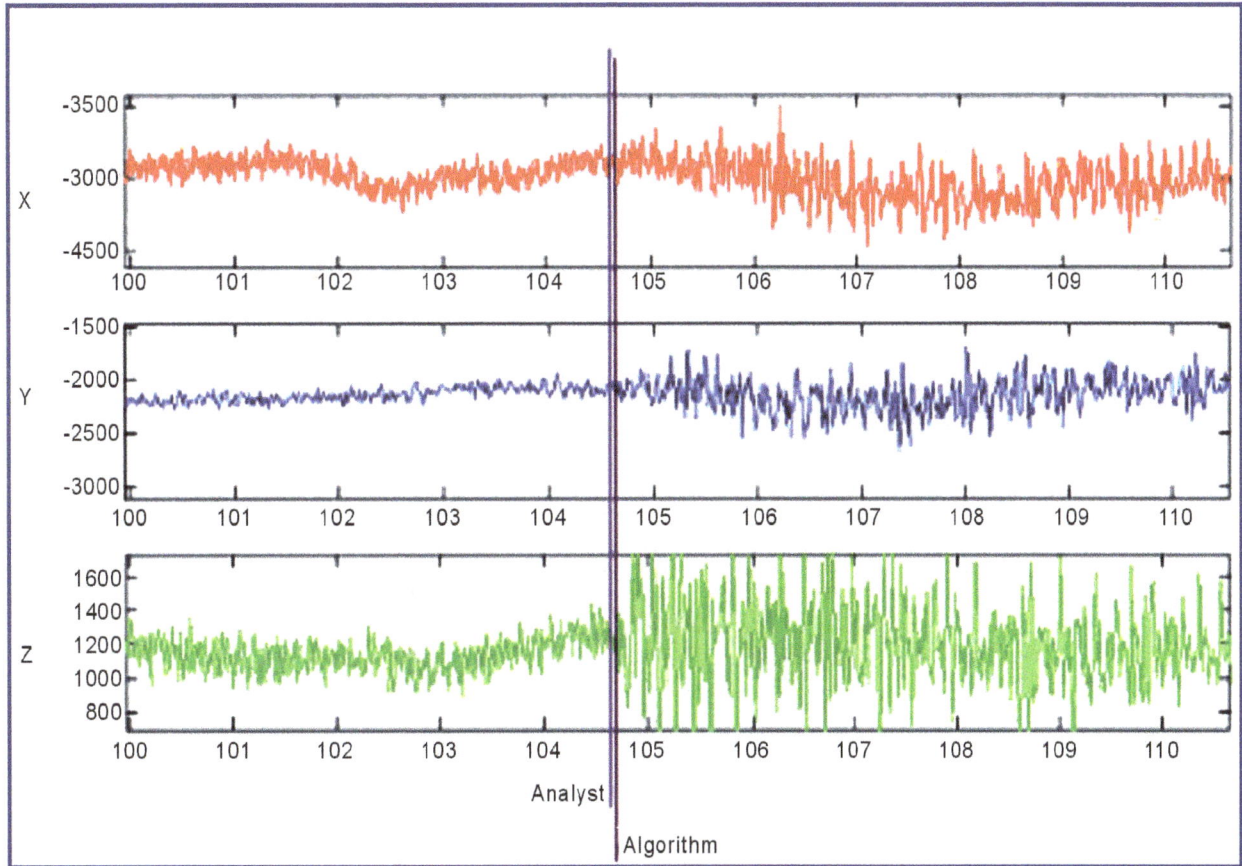

Figure 14. The three component signals of event 17 zoomed around P-arrival time to compare the analyst result with the algorithm result.

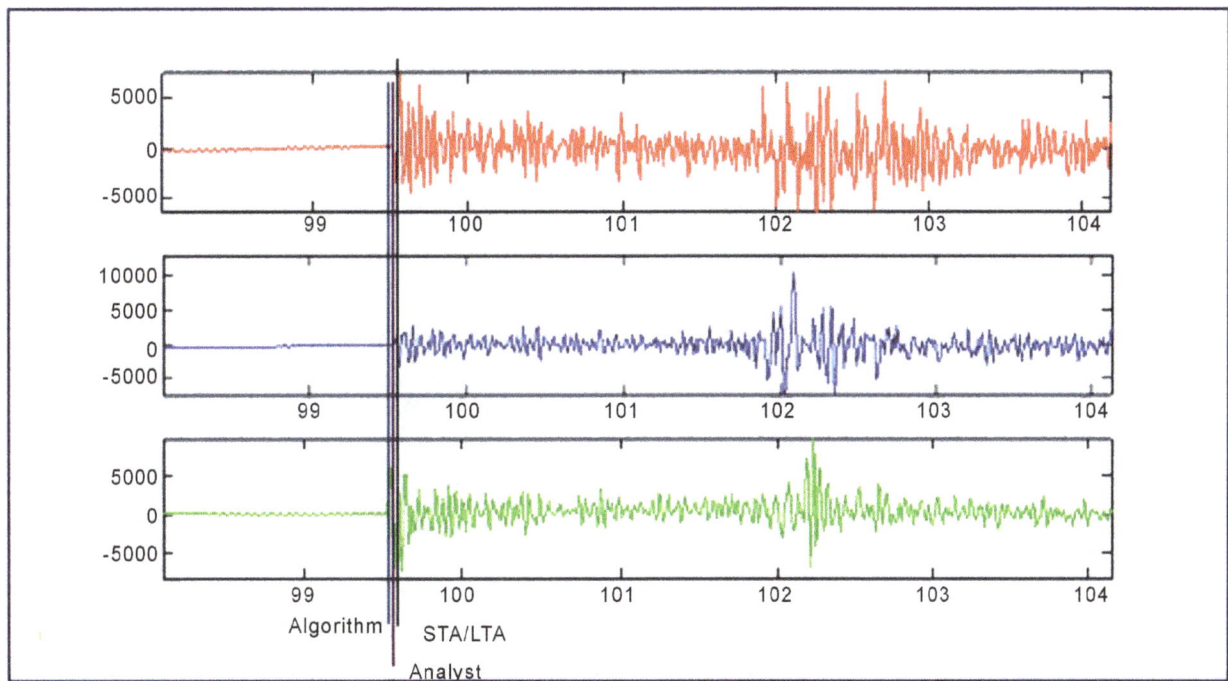

Figure 15. The result of comparing the algorithm with the analyst result and STA/LTA result when testing it by event 17 recorded by MONP station(Anza network) with epicentral distance of 10.656 km and backazimuth of 255.49 degrees.

Table 4. List of tested events.

Event No.	Date of occurrence	Origin time	Latitude	Longitude	Depth	Location
1	01/12/2002	03:43:37	22.11	63.41	30	Arabian Sea
2	02/12/2002	05:00:28	33.8	32.22	60	Eastern Mediterranean
3	03/12/2002	21:42:48	28.26	52.77	50	Southern Iran
4	03/12/2002	23:08:47	17.21	59.56	50	Arabian Sea
5	09/05/2003	00:03:38	26.7817	54.047	50	Southern Iran
6	09/05/2003	00:47:03	27.2357	54.1686	20	Southern Iran
7	09/05/2003	05:54:43	27.4024	55.9109	20	Southern Iran
8	09/05/2003	10:34:07	27.2866	54.2007	35.4	Southern Iran
9	09/05/2003	12:01:30	27.0192	54.0923	30	Southern Iran
10	09/05/2003	16:42:17	28.0641	58.1985	50	Southern Iran
11	09/05/2003	19:17:03	28.6498	52.124	20	Southern Iran
12	10/05/2003	06:39:31	27.6593	60.5815	20	Southern Iran
13	10/05/2003	15:47:46	27.7768	50.1192	30	Persian Gulf
14	11/05/2003	10:17:28	28.9479	51.5855	20	Southern Iran
15	11/05/2003	10:56:14	24.1477	58.8162	36.7	Gulf of Oman
16	11/05/2003	18:01:53	19.1998	59.1707	20	Arabian Sea
17	26/07/2003	05:56:37	15.4463	116.3117	32.92	California-Mexico Border Region
18	24/06/2004	21:56:05	28.0082	58.2898	50	Southern Iran
19	24/07/2004	00:00:01	23.1694	58.662	30	Eastern Arabian Peninsula
20	27/07/2004	05:10:44	23.3008	63.3888	30	Near Coast of Pakistan
21	05/08/2004	08:38:24	14.3831	51.4685	20	Eastern Gulf of Aden
22	29/08/2004	00:43:18	22.1723	61.7808	30	Arabian Sea
23	30/08/2004	08:19:58	26.0658	56.5246	31.8	Southern Iran
24	13/12/2004	01:36:04	21.6386	62.568	35	Arabian Sea
25	13/03/2005	03:31:27	26.6853	61.7142	30	Southern Iran
26	23/03/2005	19:49:33	22.3243	59.4517	13.95	Eastern Arabian Peninsula
27	24/03/2005	00:37:12	22.0547	61.5609	38.6	Arabian Sea
28	03/05/2005	07:20:43	35.8229	47.9509	30	Western Iran

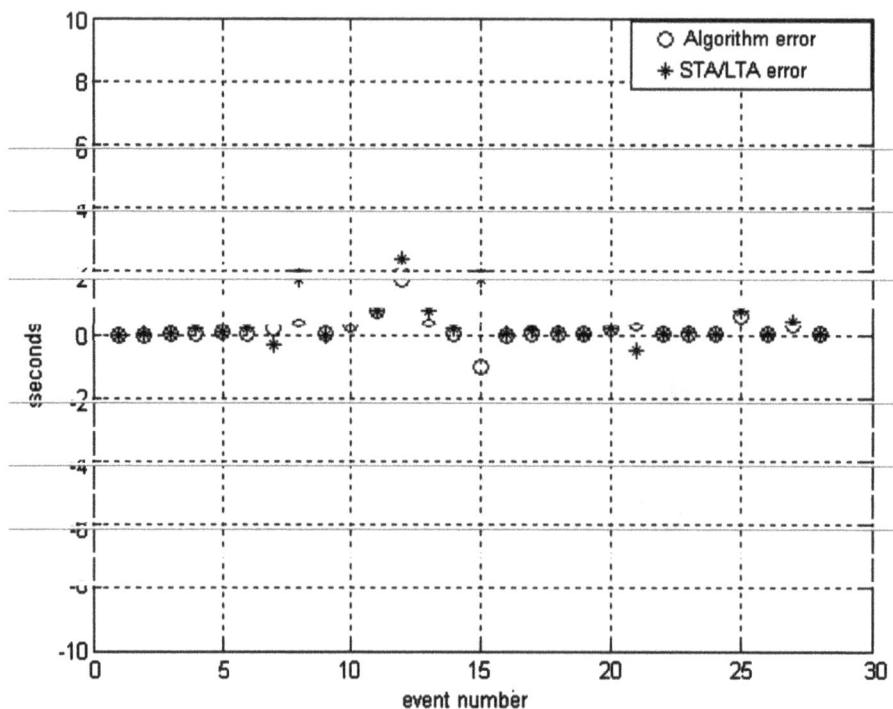

Figure 16. Residual plot of P-Phase arrival.

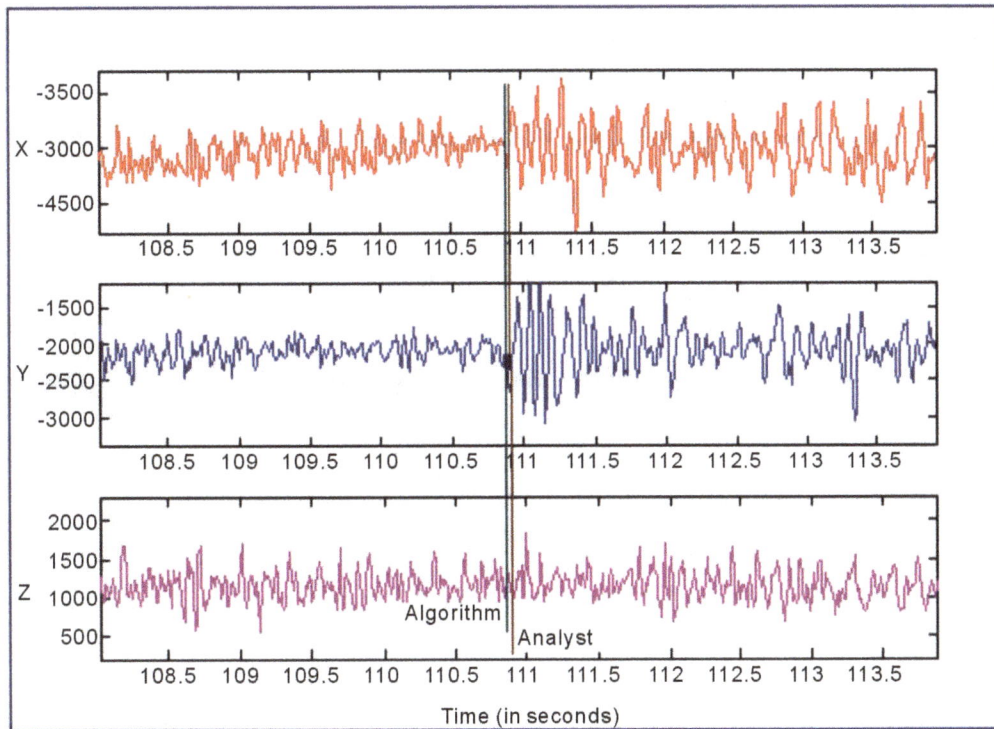

Figure 17. The three component signals of event 17 recorded by BVDA2 station with epicentral distance of 45.732 km and backazimuth of 353.73 degrees, zoomed around S-arrival time to compare the analyst result with the algorithm result.

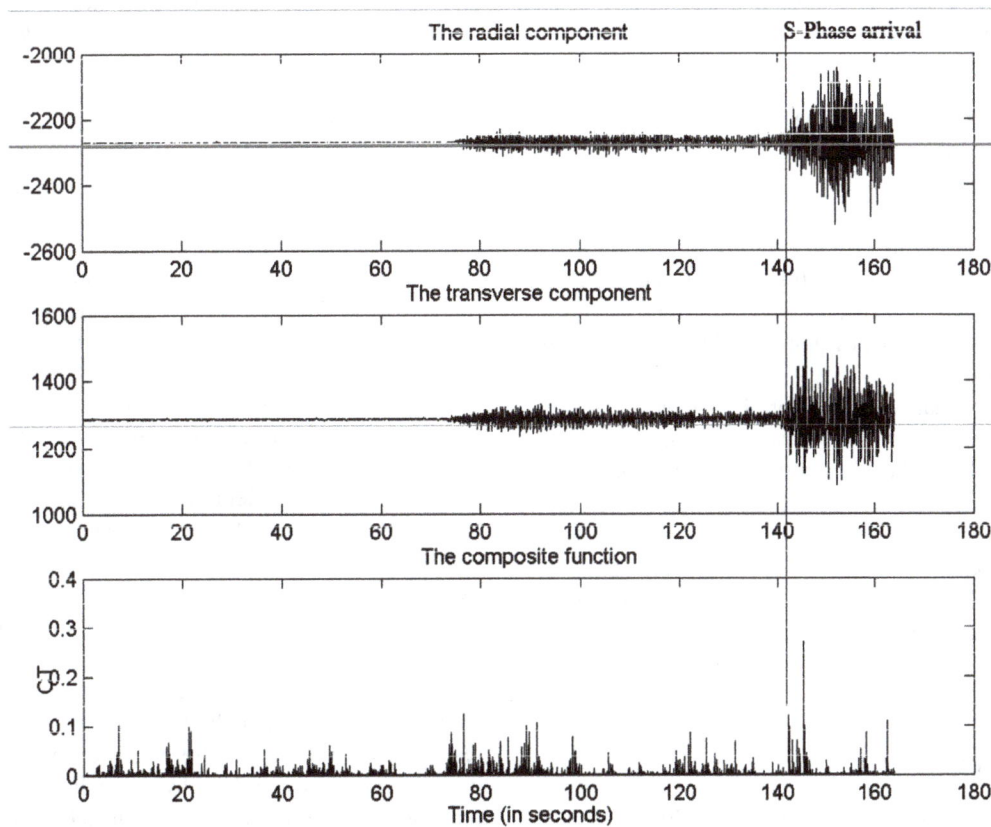

Figure 18. The radial and transverse components of event 4 recorded by ABT station with epicentral distance of 664.224 km and backazimuth of 43.9 degrees, and the composite function of the transverse to radial amplitude ratio.

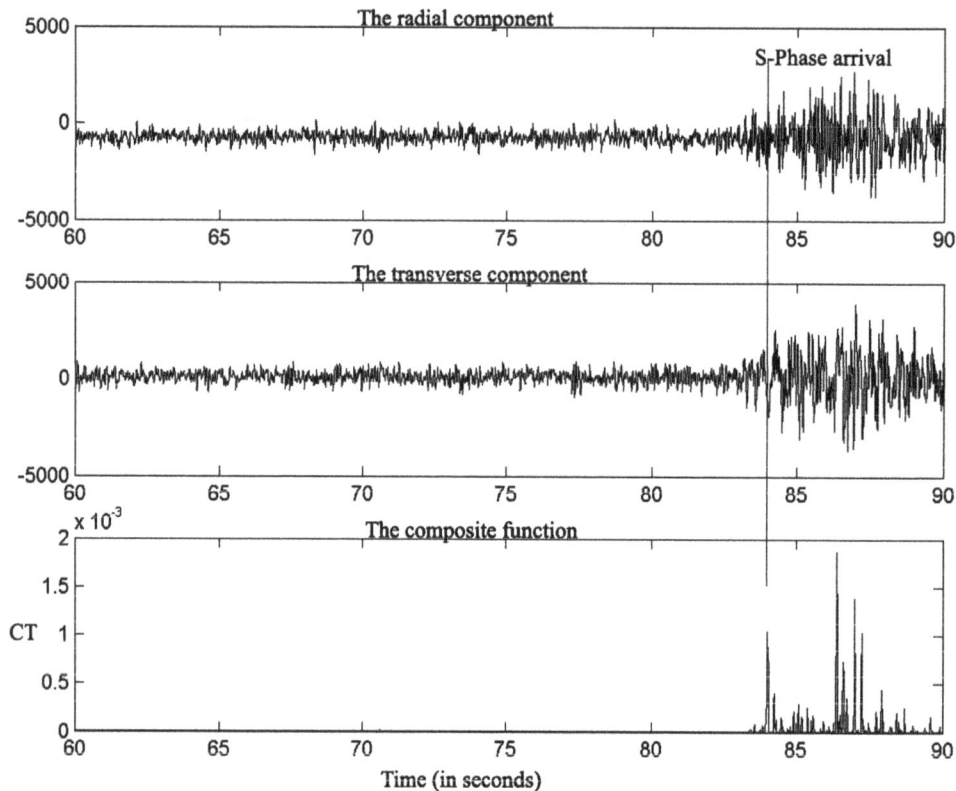

Figure 19. The radial and transverse components of event 18 recorded by BAN station (one of the broadband stations in northern Oman) with epicentral distance of 274.5 and backazimuth of 40.8 degrees and the composite function of transverse to radial amplitude ratio.

Figure 20 shows the residual plot of S-Phase arrival that illustrates the error between the algorithm and the analyst results (in circles) and the error between STA/LTA results and the analyst results (in asterisks). It can be noticed that the mean magnitude error of the algorithm error is low (0.8197 seconds). In addition, it can be observed that there is no S-pick from STA/LTA for event 17 while the algorithm has good results and consistent with the analyst results.

As mentioned earlier in Section 4.3, several wavelets are independently used in the wavelet decomposition and the resulting CT of the highest dynamic range has been chosen to determine the S-arrival time. The results show that "Daubechies 8" wavelet is a good choice for most of the tested events. While the other wavelets such as "Daubechies 12" and "Daubechies 20" give good results for some events. This wavelet selection has been done according to the expected arrival shapes as described in details in Section 2.

5. Conclusions

Since the main objective of the project is to develop an algorithm that automatically detects particular classes of seismic wave, the software has been developed to detect

P and S-phases in three component seismic data.

The algorithm has been tested by a real seismic data represented by multiple local and regional events recorded on various three components stations where it successfully detected P and S-phases.

Two important concepts have been presented in this research. The first concept is decomposing the signal into several resolutions or scales; important features in the seismic signals can be identified. Strong features in a signal can be maintained in several scales while weaker features will present in fewer scales or just one scale. The second concept is the wavelet type selection that depends on matching the wavelet to the arrival shapes.

The results of the software have been compared with a human analyst results as well as with the results obtained by software that uses STA/LTA (short term average/long term average) method. From the comparison results it can be concluded that the composite rectilinearity function and the composite function of transverse to radial amplitude ratio accurately determined P and S arrivals respectively. The results are consistent with the manual inspection results that done by an analyst as well as with other software based method in current professional use. The results showed that the STA/LTA failed to give any detection for S-phase of the tested events recorded by

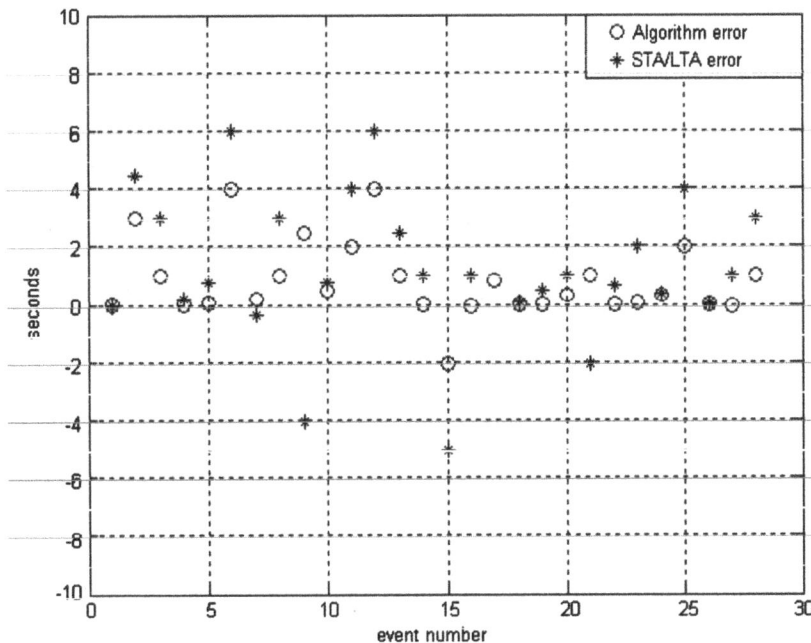

Figure 20. Residual plot of S-phase arrival.

stations with low signal to noise ratio conditions. In addition, the STA/LTA will fail to give a result for P-phase in low signal-to-noise conditions. However, the P and S-phase wavelet detection algorithms provided good results for P and S-arrivals and the results are consistent with the human analyst results.

To sum up, the developed wavelet algorithm can accurately determine the P and S arrivals in spite of the low signal-to-noise ratio. Furthermore, the software succeeded in determining back azimuths towards the earthquake sources.

The developed software can be used for a daily routine analysis in the seismological laboratories. In order to achieve this it should be tested by a huge amount of data set so that the parameters such as wavelet type that used in the DTWT analysis and the window size that used to construct the covariance matrices can be fixed.

6. Acknowledgements

Sincere thanks to Sultan Qaboos University for sponsoring this study. The first author would like to thank his colleagues at the earthquake monitoring center in Oman for sending the seismic data that are used in testing the software. In addition, many thanks to the staff at the Broadband Array data collection center, UCSD (http://eqinfo.ucsd.edu), for providing the seismic data that are used for results comparison. Partial support for this research was provided by NEHRP/USGS grants 03HQPA-0001 and 01HQAG0021. Furthermore, thanks to the University of Newcastle upon Tyne for providing computer facilities.

REFERENCES

[1] H. Dai and C. MacBeth, "Application of Back-Propagation Neural Networks to Identification of Seismic Arrival Types," *Physics of the Earth and Planetary Interiors*, Vol. 101, No. 3-4, 1997, pp. 177-188.

[2] H. Dai and C. MacBeth, "The Application of Back-Propagation Neural Network to Automatic Picking Seismic Arrivals from Single-Component Recordings," *Journal of Geophysical Research*, Vol. 102, No. B7, 1997, pp. 15105-15113.

[3] R. Allen, "Automatic Phase Pickers: Their Present Use and Future Prospects," *Bulletin of the Seismological Society of America*, Vol. 72, No. 68, 1982, pp. S225-S242.

[4] M. Baer and U. Kradolfer, "An Automatic Phase Picker for Local and Teleseismic Events," *Bulletin of the Seismological Society of America*, Vol. 77, No. 4, 1987, pp. 1437-1445.

[5] A. Cichowicz, "An Automatic S-Phase Picker," *Bulletin of the Seismological Society of America*, Vol. 83, No. 1, 1993, pp. 180-189.

[6] P. S. Earle and P. M. Shearer, "Characterization of Global Seismograms Using an Automatic-Picking Algorithm," *Bulletin of the Seismological Society of America*, Vol. 84, No. 2, 1994, pp. 366-376.

[7] D. Patane and F. Ferrari, "ASDP: A PC-Based Program Using a Multi-Algorithm Approach for Automatic Detection and Location of Local Earthquakes," *Physics of the Earth and Planetary Interiors*, Vol. 113, No. 1-4, 1999, pp. 57-74.

[8] R. Sleeman and T. Van Eck, "Robust Automatic P-Phase Picking: An On-Line Implementation in the Analysis of Broadband Seismogram Recordings," *Physics of the*

Earth & Planetary Interiors, Vol. 113, No. 1-4, 1999, pp. 265-275.

[9] P.-J. Chung, M. L. Jost and J. F. Böhme, "Estimation of Seismic-Wave Parameters and Signal Detection Using Maximum Likelihood Methods," *Computers and Geosciences*, Vol. 27, No. 2, 2001, pp. 147-156.

[10] M. Tarvainen, "Automatic Seismogram Analysis: Statistical Phase Picking and Locating Methods Using One-Station Three-Component Data," *Bulletin of the Seismological Society of America*, Vol. 82, No. 2, 1992, pp. 860-869.

[11] K. S. Anant and F. U. Dowla, "Wavelet Transform Methods for Phase Identification in Three-Component Seismograms," *Bulletin of the Seismological Society of America*, Vol. 87, No. 6, 1997, pp. 1598-1612.

[12] P. J. Oonincx, "A Wavelet Method for Detecting S-Waves in Seismic Data," *Computational Geosciences*, Vol. 3, No. 2, 1999, pp. 111-134.

[13] H. Zhang, C. Thurber and C. Rowe, "Automatic P-Wave Arrival Detection and Picking with Multiscale Wavelet Analysis for Single-Component Recordings," *Bulletin of the Seismological Society of America*, Vol. 93, No. 5, 2003, pp. 1904-1912.

[14] I. Daubechies, "Ten Lectures on Wavelets," The Society of Industrial and Applied Mathematics, 1992.

[15] E. J. Stollnitz, T. D. DeRose and D. H. Salesin, "Wavelets for Computer Graphics, Theory and Applications," Morgan Kaufmann, Burlington, 1996.

[16] G. Strang and T. Q. Nguyen, "Wavelets and Filter Banks," Revised Edition, Wellesley-Cambridge Press,

Wellesley, 1998.

[17] S. Mallat, "A Wavelet Tour of Signal Processing," Academic Press, Waltham, 1999.

[18] S. Mallat, "A Theory for Multiresolution Signal Decomposition: The Wavelet Representation," *IEEE Transactions on Pattern Analysis and Machine Intelligence*, Vol. 11, No. 7, 1989, pp. 674-493.

[19] B. Porat, "A Course in Digital Signal Processing," John Wiley & Sons, Inc., Canada, 1997.

[20] M. Vetterli and C. Herley, "Wavelets and Filter Banks: Theory and Design," *IEEE Transactions on Signal Processing*, Vol. 40, No. 9. 1992, pp. 2207-2232.

[21] M. Vetterli and D. Le Gall, "Perfect Reconstruction FIR Filter Banks: Some Properties and Factorizations," *IEEE Transactions on Acoustics, Speech and Signal Processing*, Vol. 37, No. 7. 1989, pp. 1057-1071.

[22] E. R. Kanasewich, "Time Sequence Analysis in Geophysics," University of Alberta Press, Edmonton, 1981.

[23] A. Jurkevics, "Polarization Analysis of Three-Component Array Data," *Bulletin of the Seismological Society of America*, Vol. 78, No. 5, 1988, pp. 1725-1743.

[24] Stein, S. and M. Wysession, "An Introduction to Seismology, Earthquakes and Earth Structure," Blackwell Publishing Ltd., Hoboken, 2003.

[25] K. Aki and P. G. Richards, "Quantitative Seismology," University Science Books, Sausalito, 2002.

[26] R. Wiggins, "Minimum Entropy Deconvolution," *Geoexploration*, Vol. 16, No. 1-2, 1978, pp. 21-35.

Appendix A

As mentioned in Section 3, a covariance matrix $M[i]$ is calculated at each point I of x_1^j, x_2^j and x_3^j over a T point window. The size of window affects the eventual output, Cf. The aim is to obtain a function Cf that has a one main spike with a minimum of secondary spikes. There is a measure called the varimax norm [26] can be used for this purpose. The higher the varimax norm of a signal, the fewer spikes the signal has.

The varimax norm of the composite function Cf is calculated as:

$$V_c = \left(\sum_i Cf^4[i] \right) \Big/ \left(\left(\sum_i Cf^2[i] \right)^2 \right)$$

Thus, before selecting the size of the window (Nwin), several size windows are tested in order to maximize V_c. So, the window that gives a Cf with the highest varimax norm is selected as the window for a particular event.

Figure A1 and **A2** show three component data of one of the tested events and a plot of the varimax norm versus the window size used to construct the covariance matrix. For this particular event, 6 seconds is chosen as the window length.

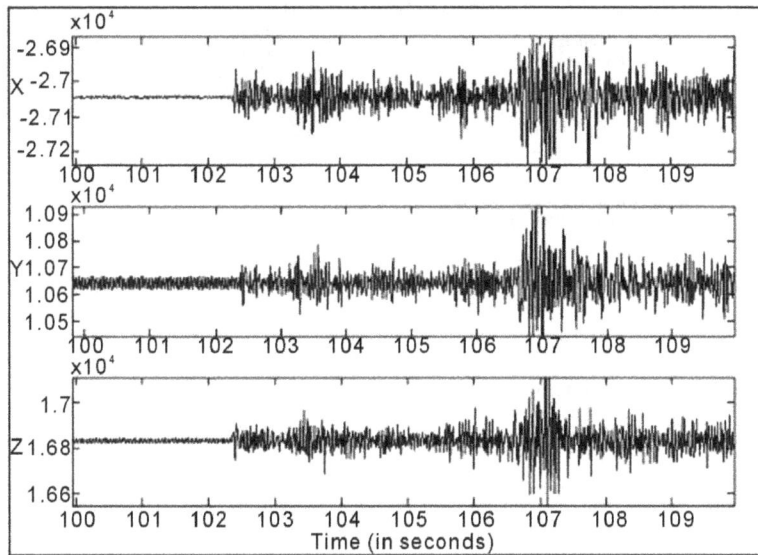

Figure A1. Three component data of an event.

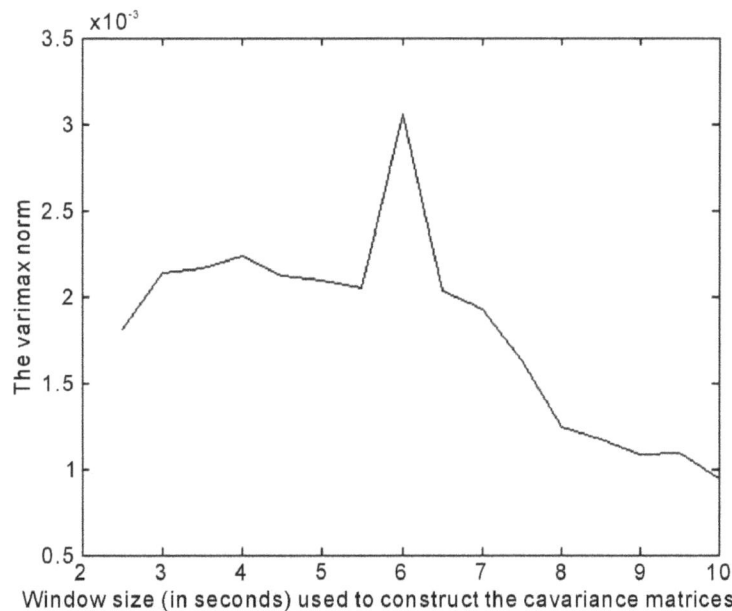

Figure A2. Demonstrating the effect of window size used to construct covariance matrices on the varimax norm of the composite rectilinearity function Cf.

Appendix B

At each scale j, a 3*3 Covariance Matrix $M^j[i]$ is calculated over a T point window centred at point i

Wavelet Coefficients (WC) $(x_1^j, x_2^j$ and $x_3^j)$ where, j: scaling number

Processing Using Discrete Time Wavelet Transform (DTWT)

Three Components data (X, Y and Z)

At each scale j, a rectilinearity function is calculated:

$F^j=1-(\lambda_2/\lambda_1)$
Where,
λ_1: The largest eigenvalue
λ_2: The second largest eigenvalue

A Composite rectilinearity function is constructed:

$C_F = \prod_j F^j$

P-arrival time is the position at which this function is a maximum

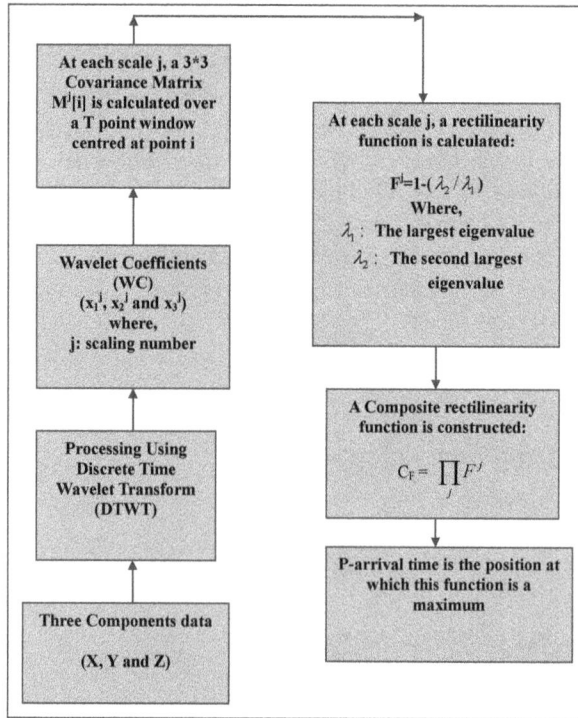

Figure B1. Flow Chart of P-wave detection algorithm.

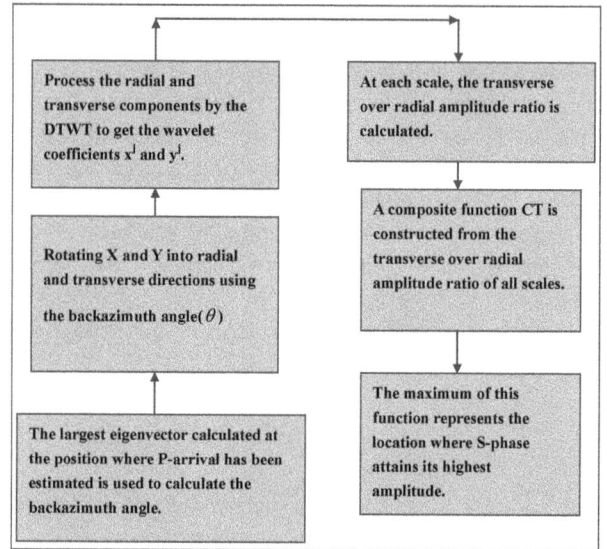

Process the radial and transverse components by the DTWT to get the wavelet coefficients x^j and y^j.

Rotating X and Y into radial and transverse directions using the backazimuth angle(θ)

The largest eigenvector calculated at the position where P-arrival has been estimated is used to calculate the backazimuth angle.

At each scale, the transverse over radial amplitude ratio is calculated.

A composite function CT is constructed from the transverse over radial amplitude ratio of all scales.

The maximum of this function represents the location where S-phase attains its highest amplitude.

Figure B2. Flow Chart of S-wave detection algorithm.

12

Estimating the Recurrence Periods of Earthquake Data in Turkey

Hande Konşuk, Serpil Aktaş
Department of Statistics, Hacettepe University, Ankara, Turkey

ABSTRACT

In this paper, the 231 earthquake data of magnitude 5 and higher, between north (39.00° - 42.00°) and east (26.00° - 45.00°) coordinates in Turkey from July 12, 1900 to October 23, 2011 are statistically analyzed. The probability density function and cumulative function of the magnitude are derived. It is shown that magnitude random variable is distributed as the exponential distribution. The recurrence periods is also calculated. Recurrence period is estimated approximately two times a year for an earthquake having magnitude 5.2. Using the Gutenberg-Richter function, the relation between magnitude and frequency is represented.

Keywords: Earthquake; Recurrence Estimation; Gutenberg-Richter Relation

1. Introduction

One of the most frightening and destructive disaster of nature is a severe earthquake and its destroying effects. If the earthquake occurs in a populated area, it may cause many deaths and injuries and extensive property damage regions [1]. Many studies have been presented to develop reliable estimates, given the large uncertainties in the pattern of earthquake occurrence. Abe and Suzuki [2] analyzed the seismic data from the viewpoint of science of complexity. These are power laws and represent the scale-free nature of the earthquake phenomenon. One of the main goals of seismology is to predict when and where the next main shock will occur after an earlier main shock.

Turkey is located on one of the important earthquake belt and, is a seismically active region. A large number of great historical earthquakes have been appeared in Turkey.

Particularly, over the past two decades Turkey faced several moderate and large earthquakes that resulted in significant loss of life and property. Since 1992, many massive earthquakes have hit the populated areas in Turkey. One of the most important earthquake, whose magnitude (M) is estimated to be about 7.4, occurred in August 17, 1998 in Marmara region.

Following the Marmara Earthquake, an earthquake having a magnitude 7.2 occurred in Düzce-Bolu region in November 12, 1999. The last massive earthquake hit Van province in October 23, 2011. All of these earth-quakes caused destructive damage and more than 16,000 people were died.

When considering the earthquake, we have to answer the four questions: Where? How often? How big? and When? The goal of the earthquake prediction is to give warning of potentially damaging earthquakes early enough to allow appropriate response to the disaster, enabling people to minimize loss of life and property [1]. The rate of recurrence of earthquakes on a seismic source can be represented with the Gutenberg-Richter relation [3].

As indicated by **Figure 1**, the 92% region of Turkey is on the seismic active region, therefore there have been a lot of studies on earthquake topic. Kasap and Gürlen [4] studied the return periods of earthquakes. Ogata [5] investigated the statistical models for earthquake occurrences. Utsu [6] applied gamma, log-normal, Weibull and exponential distributions to describe the probability distribution of interoccurrence time of large earthquakes in Japan. Aktaş et al. [7], used Poisson distribution to describe the recurrence times, and estimated the expected value and variance computed for the loss of life and damaged buildings after the change point using the compound Poisson process.

Bayrak et al. [8] evaluated the seismicity and earthquake hazard parameters of Turkey based on maximum regional magnitude. Authors divided Turkey into 24 seismic regions and used data between 1900 and 2005.

Öztürk et al. [9] estimated the mean return periods, the most probable magnitude in a time period of t-years, and the probability of earthquake occurrence for a given magnitude during a time span of t-years, they also

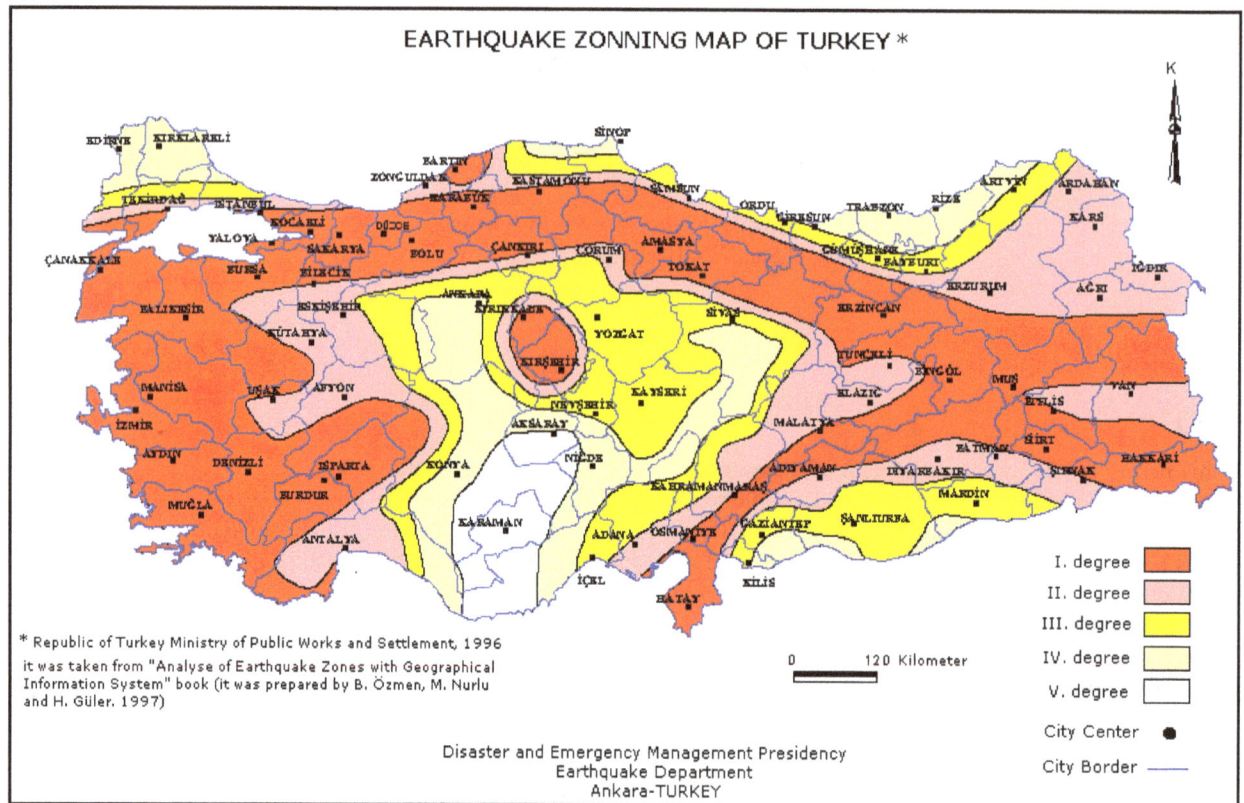

Figure 1. Seismic map of Turkey (Source: http://www.deprem.gov.tr/depbolge/).

showed that in the specific region, the most probable earthquake magnitude in the next 100 years would be over 7.5.

Bayrak *et al.* [10] calculated the seismicity parameters for the 24 seismic regions of Turkey according to Gumbel and Gutenberg-Richter methods and concluded that b-values obtained from the maximum likelihood approach gives better results for the tectonics of the examined area.

In this study, to find the probability distribution of magnitude is attempted and the statistical models are taken to interpret the observed frequency distribution. The statistical tools and methodologies used in the study are summarized and then the appropriate statistical models are developed.

Earthquake prediction can be considered into two types. First is the statistical prediction which is based on previous events; Data are collected from the records. Second is deterministic prediction which is made from the earthquake signs. In this study, 231 earthquake data of magnitude 5 and higher, between north (39.00° - 42.00°) and east (26.00° - 45.00°) coordinates in Turkey from July 12, 1900 to July 25, 2011 are analyzed. We study the records from the Turkish General Directorate of Disaster Affairs Earthquake Research Department [11].

The earthquake zones determined by using the acceleration contour map of Turkey is illustrated in **Figure 1.**

2. The Distribution of Magnitude

The distribution of magnitude has an important matter. Magnitude is defined as a continuous random variable having lower bound θ, but it has no defined theoretical upper bound. The probability density function of magnitude random variable is defined as an exponential function,

$$f_M(m) = \lambda\, e^{-\lambda(m-\theta)}, \theta \leq m \leq \infty \qquad (1)$$

where λ is calculated as,

$$\hat{\lambda} = \frac{1}{\left(E(M)-\theta\right)}. \qquad (2)$$

where, $E(M)$ is the expected value of M.

Frequency distribution of data is displayed in **Table 1**.

From the frequencies, it is noted that during the last century, total 231 earthquakes for magnitude 5 and higher occurred in Turkey.

The mean value of m can be used as an expected. Number of 231 earthquakes of magnitude 5 and higher ($M \geq 5$) data is investigated and average magnitude

Table 1. Total earthquake frequencies by magnitude.

Magnitude	5.0 - 5.4	5.5 - 5.9	6.0 - 6.4	6.5 - 6.9	7.0 - 7.4	7.5 -7.9	Total
Earthquake frequency	109	55	24	20	22	1	231

Table 2. The frequency distribution of earthquakes.

Class	Lower bound	Upper bound	Class midpoint	Frequency	Percent (%)
5.0 - 5.4	5.0	5.4	5.2	109	0.475
5.5 - 5.9	5.5	5.9	5.7	55	0.239
6.0 - 6.4	6.0	6.4	6.2	24	0.104
6.5 - 6.9	6.5	6.9	6.7	20	0.087
7.0 - 7.4	7.0	7.4	7.2	22	0.095
7.5 - 7.9	7.5	7.9	7.7	1	0.004
Total				**231**	**1**

$\overline{m} = 5.75$ is calculated. $\theta = 5.0$ is the lower bound of the first class. Then, $\hat{\lambda}$ is calculated using the Equation (2),

$$\hat{\lambda} = \frac{1}{(5.75 - 5.0)} = 1.33 .$$

Therefore, the density function can be defined as,

$$f_M(m) = 1.33 e^{-1.33(m-5.0)}, \quad 5.0 \le m \le \infty \quad (3)$$

By integrating the Equation (1), the probability density function of random variable M is obtained as,

$$F_M(m) = 1 - e^{-1.33(m-5.0)}, \quad m \ge 5.0$$
$$= 0, \quad m < 5.0 \quad (4)$$
$$= 1, \quad m \to \infty .$$

The frequency distribution of earthquakes which occurred between the dates July 12, 1900 and October 23, 2011 with respect to the magnitude is given in **Table 2**. Data are classified into six categories.

The goodness of fit test is performed to compare the observed frequency distribution with the theoretical exponential distribution of earthquake data using the chi-square distribution. Observed frequencies were calculated with summing the percent values as cumulative. Expected frequencies are calculated from the Equation (1) by integrating the probability density function between θ and the upper bounds of every class.

The following null hypothesis is tested against the alternative hypothesis,

Ho: There is no difference between experimental distribution and theoretical exponential distribution;

H_A: There is difference between experimental distribution and theoretical exponential distribution.

Table 3. The values which are related to experimental and theoretical distribution.

Class point	Frequency	Percent	Observed	Expected
5.2	109	0.454	0.454	0.233
5.7	55	0.238	0.692	0.605
6.2	24	0.103	0.795	0.797
6.7	20	0.086	0.881	0.895
7.2	22	0.095	0.976	0.946
7.7	1	0.004	1.000	0.972

Table 4. The estimation of recurrence periods for certain earthquakes.

Class point	$F_M(m)$ observed	$F_M(m)$ expected	$fm(m)$	Annual expected earthquake frequency	Average recurrence period (year)
5.2	0.454	0.233	0.233	0.538	1.85
5.7	0.238	0.605	0.372	0.859	1.16
6.2	0.103	0.797	0.192	0.443	2.26
6.7	0.086	0.895	0.098	0.226	4.43
7.2	0.095	0.946	0.051	0.118	8.47
7.7	0.004	0.972	0.026	0.060	16.67

The expected and observed probabilities are displayed in **Table 3**.

Chi-square value is calculated as 0.222. ($P > 0.05$). Therefore, the null hypothesis is not rejected at the level of significance 0.05. Magnitude random variable is distributed as the exponential distribution.

3. Estimation the Recurrence Periods of Earthquakes

Recurrence time is widely used for hazard assessment in seismology.

In this section, the recurrence periods of earthquakes are estimated as annually. The average recurrence time of an earthquake is usually defined as the number of years between occurrences of an earthquake of a given magnitude in a particular area. The estimations are given in **Table 4**.

Column $f_M(m)$ shows the occurrence probabilities of earthquakes for certain magnitudes. Fourth column of **Table 4** shows that annual expected earthquake frequencies which are obtained by multiplying probability at column $f_M(m)$ by annual average observed earthquake frequencies whose magnitude $M \geq 5$. Last column of **Table 4** shows that recurrence periods of earthquakes for certain magnitudes.

The recurrence interval is defined as the average time span between earthquake occurrences on a fault or in a source zone. The frequency or probability distributions of intercurrence times of earthquakes are of interest as well [12].

Recurrence year can be calculated using the following formula

Recurrence year

$$= \frac{1}{\text{Expected earthquake frequency for a year}}$$

For example, recurrence period is estimated approximately 1.85 times a year for an earthquake having magnitude 5.2. Similarly, the recurrence period of a 7.2 magnitude earthquake is estimated as 8.47 years.

4. Modeling Approach

Although seismicity is characterized by complex incident that makes it difficult to develop coherent explanation and prediction of earthquakes, there are some wellknown empirical laws. Omori law for temporal pattern of aftershocks and the Gutenberg-Richter law for frequency and magnitude are the examples.

The rate of recurrence of earthquakes on a seismic source is assumed to follow the Gutenberg-Richter relation [3,13],

$$\log_{10}^{N} = b_0 + b_1 M . \tag{5}$$

where,
N: number of earthquakes in magnitude range;
M: earthquake magnitude;
b_0: intercept;
b_1: slope.

There is a tendency for the b-value to decrease for smaller magnitude events and there is systematic decrease in b-value with increasing depth of earthquake.

In Gutenberg-Richter function, that the number of earthquakes greater than magnitude 6 that would occur in a given area over, say, 10 years, is proportional to the number of earthquakes greater than magnitude 5 in that area, which is proportional to the number greater than magnitude 4.

The relation between magnitude and frequency can be represented by the linear function as

$$\ln N = 12.52 - 1.48 \times M .$$

Model has R^2 = 77% and is statistically significant with the p value = 0.0221. A big value of intercept represents the lower magnitude earthquakes.

For instance, for M = 7.5, N is predicted as 4.14.

5. Results

Geological and Statistical surveys conduct and support research on the likelihood of future earthquakes. Earthquake researches mostly include field, laboratory, and theoretical investigations of earthquake facts. A primary goal of earthquake research is to increase the reliability of earthquake probability estimates. With a greater understanding of the causes and effects of earthquakes, we may be able to reduce damage and loss of life from this destructive event. Statistics help us to predict the future events based on previous events. The significance of the results to the problem of statistical prediction of earthquakes if of interest. In this paper, the magnitude is considered as a continuous random variable and the density function of magnitude random variable is defined as an exponential function. It is shown thatmagnitude random variable of Turkish earthquake data is distributed as the exponential distribution. The estimations of recurrence periods of earthquakes for some magnitudes are estimated. The number of earthquakes having magnitude in the Richter scale greater than or equal to M, is observed to decrease with M exponentially. For instance, the recurrence period of a 7.2 magnitude earthquake is estimated as 8.47 years. The relation between magnitude and frequency can be represented by the linear function as $\ln N = 12.52 - (1.48 \times M)$. This model gives a linear relation between magnitude and earthquake frequency.

Consequently, earthquake is an unavoidable natural disaster for Turkey. Hence, to take precautions for the future by utilizing the past experiences are very substantial.

We might conclude from these results about the distribution of earthquakes and statistical models. Models can be extended into the future research to make forecasts or predictions of the future seismicity.

REFERENCES

[1] D. R. Brillinger, "Earthquake Risk and Insurance," *Environmetrics*, Vol. 4, No. 1, 1993, pp. 1-21.

[2] S. Abe and N. Suzuki, "Scale-Free Statistics of Time Interval between Successive Earthquakes," *Physica A: Statistical Mechanics and Its Applications*, Vol. 350, No. 2-4, 2005, pp. 588-596.

[3] R. Gutenberg and C. F. Richter,"Frequencies of Earthquakes in California," *Bulletin of the Seismological Society of America*, Vol. 34, No. 4, 1944, pp. 185-188 .

[4] R. Kasap and Ü. Gürlen, "Deprem Magnitüdleri İçin Tekrarlanma Yıllarının Elde Edilmesi: Marmara Bölgesi Örneği," *Doğuş Üniversitesi Dergisi*, Vol. 4, No. 2, 2003, pp. 157-166.

[5] Y. Ogata,"Statistical Models for Earthquake Occurrences and Residual Analysis for Point Processes," *Journal of the American Statistical Association*, Vol. 83, No. 401,

1988, pp. 9-27.

[6]   T. Utsu, "Estimation of Parameters for Recurrence Models of Earthquakes," *Bulletin of the Earthquake Research Institute*, Vol. 59, 1984, pp. 53-66.

[7]   S. Aktaş, H. Konşuk and A. Yiğiter, "Estimation of Change Point and Compound Poisson Process Parameters for the Earthquake Data in Turkey," *Environmetrics*, Vol. 20, No. 4, 2009, pp. 416-427.

[8]   Y. Bayrak, S. Öztürk, H Cinar, D. Kalafat, T. M. Tsapano, G. Ch. Koravos and G. A. Leventakis, "Estimating Earthquake Hazard Parameters from İnstrumental Data for Different Regions in and around Turkey," *Engineering Geology*, Vol. 105, No. 3-4, 2009, pp. 200-210.

[9]   S. Öztürk, Y. Bayrak, G. Ch. Koravos and T. M. Tsapanos, "A Quantitative Appraisal of Earthquake Hazard Parameters Computed from Gumbel I Method for Different Regions in and around Turkey," *Natural Hazards*, Vol. 47, No. 3, 2008, pp. 471-495.

[10] Y. Bayrak, S. Öztürk, G. Ch.Koravo, G. A. Leventakis and T. M. Tsapanos, "Seismicity Assessment for the Different Regions in and around Tyrkey Based on İnstrumental Data: Gumbel First Asymptotic Distribution and Gutenberg-Richter Cumulative Frequency Law," *NHESS*, Vol. 8, 2008, pp. 1090-1122.

[11] http://sismo.deprem.gov.tr

[12] J. H. Wang and C. H. Kuo,"On the Frequency Distribution of Interoccurence Times of Earthquakes" *Journal of Seismology*, Vol. 2, No. 4, 1988, pp. 351-358.

[13] http://en.wikipedia.org/wiki/Gutenberg%E2%80%93Richter_law

Numerical Analysis of Seismic Elastomeric Isolation Bearing in the Base-Isolated Buildings

M. Jabbareh Asl[1], M. M. Rahman[1], A. Karbakhsh[2]

[1]Department of Mechanical Engineering, Universiti Tenaga Nasional, Selangor, Malaysia
[2]Department of Civil Engineering, Islamic Azad University, Kerman Branch, Kerman, Iran

ABSTRACT

Base isolation concept is currently accepted as a new strategy for earthquake resistance structures. According to different types of base isolation devices, laminated rubber bearing which is made by thin layers of steel shims bonded by rubber is one of the most popular devices to reduce the effects of earthquake in the buildings. Laminated rubber bearings should be protected against failure or instability because failure of isolation devices may cause serious damage on the structures. Hence, the prediction of the behaviour of the laminated rubber bearing with different properties is essential in the design of a seismic bearing. In this paper, a finite element modeling of the laminated rubber bearing is presented. The procedures of modeling the rubber bearing with finite element are described. By the comparison of the numerical and the experimental, the validities of modelling and results have been determined. The results of this study perform that there is a good agreement between finite element analysis and experimental results.

Keywords: Base-Isolated Structure; Seismic Isolation Bearing; Laminated Rubber Bearing; Finite Element Analysis

1. Introduction

The use of seismic isolation devices for buildings and bridges has been gaining worldwide acceptance and a well recognized approach to anti-seismic design. The seismic isolation method is generally preferred over more traditional methods, which rely mainly on strengthening of structural elements. This innovative technology reduces the earthquake forces transmitted to the superstructure and therefore eliminates permanent damage to the structure itself, protecting contents and secondary elements. The basic feature of a base isolation system is that the superstructure vibrates almost like a rigid body due to the combination of the flexibility and energy dissipation mechanisms of the components of the base isolation system. The seismically isolation reduces the natural frequency of the isolated structure rather than the same building, if conventionally founded, and the dominant frequencies of a strong ground motion [1,2].

The flexibility of the base-isolation structure is usually achieved by providing elastomeric bearings made of rubber reinforced with steel. The laminated rubber bearings for base isolation devices are made of alternating thin horizontal layers of natural or synthetic rubbers bonded to steel plates. In the concept of base isolation, the steel plates provide large stiffness under vertical load, while the rubber layers provide low horizontal stiffness, when the structure is subjected to lateral loads (e.g., earthquake, wind, etc.). The devices are usually subjected either to compression or to a combination of compression and shear [3]. The great advantage of elastomeric bearings is that: they have no moving parts; they are not subject to corrosion and they are reliable, cheap to manufacture and need no maintenance [4,5].

In the base-isolated buildings, the laminated rubber bearings, being as protectors of the superstructure, should sometimes be protected from failure or instability because the failure of rubber bearings may result in serious damage to superstructure. The evaluation of the collapse conditions is an essential step in designing the rubber bearing. The collapse of the device can occur either by global failure, due to buckling or roll-out of the device [6], or by local rupture, due to tensile rupture of the rubber, through detachment of the rubber from the steel or steel yielding [7]. Therefore, it is necessary to have an accurate knowledge of the global characteristics and behaviour of the device under maximum lateral displacement in compression and shear loads.

A finite element model has the potential for being a powerful tool to be used for improving the knowledge of the local behaviour of the isolation devices. Simo and Kelly, [4] used the finite element modelling to study the variation of lateral load-displacement behaviour under increasing axial load by consideration about 2-D model. Finite element analyses of a circular elastomeric bearing subjected to vertical loads have been performed by Imbimbo and Luca, [7]. Doudoumis *et al.*, [8] have done the numerical modelling of Lead-Rubber Bearings with the use of finite element micromodels and it has shown that the finite element micromodels provide increased possibilities for a more detailed study of the stress, strain and available strength of Lead-Rubber Bearings. However, at previous finite element modelling, there is no brilliant comparison between theoretical and numerical model of laminated rubber bearing.

In this research, firstly, the process of modelling of 3-D finite element analysis of a laminated rubber bearing has been explained. Then, by comparing the results of finite element analysis with experimental results, the validity of models has been shown.

2. Numerical Modelling of Laminated Rubber Bearing

The objective of the numerical analysis is to examine and verify the behaviour of laminated rubber bearing with different compression loads and shear forces. In order to verify the behaviour of rubber bearing predicted by the analytical theory, a computer program has been written for calculating the above mentioned formulation. In other hands, to verify the numerical modelling, the general purpose of finite element ABAQUS application [9] is used to model a multilayer rubber bearing.

2.1. Determination of Material Parameters and Dimensions

Rubbers are usually considered as almost incompressible material including nonlinear geometric effects. The mechanical rubber behaviour is modelled by a homogeneous, isotropic and hyperelastic model and it is usually described in terms of a strain energy potential U, which defines the strain energy stored in the material per unit of reference volume in the initial configuration as a function of the strain at that point in the material [10]. The rubber is selected as a polynomial hyperelastic material of the second order. In this case, the strain energy potential has the form as follows:

$$U = \sum_{i+j=1}^{2} C_{ij} \left(I_1 - 3\right)^i \left(I_2 - 3\right)^j + \frac{\left(J^{el} - 3\right)^2}{D_1} \quad (1)$$

where C_{ij} and D_1 are the material parameters, I_1 and I_2 are the first and the second invariants of the deviatoric strain, and J^{el} is the elastic volume ratio.

An almost incompressible (neo-Hookean) model has been considered. The material parameters of the rubber can be expressed in terms of initial shear modulus G, and initial bulk modulus K via $G = 2(C_{10})$, $K = 2/D_1$ [9]. Since D_1 is not equal to zero, this model allows some compressibility in the rubber material. The values of the neo-Hookean coefficients and material properties of the laminated rubber bearing and steel shims are shown in **Table 1**. The steel material of the top and bottom plates and the shims was assumed to be mild steel, bilinear elastoplastic constitutive law. However, in the simulation of the laminated rubber bearings yielding of steel plate did not occur.

2.2. Numerical Example and Bearing Properties

The experimental data provided by Tsai and Hsueh, [11] are used to validate the comparison between theoretical and numerical modelling. **Figure 1(a)** shows the laminated rubber bearing with the following dimensions: The steel shim radius R_1 = 140 mm, with an additional 10 mm of protective rubber covering, for a total radius of R_2 = 150 mm, the number of rubber layers n_r = 20 with thickness t_r = 10 mm, the total rubber thickness is t = 200 mm, the number of steel shims n_s = 19 with thickness t_s = 2 *mm*, and each one of the top and bottom end steel plates is 21 mm. The shear modulus of the rubber material G is 0.611 *MPa*.

2.3. Finite Element Modelling

The finite element models of the above-mentioned elastomeric isolators have been developed and implemented

Table 1. Material properties of the selected laminated rubber bearing.

Material	Properties	Explanation
Rubber	C_{10} = 0.3055 MPa	*Neo-Hookean* *(Finite element modelling)*
	D_1 = 9.7 × 10^{-4} MPa^{-1}	
	G = 0.611 MPa	*Shear modulus* *(Theoretical approach)*
Steel shims	E = 2.1 × 10^5 MPa	*Young's modulus*
	V = 0.3	*Poisson's ratio*
	σ_y = 240 MPa	*Von Mises yield criterion*

in computer code ABAQUS [9]. The hybrid C3D8H element and reduced integration element, C3D8R of ABAQUS have been used to model rubber and steel shims respectively. The finite element model consists of 49,813 nodes and 27,049 elements. **Figure 1(b)** shows the meshing of the rubber bearing. Each layer of rubber and steel shim plate is divided into 3 and 2 layers respectively. In order to simulate the real rubber bearing, the central hole (20 mm diameter) is modelled. The hole is used during manufacturing phase and for a better heat exchange during vulcanisation process. The rubber cover is also included in the consideration which is not considered in the previous finite element modelling of laminated rubber bearings [8,12].

The laminated rubber bearing is first compressed with vertical load. Subsequently, the bearing is sheared by applying shear force through the effective area of top end plate keeping the vertical force constant. The applied shear force is increased until the required level of deformation is reached. A typical deformed shape of half of laminated rubber bearing under these loading conditions is given in **Figure 2**. At the next step, the vertical load is increased and keeping constant then the shear forces are applied frequently. At each step of loading, the maximum lateral displacement of the top end plate to measure the horizontal stiffness is extracted with respect to applied shear force. Other requirements like internal rotation of steel shims and stress concentration at rubber bearing and steel shims are measured as well.

3. Results and Discussions

In the past, researches in the stability issue of laminated rubber bearings show that the P-Δ effect causes the horizontal stiffness to decrease with an increase in the compression force [11,13,14]. A theoretical curve, calculated from Haringx's theory, results of finite element analysis, and experimental data presented by Tsai and Hsueh, [11] is shown in **Figure 3**.

As shown in **Figure 3**, the experimental work has been

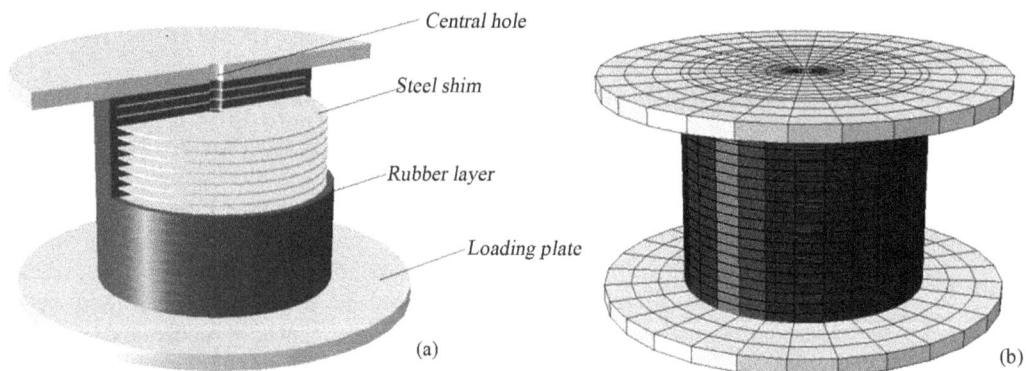

Figure 1. (a) Laminated rubber bearing (b) Finite element mesh.

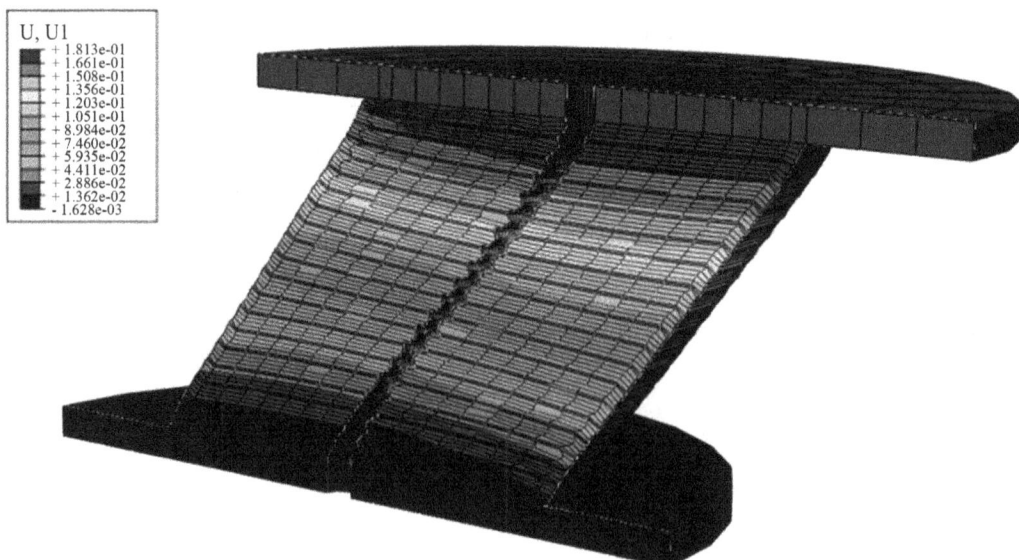

Figure 2. Deformed shape of half part of rubber bearing mesh.

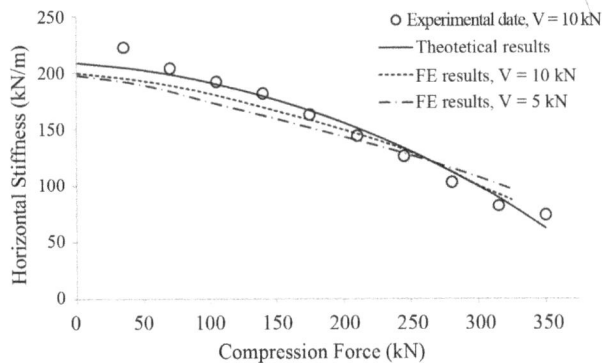

Figure 3. Comparisons of theoretical, finite element and experimental results of horizontal stiffness with compression force.

done with consideration of shear force equal to 10 kN. However, the results of theoretical method are independent of shear force, whereas; the finite element modelling was done with various values of shear forces. **Figure 3** shows the results of finite element analysis with consideration of shear forces equal to 5 kN and 10 kN and it shows that the shear force affected the variation of horizontal stiffness of laminated rubber bearings slightly. The slight difference between theoretical and numerical is because the central hole and rubber cover were not considered in the theoretical study, whereas; in the finite element analysis, whole specifications of the laminated rubber bearing were modelled. **Figure 3** shows good agreement between the experimental, theoretical and numerical analysis and the validity of finite element results are shown appropriately by this figure.

4. Conclusions

In this paper, the comparison between the finite element analysis and experimental results of a laminated rubber bearing has been performed. Using computer code ABAQUS, numerical simulation of rubber bearing test is compared with theoretical formulation. The comparison of variations of horizontal stiffness of rubber bearing confirms that simulation results have satisfactory agreement with the test results and analytical solution. Concerning the numerical results, it can be noted that varying the values of shear forces affects the horizontal stiffness and maximum lateral displacement which cannot be observed in analytical solution.

Hence, the finite element modelling provides more detailed study of the rubber bearing and a better understanding of their mechanical behaviour. Therefore, the use of finite element modelling in the study of laminated rubber bearings would be advisable.

REFERENCES

[1]   F. Naeim and J. M. Kelly, "Design of Seismic Isolated Structures from Theory to Practice," John Wiley & Sons, Hoboken, 1999.

[2]   X. Y. Zhou, M. Han and I. Yang, "Study on Protection Measures for Seismic Isolation Rubber Bearings," *Earthquake Technology*, Vol. 40, 2003, pp. 137-160.

[3]   A. F. M. S. Amin, S. I. Wiraguna, A. R. Bhuiyan and Y. Okui, "Hyperelasticity Model for Finite Element Analysis of Natural and High Damping Rubbers in Compression and Shear," *Engineering Mechanics*, Vol. 132, No. 1, 2006, pp. 54-64.

[4]   J. C. Simo and J. M. Kelly, "Finite Element Analysis of the Stability of Multilayer Elastomeric Bearings," *Engineering Structures*, Vol. 6, No. 3, 1984, pp. 162-174.

[5]   A. Karbakhsh Ravari, I. B. Othman, Z. B. Ibrahim and K. Ab-Malek, "P-Δ and End Rotation Effects on the Influence of Mechanical Properties of Elastomeric Isolation Bearings," *Journal of Structural Engineering-ASCE*, Vol. 138, No. 6, 2012, pp. 669-675.

[6]   J. M. Kelly, "Earthquake-Resistant Design with Rubber," Springer, Berlin, 1997.

[7]   M. Imbimbo and A. D. Luca, "F. E. Stress Analysis of Rubber Bearings under Axial Loads," *Computers & Structures*, Vol. 68, No. 1-3, 1998, pp. 31-39.

[8]   I. N. Doudoumis, F. Gravalas and N. I. Doudoumis, "Analytical Modeling of Elastomeric Lead-Rubber Bearings with the Use of Finite Element Micromodels," *Proceedings 5th GRACM International Congress on Computational Mechanics*, Limassol, 2005.

[9]   "ABAQUS 6.2 User Manual, Version 6.2," HbbitKarlsson Sorensen Inc., Pawtucket, 2001.

[10]   J. Yan and J. S. Strenkowski, "A Finite Element Analysing of Orthogonal Rubber Cutting," *Materials Processing Technology*, Vol. 174, No. 1-3, 2006, pp. 102-108.

[11]   H. C. Tsai and S. J. Hsueh, "Mechanical Properties of Isolation Bearings Identified by a Viscoelastic Model," *Solids and Structures*, Vol. 38, No. 1, 2001, pp. 53-47.

[12]   J. M. Kelly and S. M. Takhirov, "Tension Buckling in Multilayer Elastomeric Isolation Bearings," *Mechanics of Materials and Structures*, Vol. 2, No. 8, 2007, pp. 1591-1605.

[13]   A. Karbakhsh Ravari, I. Othman, Z. Ibrahim and H. Hashamdar, "Variations of Horizontal Stiffness of Laminated Rubber Bearings Using New Boundary Conditions," *Scientific Research and Essays*, Vol. 6, No. 14, 2011, pp. 3065-3071.

[14]   C.-H. Chang, "Modeling of Laminated Rubber Bearings Using an Analytical Stiffness Matrix," *International Journal of Solids and Structures*, Vol. 39, No. 24, 2002, pp. 6055-6078.

14

Seismicity in the Antarctic Continent and Surrounding Ocean

Masaki Kanao

National Institute of Polar Research, Midori-cho, Tachikawa-shi, Tokyo, Japan

ABSTRACT

Seismicity in the Antarctic and surrounding ocean is evaluated based on the compiling data by the International Seismological Centre (ISC). The Antarctic continent and surrounding ocean have been believed to be one of the aseismic regions of the Earth for many decades. However, according to the development of Global Seismic Networks and local seismic arrays, the number of tectonic earthquakes detected in and around the Antarctic continent has been increased. A total of 13 seismicity areas are classified into the Antarctic continent (3 areas) and oceanic regions within the Antarctic Plate (10 areas). In general, seismic activity in the continental areas is very low in Antarctica. However, a few small earthquakes are identified. Wilkes Land in East Antarctica is the most tectonically active area in the continent, with several small earthquake events having been detected during the last four decades. In the oceanic region, in contrast, seismic activity in the area of 120°-60° W sector is three times higher than that in the other oceanic areas. This may be considered to be involved in a tectonic stress concentration toward the Easter Island Triple Junction between the Antarctic Plate, the Pacific Plate and the Nazuca micro-Plate. Three volcanic areas, moreover, the Deception Island, the Mts. Erebus and Melbourn, are also found to be high seismic activities in contrast with surrounding vicinity areas.

Keywords: Seismicity; Antarctic Plate; Antarctic Continent; Southern Ocean; Tectonic Stress

1. Introduction

The Antarctic Plate has a unique tectonic setting that it is surrounded by almost completely divergent margins, with a very small amount of convergent or transformed margins. The divergent margins are characterized by the circum Antarctic seismic zones and cover about 92% of the surroundings of the Antarctic Plate. The convergent margins locate in the northwestern part of the South Shetland Islands and are less than 2% of the plate boundary. The transformed margins cover less than 7% along the boundary of the Scotia micro-Plate (**Figure 1**).

Seismicity in the Antarctic Plate is not so high compared with other plates. However, the seismicity becomes gradually clear in the last few decades by the development of Global Seismic Networks as well as the local seismic stations deployed in the Antarctic. It was a general knowledge by 1957, when the International Geo-

physical Year (IGY) started, that the seismic activity in the Antarctic continent was very low and only minor activity in the vicinity of active volcano, such as the Mount Erebus, was to be expected [1,2]. One of the other active volcanoes, the Deception Island in the South Shetland Islands was already known since the 19th century, but no seismic observation was carried out in the Island until the IGY. During the IGY period, ten seismic stations were established in the Antarctic and arrival time information on seismic phases recorded at the deployed stations had been reported to the United States Coastal Geological Survey (USGS), significantly contributed to the determination of earthquake locations globally. However, no earthquakes were determined by their hypocenters in the Antarctic at the time. USCGS had published "Antarctic Seismological Bulletin" based on travel time information received telegraphically from the Antarctic seismic stations during 15 years since IGY.

As the number of seismic stations in the Antarctic has been increased, local and regional earthquakes came to be detected within the Antarctic continent, even though their seismic activities were very low [3,4]. It became clear that an earthquake with magnitude around 4 occurred every few years in the Antarctic continent. However, no earthquakes with magnitude larger than 5 have occurred during the last four decades. [5] proposed an explanation for the lack of seismic activity of the Antarctic continent in terms of loading effects by overlying continental ice sheet. As the number of seismic stations of the globe increased, several tectonic earthquakes have been detected in the Antarctic. Inside the Southern Ocean, moreover, the intra-plate seismicity has been very low. However, one great earthquake with Mw 8.1 struck the Balley Islands region on March 25, 1998. This intra-plate earthquake is the largest one ever recorded in the Antarctic Plate [6,7].

[8] compiled the seismicity in and around the Antarctic continent for the first time, and also mentioned the existence of the ice related seismic signals (i.e., icequakes, ice shocks) recorded at several seismological stations. [9] also evaluated the seismicity in the Antarctic from the viewpoint of seismotectonics and neotectonics, by dividing the continent into several tectonic brocks. [10] presented a hypocentral distribution map of Antarctic intra-plate earthquakes in the period of 1900-99 including the records by Global Seismic Network, combined with

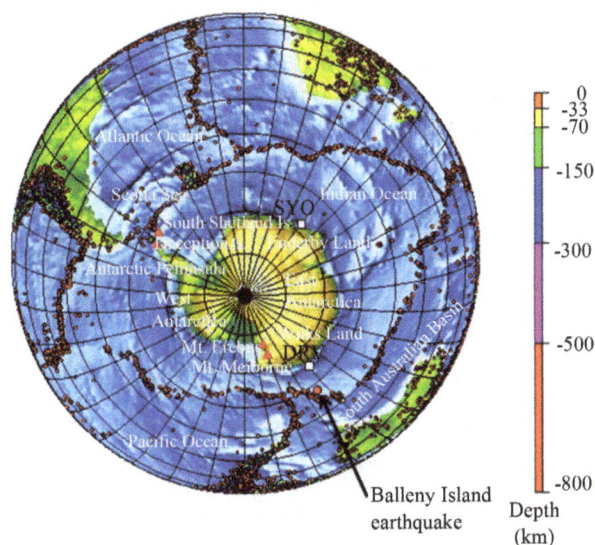

Figure 1. Seismotectonic regions discussed in this paper. Earthquake distribution in and around the Antarctic plate was compiled by NEIC (National Earthquake Information Center) of USGS. Three volcanoes and South Shetland Islands are also shown by solid (red) triangles, and the Balleny Island earthquake is shown by solid (red) circle, respectively.

those by permanent Antarctic stations in addition to temporary field stations.

In this paper, seismicity in the Antarctic continent and surrounding ocean is overviewed by using the data from the International Seismological Centre (ISC). Seismicity in terms of each tectonic setting is discussed by being divided into the following five regions: intra-plate low-seismic regions; high-seismic region around the Antarctic Peninsula; the Antarctic Continent aseismic region; low-seismic region at the edge of the Continent; and the volcanic regions.

2. Antarctic Continent

A general seismicity map in and around the Antarctic Plate is shown in **Figure 1**. The earthquake locations are compiled by the National Earthquake Information Center (NEIC) of USGS. The Antarctic continent is roughly divided into two large tectonic provinces; East Antarctica (eastern hemisphere part from the Transantarctic Mountains) and the other younger province; West Antarctica. East Antarctica is characterized by a fragment of Gondwana super-continent, one of the members of adjacent Pre-Cambrian terrains in southern hemisphere (South America, Africa, India and Australia; **Figure 1**). West Antarctica, on the contrary, is attributed by a chain of many islands beneath the ice sheet of Cenozoic era, including several active volcanoes.

Four earthquakes inside the continental area of Antarctica were reported to the ISC catalog in early stage of Antarctic scientific research. However, except only one event out of the four, no significant earthquakes were located in the continent before the IGY, because neither their locations nor their magnitudes were accurately determined [11]. One event which occurred on June 26, 1968 was located in Coats Land, 20°W, 80°S using the initial phase readings of five seismic stations on the Antarctic continent [3]. The magnitude of the event was 4.3 determined by the waveform amplitude of a three-component seismograph at South Pole Station (SPA). This was the first earthquake located instrumentally in the continent using the data of only seismic stations in Antarctica. However, this event was not listed in the ISC catalog. [3] used the data of five stations for hypocenter determination and two of the five stations were Byrd Seismic Array. As the focal depth was determined 1 km beneath the sea level, the event was considered to be an earthquake in the crust, not an ice-quake.

On the other hand, [4] reported an earthquake near the coast of Oates Land at 70.5°S and 161.3°E on October 15, 1974. This event was the only shock to be located on the Antarctic continent by international agency by that time. Magnitude of the event was estimated to be 5 and he concluded that the event might be an ice-quake associated with ice movement or cracking. Therefore, there are ambiguity to identify the very shallow events near the surface only from the hypocentral information, whether they are tectonic events or the ice related signals.

Regarding the wide inland area of East Antarctica, seismicity in Wilkes Land ("G" in **Figures 2** and **3**) is found to be one order higher than that in vicinity of Syowa Station (69.0°S, 39.0°E), the Lützow-Holm Bay (LHB), East Antarctica [9,12]. Seven earthquakes with Mb 4.0 - 4.9 were located in the coastal area of 100° - 170°E and 66° - 82°S in Wilkes Land in 1964-1996. Another nine earthquakes were located in the inland area and two were located offshore. The magnitudes of these eleven earthquakes were not accurately determined. Not only micro-seismic activity, but also small earthquake activity in Wilkes Land and surrounding coasts are higher than that in other area on the East Antarctic continent. The sub-glacial topography in Wilkes Land is characterized by a sub-glacial basin with 1000 m below sea level in minimum elevation of the bedrock [13]. The maximum thickness of the ice sheet in the area is over 4000 m and the surface elevation of the ice sheet is mostly over 2000 m. The formation, distribution and stability of these sub-glacial lakes might affect on tectonic processes involving relatively high seismicity in this area.

3. Southern Ocean

Seismic activity inside the Antarctic Plate has been known as very few distributions of their epicenters both in ocean and continent areas. As the space distribution and time variations in seismicity around the Antarctic Plate, in particular for the Indian Ocean sector (0° - 160°E, 20° - 80°S), was previously evaluated and intra-plate seismicity was discussed associated with far-field tectonic stress in the oceanic lithosphere [14]. Therefore in this study, a detail hypocentral distribution within the whole Antarctic Plate is investigated by using the data from ISC in 1964-2002. **Figure 2** represents the earthquake locations compiled by ISC in the area of 20° - 80°S and every 60° of longitude. Among the individual 60° longitude sectors and South Pole area in 80° - 90°S, the earthquake activities are divided into 13 regions (from "A" to "M", **Table 1**). Since the seismic activities are extremely high along the plate boundaries around the Antarctic Plate between the surrounding plates, a criterion of the area for intra-plate was selected very carefully not to include the events associated with the plate boundaries.

4. Continental Margins

[10] pointed out the considerable number of intra plate earthquakes in the 90° - 180°E quadrant and divided the earthquakes into two groups as poorly located earthquakes and well located ones. Inside the continental area in Antarctica, seismicitiy is almost very few in "D", "G"

and "M" areas of **Figure 2**. However, the Wilkes Land ("M" in **Figures 2** and **3**) had been identified as the most active within the Antarctic continent at IGY period. Poorly located earthquakes were lined from north to south along the 140°E longitude [9]. In the earthquake locating area, the Resolution sub-glacial highland, the Adventure sub-glacial trench, and the Belgica sub-glacial highland are existed along the longitudinal direction from east to west [13]. There is a possibility that the poorly located earthquakes were ice-quakes, because the thick ice sheet and complicated sub-glacial topography must cause ice-shocks. Generally the edge of the continent, the coast area, is aseismic region.

Over the past few decades, more seismic observations in polar region have detected local seismicity by both temporary seismic networks and permanent stations. [16] found the majority of seismicity near the Scott Base (SBA, 167°E, 78°S) and Wright Valley area (VNDA, 162°E, 78°S) located along the coast, particularly near large glaciers. They suggested a few generation mechanisms for these events, distinguishable by their focal mechanism and depth: basal sliding of the continental ice sheet, movement of ice streams associated with several scales of glaciers, movement of sea-ice, and tectonic earthquakes.

[17] deployed a local seismic network around the Neumayer Station (08°W, 71°S), and determined hypocenters of local tectonic events, located along the coast and the mid area of the surrounding bay. A seismic array has been operated more than one decade at the Neumayer Station. Since the deployment of the seismic network/array, several local events could be detected. Two seismic active regions were figured out at inland area and offshore of the continent. In addition, a broadband seismic network had been developed in the large region between Mawson and Casey stations and inland as far as 75°S by Australia [18]. The aim to establish the seismic network is to discover the" seismic structure of the continent under Antarctica (SSCUA)". Moreover, India has also been carried out seismic observation at the Maitri Station (12°E, 71°S) since 1997. The seismic data have already been contributed to earthquake locations by ISC. India also has published "Seismological Bulletin of Maitri Station, Antarctica" in every year [19].

5. Lützow-Holm Bay

Once a denser seismic network was established, small/micro earthquake activities became gradually clear. The LHB area around Japanese Syowa Station, East Antarctica, is one of the area where seismic activity has been well studied since 1980s [20]. Since seismic observations started by a tripartite network in 1987, seismicity for relatively small events became clear in and around Syowa Station. A total of 18 local earthquakes were located

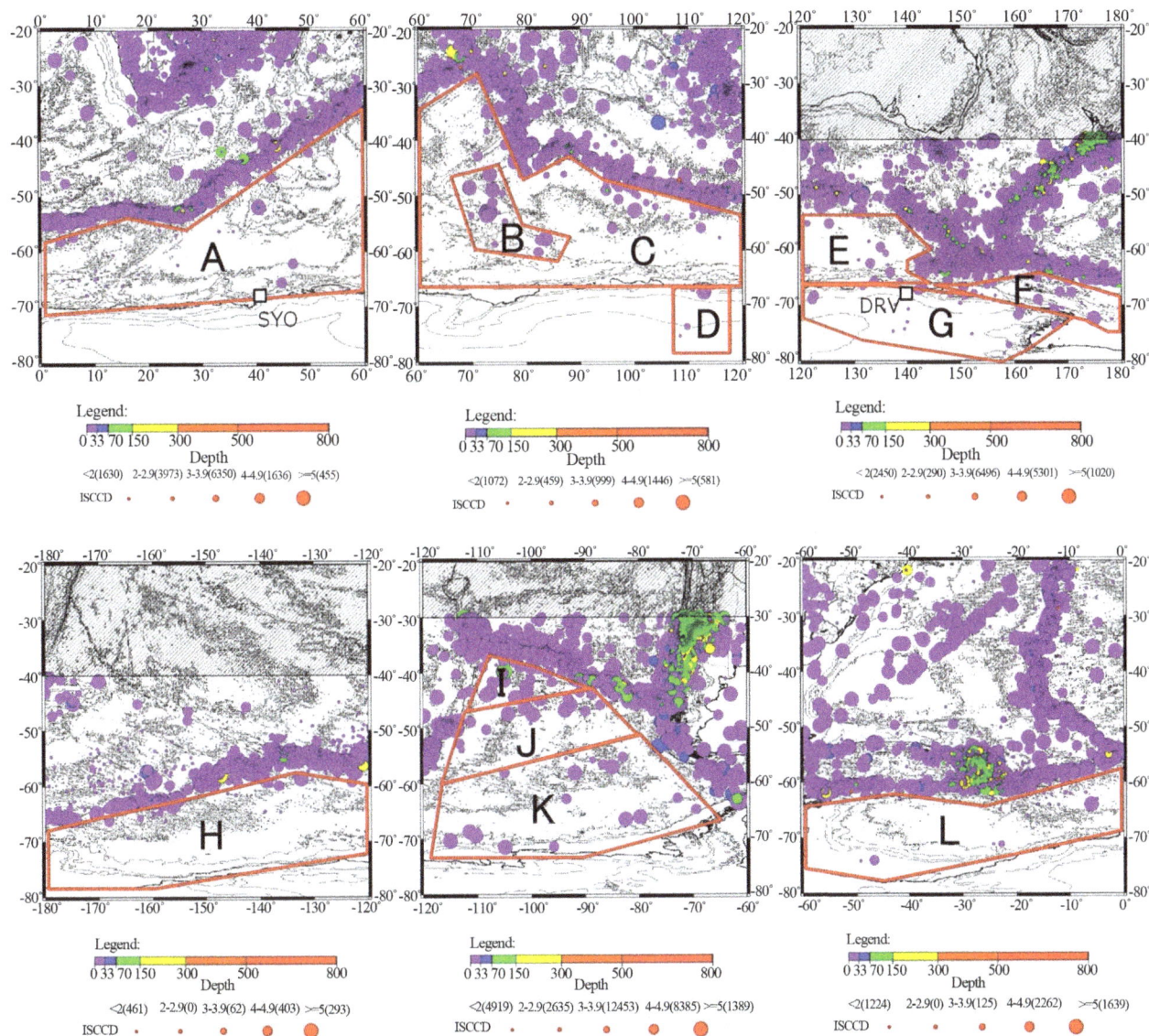

Figure 2. Earthquake locations determined by ISC in 20° - 80°S and every 60° of longitude. Surrounded areas by solid red lines indicate individual blocks ("A" - "L") discussing seismicity in this paper. These areas were classified into Antarctic continent ("D" and "G") and oceanic region within Antarctic Plate (other regions). Numerous denoted in the brackets correspond to magnitude ranges representing the number of included events for each 60° longitude sector.

during the 15 years in 1987-2003 ([12]; **Figure 4**).

Characteristic features in time variations of seismic activity are summarized as follows; Seismicity in 1987-1989: A three-station seismic network was operated around the Syowa Station. Epicenters of ten local earthquakes were determined during these three years. Many different types of earthquakes, such as a main-shock-aftershock, twin earthquake, earthquake swarms, were detected and identified at that time. The seismic activity during this period was higher than that of the following decade. In 1990-1996, nine local earthquakes were recorded with many different types of events. The seismic activity during this period was very low and the magnitudes of the earthquakes ranged from −0.5 to 1.4. One local event

was detected in 1997, two events in 1998 and one event in 2001 and 2003, respectively. The low seismic activity continues to the present day in 2004.

The seventeen events were only detected by local seismic network deployed around the LHB, except for the September 1996 Mb = 4.6 earthquake in the southern Indian Ocean. Almost all the hypocenters were located along the coast, apart from a few on the northern edge of the continental shelf. Local earthquakes in and around Syowa Station were presumably caused by tectonic stress accumulated with crustal uplift after deglaciation [21]. The effect of ice sheet changes may have caused phenomena such as crustal deformation, earthquake occurrence, faulting systems in the shallow part of the lithosphere

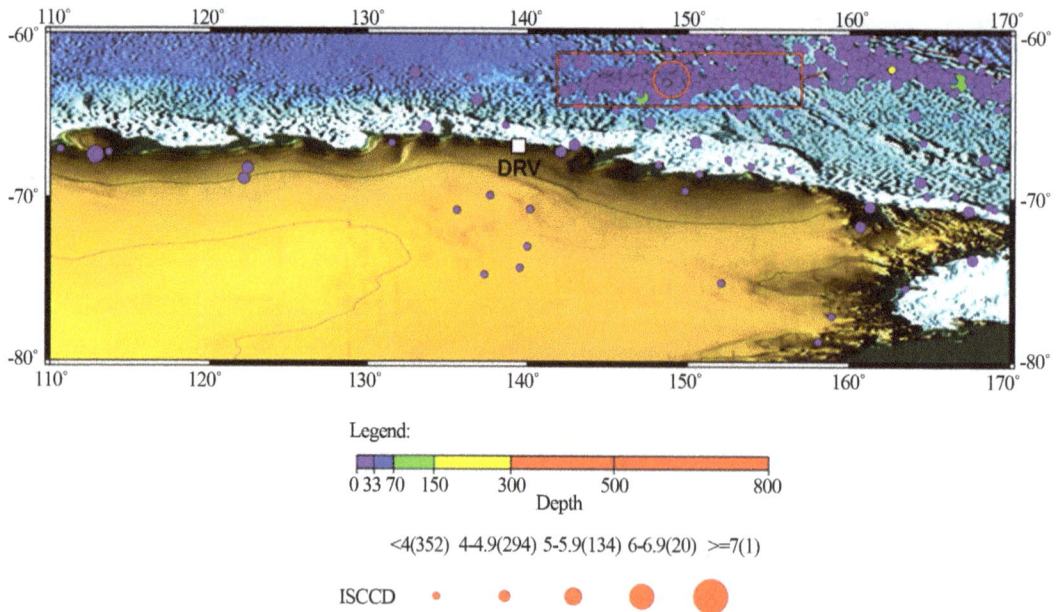

Figure 3. Earthquake locations in 60°-80°S and 110°-170°E, including both the Wilkes Land and aftershock area (red square) of the March 25, 1998 Balleny Island earthquake (Mw = 8.1, red circle). Numerous denoted in the brackets correspond to magnitude ranges representing the number of included events.

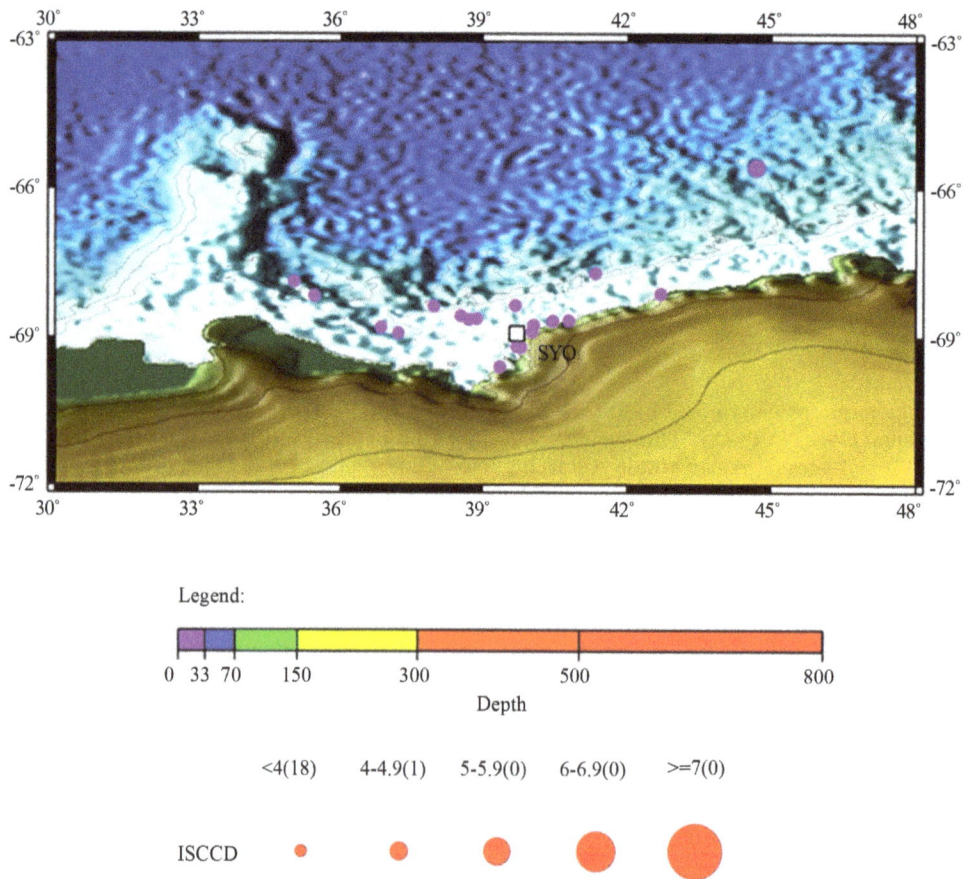

Figure 4. Earthquake locations in and around the Lützow-Holm Bay (LHB), East Antarctica detected at Syowa Station. Total number of 19 events were determined their hypocenters. Numerous denoted in the brackets correspond to magnitude ranges representing the number of included events.

Table 1. Classification of the Antarctic intra-plate earthquakes in 1964-2002 period compiled by ISC, corresponding the individual blocks in Figure 2.

Block	LAT	LON	Event number	Comment
A	0°-60°E	20°-80°S	33	Ocean
B	60°-120°E	20°-80°S	58	Ocean, Kerguelen Plateau
C	60°-120°E	20°-80°S	26	Ocean
D	60°-120°E	20°-80°S	5	Continent
E	120°-180°E	20°-80°S	33	Ocean
F	120°-180°E	20°-80°S	26	Ocean
G	120°-180°E	20°-80°S	23	Continent, Wilks Land
H	180°-120°W	20°-80°S	22	Ocean
I	120°-60°W	20°-80°S	19	Ocean, Easter Is. Triple Junct.
J	120°-60°W	20°-80°S	20	Ocean
K	120°-60°W	20°-80°S	17	Ocean
L	60°-0°W	20°-80°S	33	Ocean
M		80°-90°S	8	Continent, South Pole

beneath Antarctica [16]. Several of these events, moreover, could be large ice-quakes associated with the sea-ice dynamics around the LHB or in the Southern Ocean.

6. Antarctic Peninsula

The epicentral locations in the Antarctic Peninsula of 20° - 80°W, 50° - 70°S are shown in **Figure 5**, on the basis of the ISC data catalog in 1964-2002. The earthquake activity in this region is the highest in Antarctic including active volcanic areas in Deception Island, together with subduction zone in the Bransfield Strait. The focal depths of the earthquakes are mostly shallower than 40 km. During two decades in 1971-1989, only four earthquakes occurred with focal depths between 40 and 100 km [9]. One big event occurred on Feb. 8, 1971, for which Mb and Ms were determined to be 6.3 and 7.0 respectively. In addition that the event is the largest earthquake recorded in the region, the only recorded event with magnitude lager than 7.0 in Antarctica. This earthquake was the first tectonic earthquake to be felt in the Antarctic, observed at the Farady Station (65°S, 64°W) of United Kingdom [22].

According to the report of seismic events recorded at King Sejoung Station (60°S, 59°W) of South Korea, King George Island of the South Shetland Islands, some earthquake swarms occurred around Bridgeman Island [23]. [24] evaluate these earthquake swarms in view from physical volcanology. The earthquake swarms appeared to represent typical volcanic earthquake waveforms. The

earthquake swarm might be originated around ORCA Sea Mount about 20 km southeast of King Sejoung Station. Many earthquakes are located in the back-arc area, either on large submarine volcanoes or rifting region along the center of the Bransfield Strait. Earthquakes concentrated around ORCA Sea Mount and Bridgeman Island. These earthquake locations indicate that the seismicity is associated with active volcanism rifting, non active subduction systems.

[25] deployed seven continuously recording broadband seismometers in the South Shetland Islands region during 1997 and 1998. High level of local seismicity was identified in the first 15 months. About 90 earthquakes with magnitude mb 2 - 4 were located. Many earthquakes with 10 - 30 km depth are located in the forearc region extending from the South Shetland Trench axis toward and beneath the South Shetland Islands. The earthquake locations suggest active convergence along the South Shetland subduction zone, however, there is no significant subduction zone extending into mantle.

7. Volcanic Regions

Seismic observations in volcanic regions have been carried out only three volcanoes as shown in **Figure 1**. Eruptions were recognized at Deception Island (63°S, 61°W) and Mount Erebus (78°S, 167°E, 3974 m), on the contrary, no eruption was recognized at Mount Melbourne (74°S, 165°E, 2732 m). The first felt shock in the Antarctic was the magnitude 4.7 earthquake accompanied by the volcanic eruptions of Deception Island on December 4, 1967 [22]. This eruption was the first recognized explosion of Deception Island by humankind. The buildings of the stations on Deception Island were destroyed by the eruptions and all members in the stations of Argentina, Chile and the UK evacuated safely after the eruption. All of the stations at Deception Island have been closed since that time.

Since 1980's Spanish established a summer stations at Deception Island and started geophysical research on Deception Volcano. [26] have carried out seismic observations with five stations during the austral summer since 1987, and observed approximately 1000 local events per month. Earthquake locations seem to be concentrated along the NE-SW direction crossing the central part of the island. In the 1991-92 survey, the seismic activity was significantly increased with total number of recorded earthquakes more than 700, an extending of the areas with geothermal anomalies and major activity of the fumaroles. The seismic activity was low in 1992-93 and 1993-94 surveys, and no other volcanic anomalies were observed. In the 1994-95 survey, a new increase of seismic activity was recorded with more than 800 seismic events [27].

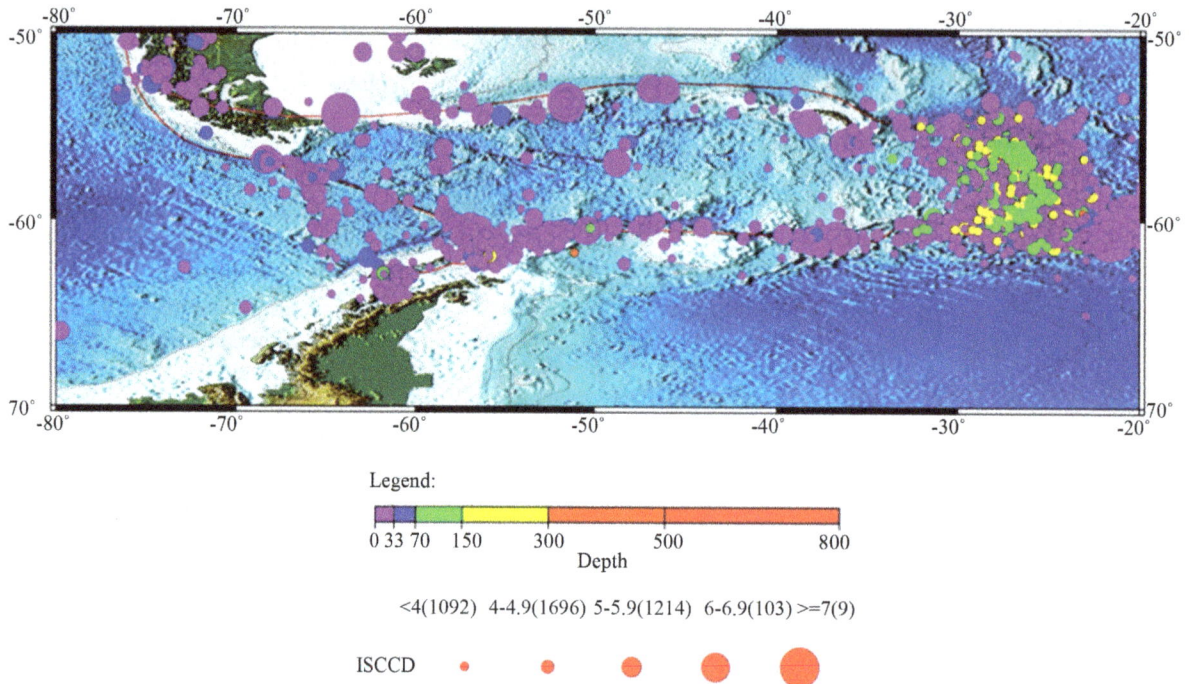

Figure 5. Earthquake location around Antarctic Peninsula. Numerous denoted in the brackets correspond to magnitude ranges representing the number of included events.

Seismic observations by radio-telemetry continued in the summit area and on the slope of Erebus Volcano on Ross Island by the international cooperation among Japan, New Zealand and US during 1981-90 [28]. Seismic activity around Mount Erebus became clear and remarkable change of seismic activity were recognized before and after the new phase of volcanic activity started on September 13, 1984. It was characterized by larger and more frequent strombolian eruptions. Significant changes in seismicity within Ross Island and Mount Erebus were recognized both before and after the increased eruptive activity. The seismic activity of Mount Erebus in 1980-1990 is divided into the following four stages: 1) normal high activity, 2) preceding the new phase, 3) new phase in volcanic activity and 4) low seismic activity. These stages suggest a general pattern of volcanic activity and give a fundamental information concerning production of volcanic eruption.

US scientists have still been carrying on the monitor of seismic activity around Mount Erebus [29]. A research program on physical volcanology has been conducted around Mount Melbourne by Italian scientists since the end of the 1980's. Four seismic stations were installed on Mount Melbourne since 1990 and many local earthquakes have been reported together with other geophysical data [30]. Local seismicity clustered along the eastern flank of Mount Melbourne, possibly in spatial association with the local north-south magnetic trend. One event occurred on December 10, 1990 is the largest earthquake with magnitude 1.9 around Mount Melbourune [31].

8. Ice-Quakes, Glacial Earthquakes

Seismic signals involving ice-related phenomena are called "ice-quakes (ice-shocks for smaller ones)", and are most frequently reported in association with glacially related mass movements of ice-sheets, or with sea-ice, tidecracks and icebergs in the other polar areas [32,12]. The so-called "ice-micity" detected around the Bransfield Strait and Drake Passage by a local network of hydrophone arrays in 2006-2007 illustrate the dynamic behavior of sea ice in the Bransfield and Antarctic Peninsula regions [33].

Dynamics of sea-ice and icebergs also affect on the seismic signals. For example, a large volume of sea ice was discharged from LHB during the 1997 austral winter, and clearly imaged by the NOAA satellite. The broadband seismographs at Syowa Station recorded distinct waveforms associated with the discharge events [34]. The long-duration sea-ice tremors had very distinct spectral characteristics that distinguished them clearly from ordinary teleseismic and/or local tectonic events. Several sequences of harmonic over-toned signals, presumably associated with the merging of multiple ice volumes, appeared on the seismic signals [34]. The characteristic signals also represent the surge events that seem more closely related to the break-up process of the sea-ice mass volume. Both kinds of cryoseismic waves occurred continuously for few hours, and repeated themselves several times within a few days during late July, 1997. Identification of the exact sources that produced these characteristic signals has not yet been completed, and theoretical modeling will most likely be required to explain the physical processes.

Similar cryoseismic phenomena have also been reported around the Ross Sea region (MacAyeal et al., 2009), the marginal sea of the Antarctic Peninsula [33], as well as the continental margin of Dronning Maud Land [17]. In particular, iceberg-originated harmonic tremor emanating from tabular icebergs was observed by both seismo-acoustic and local broadband seismic signals [35]. The tremor signals consisted of extended episodes of stick-slip ice-quakes generated when the ice-cliff edges of two tabular icebergs rubbed together during glancing, "strike-slip" type iceberg collisions. Source mechanisms of such harmonic tremors might provide useful information for the study of iceberg behavior, and a possible method for remotely monitoring iceberg activity.

In addition to the short-period cryoseismic signals mentioned above, a new class of seismic events associated with melting of large ice cap was discovered in particular around Greenland [36,37]. These large events were called "glacial earthquakes", generating long-period ($T > 25$ s) surface waves equivalent in strength to those radiated by standard magnitude five earthquakes, and were observable worldwide. The glacial earthquakes radiated little high-frequency energy, which explains why they were not detected or located by traditional earthquake-monitoring systems. These events are two magnitude units larger than previously reported seismic phenomena associated with glaciers, a size difference corresponding to a factor of 1000 in a seismic moment.

9. Discussion

For many decades, the Antarctic continent and surrounding area have been believed as aseismic. Several earthquakes were found to be located in the Antarctic continent since IGY, as mentioned in this paper. It became general knowledge that there are small earthquake activities detected in the Antarctic Continent even though the activity itself is not so high as other continents. However the event of which magnitude lager than 5 is not detected during last four decades. The number of tectonic earthquakes located in the Antarctic has increased according to the development of Global Seismic Networks and local seismic arrays. The most of the tectonic earthquakes along the coast of the Antarctic continent are caused by tectonic stress accumulated by crustal deformation after deglaciation.

The micro-earthquake activity around Syowa Station as mentioned the previous section might be a typical activity on the coastal area of the Antarctic continent where crustal uplift after deglaciation is now going on. As seismological observation like the local array deployment around Syowa Station have been developing at several stations in the coastal area, high activity of micro-earthquakes will be detected on the continental margin of Antarctica. If the other geophysical observations such as tide gauges, gravimeters and GPS have been carried out at coastal stations, the relation between micro-earthquake activity and crustal movement will become clear.

Around continental area in East Antarctica from 90°-180°E, in generally, seismicity is low in an average. As already treated in Chapter 2, the Wilkes Land, in contrast, has been identified as the most tectonically active region within the Antarctic. It is also noticed that there is a significant number of sub-glacial lakes in this area, then can be considered a relation to produce the occurrence of local seismic activities. The formation, distribution and stability of sub-glacial lakes could give rise to the fundamental question how the tectonic processes control the existence of the lakes [38]. The poorly located earthquakes, accordingly, could be a kind of large ice-quakes, because complicated sub-glacial topography with overlying ice-sheet in this area may efficiently cause ice related seismic phenomena.

A complexity of the surface bedrock structure of Wilkes Land, moreover, might be influenced by tectonically weakened upper crust along with the present mobile belts, forced by the Tasmania micro plate [7,12]. The aftershock area of the large Balleny Island Earthquake in the Antarctic Oceanic Plate appears to be continued with the continental area of Wilkes Land (**Figure 3**). It might be supposed that the large earthquakes generate in this area, by assuming thermal stress of the young lithosphere, associated with unusual deformation due to Macquarie triple junction [39]. Recently several intra-plate earthquakes became to be located in the southern ocean. In contrast, there is a possibility that some earthquakes off of the Antarctic Plate are also caused by crustal deformation and tectonic stress involving deglaciation [6]. It is strongly suggested that the effect of volume and shape change of ice sheet causes some phenomena in the crust and upper mantle such as crustal uplift, earthquake occurrence, lithospheric deformation.

10. Conclusion

Seismicity in the Antarctic and surrounding ocean is evaluated based on the compiled data in 1964-2002 by International Seismological Centre. The Antarctic continent and surrounding ocean were believed to be one of the aseismic regions of the Earth for many decades. However, according to the development of Global Seismic Networks and local seismic arrays, the number of tectonic earthquakes detected in and around the Antarctic continent has increased. The total of 13 seismicity areas were classified into the Antarctic continent (three areas) and oceanic regions within Antarctic Plate (the other 10 areas). Generally, seismic activity in the continental areas is very low; few small earthquakes located. The Wilkes Land is the most active area in the Antarctic continent; several small earthquakes were detected during these four decades. In the ocean area, in contrast, the seismic activity in the area of 120° - 60°W sector is three times larger than that in the other areas. This evidence is considered to be involved in a stress concentration toward the Easter Island Triple Junction among Antarctic Plate, Pacific Plate and Nazuca micro-Plate. Three volcanic areas, Deception Island, Mt. Erebus and Mt. Melbourn, are also high seismic activity areas. Finally, there are a lot of undefined origin events which include both tectonic earthquakes involving deglaciation, as well as cryoseismic signals associated with variations in the surface environment.

Acknowledgements

The authors express their sincere thankfulness to Ms. A. Ibaraki of NIPR, for her great efforts in re-scaling the Syowa Station data in last two decades. The authors also appreciate to Profs. K. Shibuya, K. Doi, Y. Nogi and Y. Aoyama of NIPR for their valuable advice and discussions for manuscript. The authors would express their sincere appreciation to the International Seismological Centre (ISC), for the utilization of the compiled data.

REFERENCES

[1]   B. Gutenberg and C. F. Richter, "Seismicity of the Earth and Associated Phenomena," Princeton University Press, Princeton, 1954.

[2]   C. F. Richter, "Elementary Seismology," W. H. Freeman (Ed.), San Francisco, 1958, p. 768.

[3]   K. Kaminuma and M. Ishida, "Earthquake Activity in Antarctica," *Antarctica Record*, Vol. 42, 1971, pp. 53-60.

[4]   R. D. Adams, "Source Properties of the Oates Land Earthquake," In: C. Craddock, Ed., *Antarctic Geoscience*, The University of Wisconsin Press, Wisconsin, 1982, pp. 955-958.

[5]   A. C. Johonstone, "Suppression of Earthquakes by Large Continental Ice Sheets," *Nature*, Vol. 330, 1987, pp. 467-469.

[6]   S. Tsuboi, M. Kikuchi, Y. Yamanaka and M. Kanao, "The March 25, 1998 Antarctic Earthquake: Great Earthquake Caused by Postglacial Rebound," *Earth Planets Space*, Vol. 52, 2000, pp. 133-136.

[7]   T. Himeno, M. Kanao and Y. Ogata, "Statistical Analysis of Seismicity in a Wide Region around the 1998 Mw 8.1 Balleny Islands Earthquake in the Antarctic Plate," *Polar Science*, Vol. 5, No. 4, 2011, pp. 421-431.

[8]   K. Kaminuma, "Seismic Activity in and around Antarctic Continent," *Terra Antartica*, Vol. 1, 1994, pp. 423-426.

[9]   K. Kaminuma, "A Revaluation of the Seismicity in the Antarctic," *Polar Geoscience*, Vol. 13, 2000, pp. 145-157.

[10] A. M. Reading, "Antarctic Seismicity and Neotectonics," In: J. A. Gamble, *et al.*, Eds., *Antarctica at the Close of a Millennium*, The Royal Society of New Zealand Bullutin, Wellington, Vol. 35, 2002, pp. 479-484.

[11] R. D. Adams, A. A. Hughes and B. M. Zhang, "A Confirmed Earthquake in Continental Antarctica," *Geophysical Journal International*, Vol. 81, No. 2, 1985, pp. 489-492.

[12] M. Kanao and K. Kaminuma, "Seismic Activity Associated with Surface Environmental Changes of the Earth System, around Syowa Station, East Antarctica," In: D. K. Futterer, *et al.*, Eds., *Antarctica: Contributions to Global Earth Sciences*, Springer-Verlag, Berlin, Heidelberg, New York, 2006, pp. 361-368.

[13] D. J. Drewry, "Antarctica: Glaciological and Geophysical Folio, 9 Sheets," Scottish Polar Research Institute, University of Cambridge, Cambridge, 1983.

[14] M. Kanao, Y. Nogi and S. Tsuboi, "Spacial Distribution and Time Variation in Seismicity around Antarctic Plate —Indian Ocean," *Polar Geoscience*, Vol. 19, 2006, pp. 202-223.

[15] D. L. Anderson, "Superplumes or Supercontinents?" *Geology*, Vol. 22, No. 1, 1994, pp. 39-42.

[16] S. Bannister and B. L. N. Kennett, "Seismic Activity in the Transantarctic Mountains—Results from a Broadband Array Deployment," *Terra Antarctica*, Vol. 9, 2002, pp. 41-46.

[17] C. Muller and A. Eckstaller, "Local Seismicity Detected by the Neumayer Seismological Network, Dronning Maud Land, Antarctica: Tectonic Earthquakes and Ice-Related Seismic Phenomena," IX International Symposium on Antarctic Earth Science Programme and Abstracts, Potsudam, 2003, p. 236.

[18] A. M. Reading, "On Seismic Strain-Release within the Antarctic Plate," In: D. K. Futterer, *et al.*, Eds., *Antarctica: Contributions to Global Earth Sciences*, Springer-Verlag, Berlin, Heidelberg, New York, 2006, pp. 351-356.

[19] G. B. N. Chander, S. V. R. Ramachandra Rao, G. S. Srinivas, E. C. Malaimani and N. R. Kumar, "Seismological Bulletin of Maitri Station, Antarctica, 2002," National Geophysical Research Institute, 2003, pp. 1-71.

[20] K. Kaminuma and J. Akamatsu, "Intermittent MicroSeismic Activity in the Vicinity of Syowa Station, East Antarctica," In: Y. Yoshida *et al.* (Eds.), *Recent Progress in Antarctic Earth Science*, Terra Science Publication, Tokyo, 1992, pp. 493-497.

[21] K. Kaminuma and M. Kanao, "Local Seismicity and Crustal Uplift around Syowa Station, Antarctica," *Korean Journal of Polar Research*, Vol. 10, 1999, pp. 103-107.

[22] K. Kaminuma, "Seismicity around the Antarctic Peninsu-

la," *Polar Geoscience*, Vol. 8, 1995, pp. 35-42.

[23] Y. K. Jin, D. K. Lee, S. H. Nam, Y. Kim and K. J. Kim, "Seismic Observation at King Sejong Station, Antarctic Peninsula," *Terra Antartica*, Vol. 5, 1998, pp. 729-736.

[24] K. Kaminuma, "A Possibility of Earthquake Swarms around ORCA Sea Mount in the Bransfield Strait, the Antarctic," In: Y. Kim and B. K. Khim, Eds., *Proceedings of the Joint International Seminar: Recent Interests on Antarctic Earth Sciences of Korea and Japan*, 2001, pp. 23-34.

[25] S. D. Robertoson, D. A Wiens, P. J. Shore, G. P. Smith and E. Vera, "Seismicity and Tectonics of the South Shetland Islands and Bransfield Strait from the SEPA Broadband Seismograph Deployment," In: J. A. Gamble *et al.*, Eds., *Antarctica at the Close of a Millennium*, *The Royal Society of New Zealand Bullutin*, Wellington, Vol. 35, 2002, pp. 549-554.

[26] J. Vila, R. Ortiz, A. M. Correig and A. Garcia, "Seismic Activity on Deception Island," In: Y. Yoshida *et al.*, Eds., *Recent Progress in Antarctic Earth Science*, Terra Science Publication, Tokyo, 1992, pp. 449-456.

[27] R. Ortiz, A. Garcia, A. Aparicio, I. Branco A. Felpeto, R. Rey Del, M. T. Villegas, J. M. Ibanez, J. Morales, E. Pezzo Del, J. C. Olmedillas, M. Astiz, J. Vila, M. Ramos, J. G. Viramonte, C. Risso and A. Caselli, "Monitoring of the Volcanic Activity of Deception Island, South Shetland Islands, Antarctica (1986-1995)," In: C. A. Ricci, Ed., *The Antarctic Region: Geological Evolution and Processes*, Terra Antartica Publication, Siena, 1997, pp. 1071-1076.

[28] K. Kaminuma and R. R. Dibble, "Seismic Activity of Mount Erebus 1981-1988," *Polar Geoscience*, Vol. 4, 1990, pp. 142-148.

[29] R. Aster, W. McIntosh, P. Kyle, R. Esser, B. Bartel, N. Dunbar, B. Johns, J. Johnson, R. Karstens, C. Kurnik, M. McGowan, S. McNamara, C. Meertens, B. Pauly, M. Richmond and M. Ruiz, "Real-Time Data Received from Mount Erebus," *EOS Transactions*, Vol. 85, No. 10, 2004, p. 99.

[30] A. Banaccorso, S. Gambino and E. Privitera, "A Geophysical to the Dynamics of Mt. Melbourne (Northern Victoria Land, Antarctica)," In: C. A. Ricci, Ed., *The Antarctic Region: Geological Evolution and Processes*, Terra Antartica Publication, Siena, 1997, pp. 531-538.

[31] E. Armadillo, A. Bonaccorso, E. Bozzo, G. Caneva, A. Capra, G. Falzone, F. Ferraccioli, S. Gandolfi, F. Mancini, E. Privitera and L. Vittuari, "Geophysical Features of the Mt. Melbourne Area, Antarctica, and Preliminary Results from the Integrated Network for Monitoring the Volcano," In: J. A. Gamble *et al.*, Eds., *Antarctica at the Close of a Millennium*, The Royal Society of New Zealand Bulletin, Wellington, Vol. 35, 2002, pp. 571-577.

[32] S. Anandakrishnan and R. B. Alley, "Tidal Forcing of Basal Seismicity of Ice Stream C, West Antarctica, Observed Far Inland," *Journal of Geophysical Research*, Vol. 102, No. B7, 1997, pp. 15183-15196.

[33] R. P. Dziak, M. Parlk, W. S. Lee, H. Matsumoto, D. R. Bohnenstiehl and J. H. Haxel, "Tectono-Magmatic Activity and Ice Dynamics in the Bransfield Strait Back-Arc Basin, Antarctica," *The 16th International Symposium on Polar Science*, Incheon, 2009, pp. 59-68.

[34] M. Kanao, A. Maggi, Y. Ishihara, M.-Y. Yamamoto, K. Nawa, A. Yamada, T. Wilson, T. Himeno, G. Toyokuni, S. Tsuboi, Y. Tono and K. Anderson, "Interaction on Seismic Waves between Atmosphere—Ocean—Cryosphere and Geosphere in Polar Region," In: M. Kanao *et al.*, Eds., *Seismic Waves—Research and Analysis*, InTech. Publisher, Rijeka, 2012, pp. 1-20.

[35] D. MacAyeal, E. Okal, R. Aster and J. Bassis, "Seismic Observations of Glaciogenic Ocean Waves (Micro-Tsunamis) on Icebergs and Ice Shelves," Journal of Glaciology, Vol. 55, No. 190, 2009, pp. 193-206.

[36] G. Ekström, M. Nettles and V. C. Tsai, "Seasonality and Increasing Frequency of Greenland Glacial Earthquakes," *Science*, Vol. 311, No. 5768, 2006, pp. 1756-1758.

[37] M. Nettles and G. Ekström, "Glacial Earthquakes in Greenland and Antarctica," *Annual Review of Earth and Planetary Sciences*, Vol. 38, 2010, pp. 467-491.

[38] T. Wilson and R. Bell, "Earth Structure and Geodynamics at the Poles," Understanding Earth's Polar Challenges: International Polar Year 2007-2008, 2011, pp. 273-292.

[39] M. Nettles, T. C. Wallace and S. L. Beck, "The March 25, 1998 Antarctic Plate Earthquake," *Geophysical Research Letters*, Vol. 26, No. 14, 1999, pp. 2097-2100.

15

Numerical Study of Piles Group under Seismic Loading in Frictinal Soil—Inclination Effect

Fadi Hage Chehade[1], Marwan Sadek[2], Douaa Bachir[3]
[1]Civil Engineering Department, Doctotral School of Sciences and Technology, Lebanese University,
University Institute of Technology (Saida) & Modeling Center, Beirut, Lebanon
[2]Laboratory of Civil Engineering and GeoEnvironment,
University of Lille I Sciences and Technology, Villeneuve d'Ascq, France
[3]Numerical Center, Doctoral School of Science and Technology, Lebanese University, Beirut, Lebanon

ABSTRACT

Recent devastating earthquakes in some countries, such as Pakistan, Turkey, Algeria and China, call to the mind the high risk exposure of Lebanon which is located over an active seismic zone. Many experts shared the view that major seismic event may occur in Lebanon in the future. Moreover, many earthquakes, of low magnitudes between three and four, have been registered in Lebanon during 2008. These events have increased the anxiety of Lebanese people because of the poor quality of the constructions and their behavior under moderate or severe earthquake events. The efficient way to minimize seismic effects, material and human losses, is the prevention. The system piles-foundation is an appropriate way and widely used to ensure the stability of constructions when subjected to seismic excitation. It seems necessary to study the interaction of pile-foundation-pile-cap-structure in the case of non linear soil behavior and the interface pile-soil. The study will be also conducted by using measures recorded during real earthquakes for example in Turkey (Kocaeli, 1999). In this paper, we present a numerical modeling of the interaction of using FLAC3D software. According to soil behavior and pile inclination, parametric studies are also performed. The analysis of the results could give the better piles group configuration in order to minimize the seismic effect on the structures.

Keywords: Frictional Soil; Inclination; Interaction; Non Linear; Numerical Modeling; Piles Group

1. Introduction

Recent devastating earthquakes in some countries, such as Pakistan, Turkey, Algeria and China, call to the mind the high risk exposure of Lebanon which is located over an active seismic zone. There are some faults in the country represented respectively by major and secondary ones. Many geologic experts shared the view that major seismic event could occur in Lebanon in the future. Moreover, many earthquakes, of low magnitudes between three and four, have been registered in Lebanon during 2008. These events have increased the anxiety of Lebanese people because of the poor quality of the constructions and their behavior that could undergo in the case of moderate or severe earthquake events. The efficient way to minimize seismic effects both material and human losses, is the prevention and in particular it is very important to enhance the foundations of the constructions.

Piles are used as foundation elements for structures located in seismic areas. They provide stability, but may be acquired by efforts that exceed their carrying capacity. These efforts are dangerous for piles installed in soils with low fundamental frequencies, as they amplify the seismic motion of the soil endangering the stability of these structures and their functioning.

Analysis of the seismic response of soil-pile-structure systems constitutes a complex problem in earthquake engineering. In addition to post-earthquake investigations, analytical and numerical analyses show that the damage of piles in seismic area is mainly attributed to the kinematic interaction between piles and soils or (and) to the inertial interaction between the superstructure and the pile foundation which may cause foundation damages, in particular at the pile-cap connection.

Due to the complexity of the problem of interaction soil-pile-structure and the strong coupling between the elements of foundation and structure, it is necessary to conduct a comprehensive analysis of this problem. Most researches have been conducted within the framework of the elasticity, considering the link between piles and soil as rigid. In the case of strong seismic loading, the nonlinearities of soil can play an important role in modifying the state of the interface soil-pile causing a strong damping of the seismic energy injected into the structure. Using advanced computing resources, the consideration of the nonlinearities of the soil and the structure becomes possible in a comprehensive approach.

Non linear full 3D analyses considering the soil, piles and the superstructure are still limited. Such studies were conducted in the linear domain ([1-5]) to analyse the influence of micropiles inclination and boundaries conditions on the seismic behaviour of the soil-micropile structure system. Gerolymos *et al.* [6] used a full 3D finite element analysis to study the seismic performance of inclined piles assuming a linear behaviour of the soil and the structure.

The present paper is focused on a full 3D coupled modelling of the soil-pile-superstructure interaction under seismic loading considering the elastoplastic behaviour of the soil material. Analysis is performed using the FLAC3D [7] program under real earthquake records (Kocaeli, 1999). Soil plasticity is investigated in the case of frictional soils where the soil behaviour is described using the simple and popular non-associated Mohr-Coulomb criterion largely used in engineering practice. The last part discusses the efficiency of inclined piles in seismic zones. Using inclined piles in seismic zones is generally not recommended by international codes, especially when piles are anchored in hard substrata. However, the analysis of the Loma Prieta earthquake ([8]) and Kobe ([9,10]) showed that structures based on inclined piles were less affected or damaged than other structures.

2. Numerical Analysis of Soil-Pile-Structure System (Elastic Soil)

2.1. Problem Definition

The model that has been analyzed consists of a group of four piles of 1m diameter and 10 m length, embedded in a homogeneous soil layer of 15 m of thickness (**Figure 1**). The piles are fixed in a cap of 1m thick with no con-

tact with the ground, and supporting a superstructure. The spacing between piles is $S = 3.75D_p$ (D_p: is the diameter of the pile). The behaviour of the soil-pile-structure system is firstly assumed to be elastic with Rayleigh damping. The superstructure is modelled by a single degree of freedom system, consisting of a column of height $H_{st} = 1$ m and a concentrated mass $m_{st} = 100$ tons placed on the top of column. The fundamental frequency of the soil layer is equal to $f_1 = 0.67$ Hz. The rigidity (K_{st}) and the frequency of the superstructure (f_{st}), assumed fixed at its base, are calculated using the following expression:

$$K_{st} = \frac{3(E_{st}I_{st})}{(H_{st})^3}$$
$$f_{st} = \frac{1}{2\pi}\sqrt{\frac{K_{st}}{m_{st}}} \tag{1}$$

Then, $K_{st} = 86400$ kN/m and $f_{st} = 1.48$ Hz.

The mesh used is shown in **Figure 2**. A refinement is located around the piles and the area near the superstructure where the inertial forces induce high stresses. The soil basis is assumed rigid. The boundaries are placed far enough from the structure with the use of specific elements and absorbing boundaries ("free field") to reduce the reflection of waves at the edges of the model. The Soil, piles and superstructure mechanical properties are given in **Table 1**. The piles have an axial rigidity $E_pA_p = 18850$ MN and a flexural rigidity $E_pI_p = 1178$ MN.m^2.

2.2. Seismic Loading

The soil-structure system is subjected to two types of loading. The first one corresponds to a harmonic loading with a frequency equal to the soil natural frequency. This loading has very severe consequences and may lead to high internal forces that are not representative to a real earthquake. The second one corresponds to the 1999 Kocaeli earthquake in Turkey with a frequency contents close to the natural frequencies of the soil-structure system.

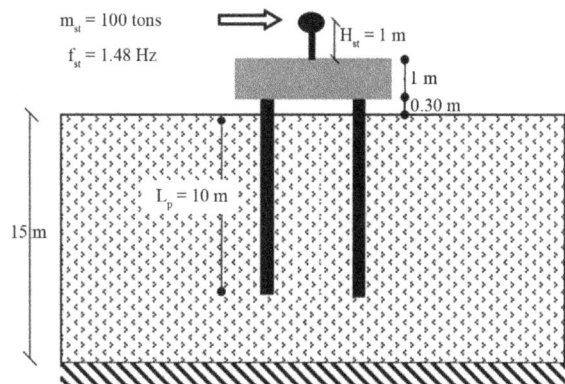

Figure 1. Problem definition.

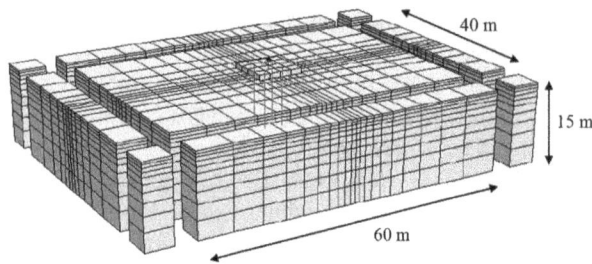

Figure 2. Mesh and boundary conditions.

Table 1. Mechanical properties of soil-pile-superstructure system.

	Mechanical properties			
	Density (Kg/m³)	Young modulus (MPa)	Poisson's ratio	Damping ratio
Soil	1700	8	0.3	$\xi = 5\%$
Pile	2500	24,000	0.3	$\xi = 5\%$
Structure	2500	24,000	0.3	$\xi = 5\%$

In the first step, numerical simulations were performed in the case of a harmonic load of 10 cycles, with a frequency equal to the fundamental frequency of the soil ($f_{loading} = f_{1,sol} = 0.67$ Hz), and an acceleration amplitude of 0.2 g ($V_g = 0.46$ m/s).

In a second step, a real loading is applied as a velocity at the base of the soil mass. The record for the base acceleration, velocity, and displacement waves are shown in **Figure 3**. It marks a maximum speed of 40 cm/s and a maximum acceleration of 0.247 g. Fourier analysis of the record of the earthquake's velocity results in a power spectrum that reveals a dominant frequency at f = 0.9 Hz (lower peaks are observed at 0.6 and 1.3 Hz) to be compared with the natural frequencies of the soil (0.67 Hz) and the superstructure (1.4 Hz). The Kocaeli earthquake has been chosen because it has frequency contents close to the natural frequencies of the soil-structure system which enhance the development of soil plasticity.

2.3. Comparison of Dynamic Forces in the Piles

The forces induced in the piles due to the real seismic loading of Kocaeli, of frequency f = 0.9 Hz, are represented in **Table 2** and **Figure 4**, compared with forces induced in piles due to the harmonic loading of frequency f = 0.67 Hz. **Figure 4** shows a significant influence of the dominant frequency loading that can lead to significant efforts values exceeding the bearing capacity of piles and especially when this frequency equals to the natural frequency of the soil. Harmonic loading is very detrimental and causes excessive forces compared to real earthquake loading. So, only real record is used in the next analysis.

Table 2. Piles response for different types of loading.

Loading	Acc mass (m/s²)	Acc cap (m/s²)	Max shear force V (KN)	Max. bending moment M (KN.m)
Sinusoidal	34.71	33.1	854.8	3137
Turkey (Kocaeli)	7.36	5.4	145	453.8

3. Effect of Soil Plasticity

This section deals with the effect of nonlinearities on the behaviour of the soil-pile-structure system, in particular, the influence of soil plasticity on the system response. Numerical simulations are performed using real seismic loading (Turkey, Kocaeli, 1999). The soil behavior is described by an elastic-perfectly plastic Mohr-Coulomb. The case relative to frictional soil is presented in this section.

3.1. Plastic Calculation

A parametric study was conducted to know the effect of plasticity for frictional soil on the seismic behaviour of the soil-pile-structure system. The friction angle is considered of 30° and a low cohesion of 2 KPa. To know the influence of dilancy angle, two values were chosen $\psi = 0°$ and $\psi = 20°$. A slight damping of Rayleigh is used for the soil. The behaviour of the cap-structure system is assumed to be elastic.

The extension of plasticity is shown in **Figure 5**. We can note that the plasticity is localized near the surface due to the low soil confinement at this zone. The seismic loading induces plasticity at the top of soil and the energy is injected into the structure. Plasticization of the soil around the pile head makes it weaker; it leads to the formation of a gap around the pile head which confirms the post seismic observations.

3.2. Effect of soil Behavior

Figure 6 shows the internal stresses induced in piles. The variation of the maximum shear force at the top is related to the change of acceleration. For the bending moment, the results at the top are not significantly affected by the change of the dilancy angle. The results are illustrated in **Table 3**. This parametric study considers two extreme cases of dilancy angles, the intermediate values between 0° and 20° lead to similar tendencies.

4. Effect of Piles Inclination

The effect on the pile inclination on the seismic answer of the system pile-soil-cap is investigated in this section. The case that has been modelled is similar to that earlier, except that the four piles are inclined outwardly. To properly analyze the influence of pile inclination on their

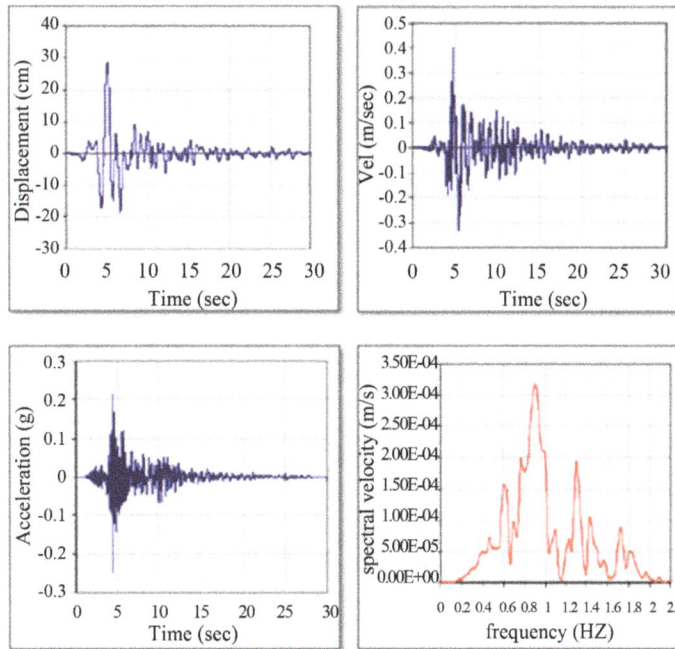

Figure 3. Data of Kocaeli earthquake (Turkey, 1999).

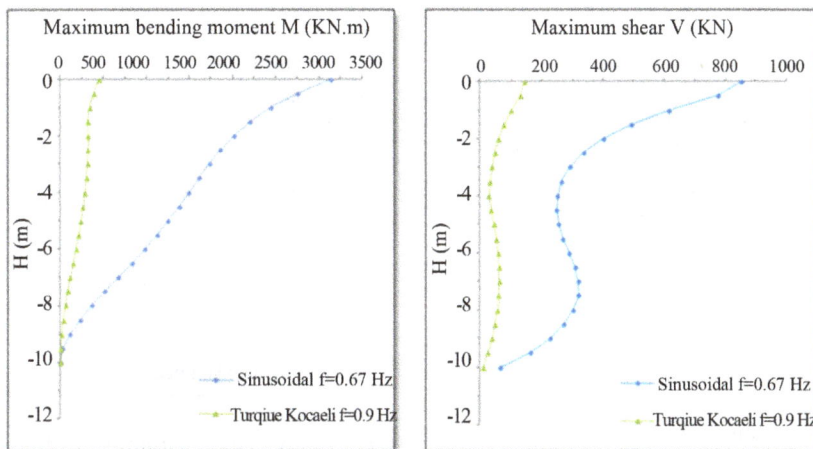

Figure 4. Maximum dynamic forces in the piles.

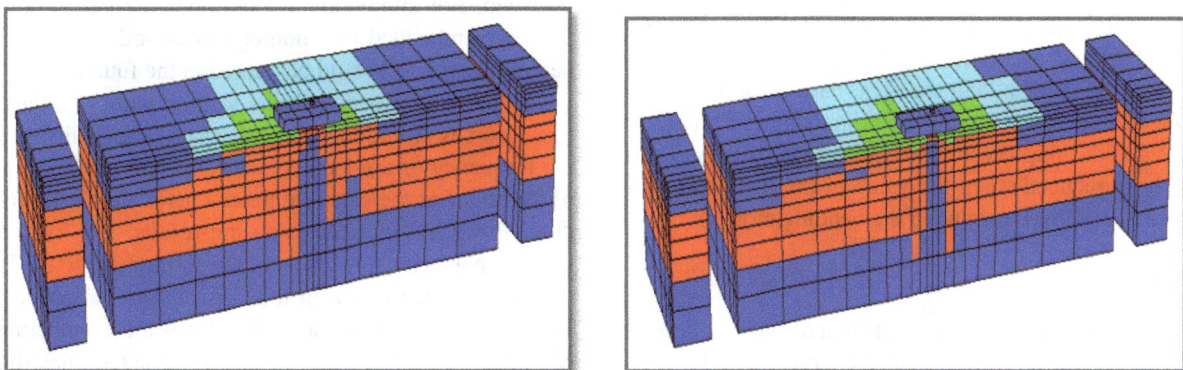

(a)   (b)

Figure 5. Plasticity extension in the case of frictional soil (C = 2 KPa, $\psi = 0°$ and $\psi = 20°$). (a) $\psi = 0°$, (b) $\psi = 20°$.

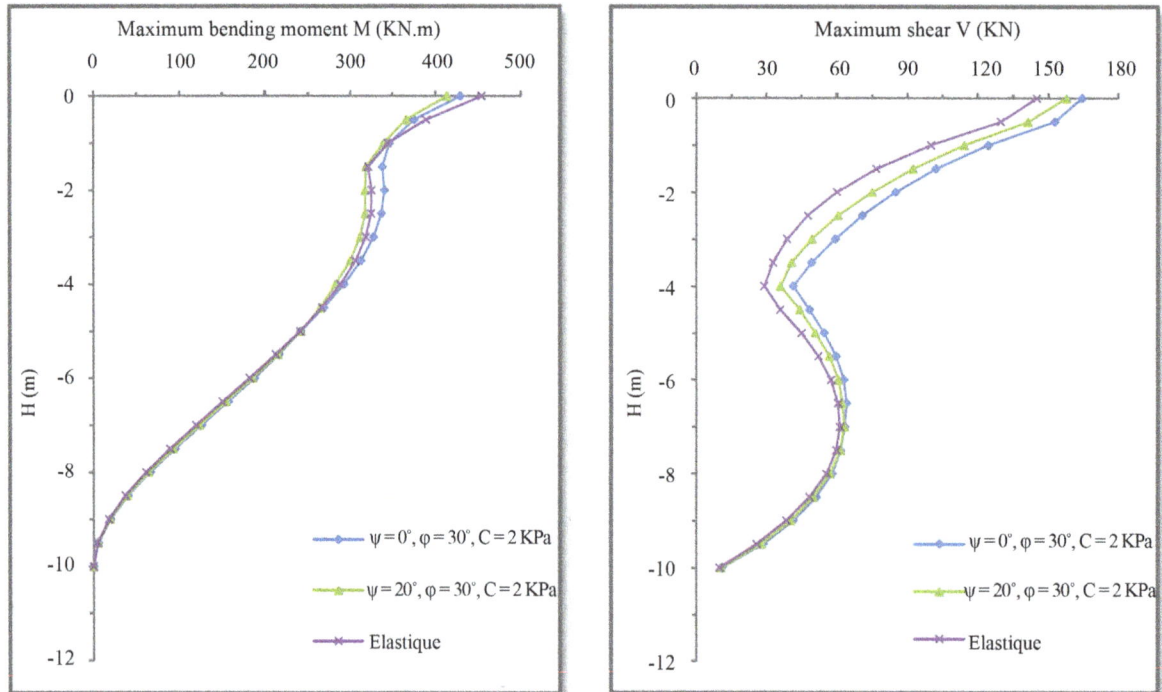

Figure 6. Influence of frictional soil plasticity on the dynamic forces in the piles (Kocaeli earthquake, Turkey, 1999).

Table 3. Influence of frictional soil plasticity on the dynamic forces in the piles.

ψ (°)	Acc mass (m/s²)	Acc cap (m/s²)	Max shear force V (KN)	Max. bending moment M (KN.m)
Elastic	7.36	5.4	145	453.8
0°	7.431	5.775	164.3	428.5
20°	7.391	6.147	157.7	413.6

Table 4. Influence of inclination on the seismic response of pile groups.

Inclination	$\alpha = 0°$	$\alpha = 10°$	$\alpha = 20°$
Amplification at the head of cap	5.40	5.217	3.445
Amplification at the head of structure	7.36	6.526	3.580
Maximum bending moment M (KN.m)	453.8	584.6	657.6
Maximum shear force T (KN)	145	122.8	116.5
Maximum axial force N (KN)	681.1	633.4	446.8

seismic response, results of calculations are presented for the two values of inclination angles respectively $\alpha = 10°$ and $\alpha = 20°$. The results are summarized in **Figure 7** and **Table 4**. For the example presented here, the inclination of the pile leads to a reduction of the numerical values of the normal load and lateral displacement of the pile group. However we can remark that along the pile, the values of the moment and the shear have been increased. **Table 4** illustrates that on the maximal values of internal forces and the amplification at cap and the structure head have been reduced when the value of the inclination increases except for the value of the bending moment.

5. Conclusions

In this paper, we present a three-dimensional numerical modeling of the soil-pile-structure interaction under seismic loading. The effect of the plasticity has been investigated in the case of a frictional soil as well as the effect of the dilancy angle. The numerical modeling has been

carried out by using harmonic excitation and real seismic loading recorded during the Kocaeli earthquake (Turkey, 1999). The effect of the pile inclination has been also analyzed. For simplicity, we consider the case that the piles are embedded in a homogeneous soil. The case of heterogeneous soil could be treated in the future.

The harmonic loading leads to high values of the internal forces (Bending moment, shear) especially when the frequency of the load is near to the proper frequency of the soil. For the example treated here, the plasticity of the soil has a minor effect on the results. For frictional soil, the plasticity spreads from the surface due to the low confinement of the soil in this area. Plasticisation of the soil around the piles head makes them more vulnerable, and the post seismic observations of damaged piles show the formation of a vacuum around the head of the piles.

The inclination of piles leads to a reduction in the lateral amplification of the superstructure resulting from an

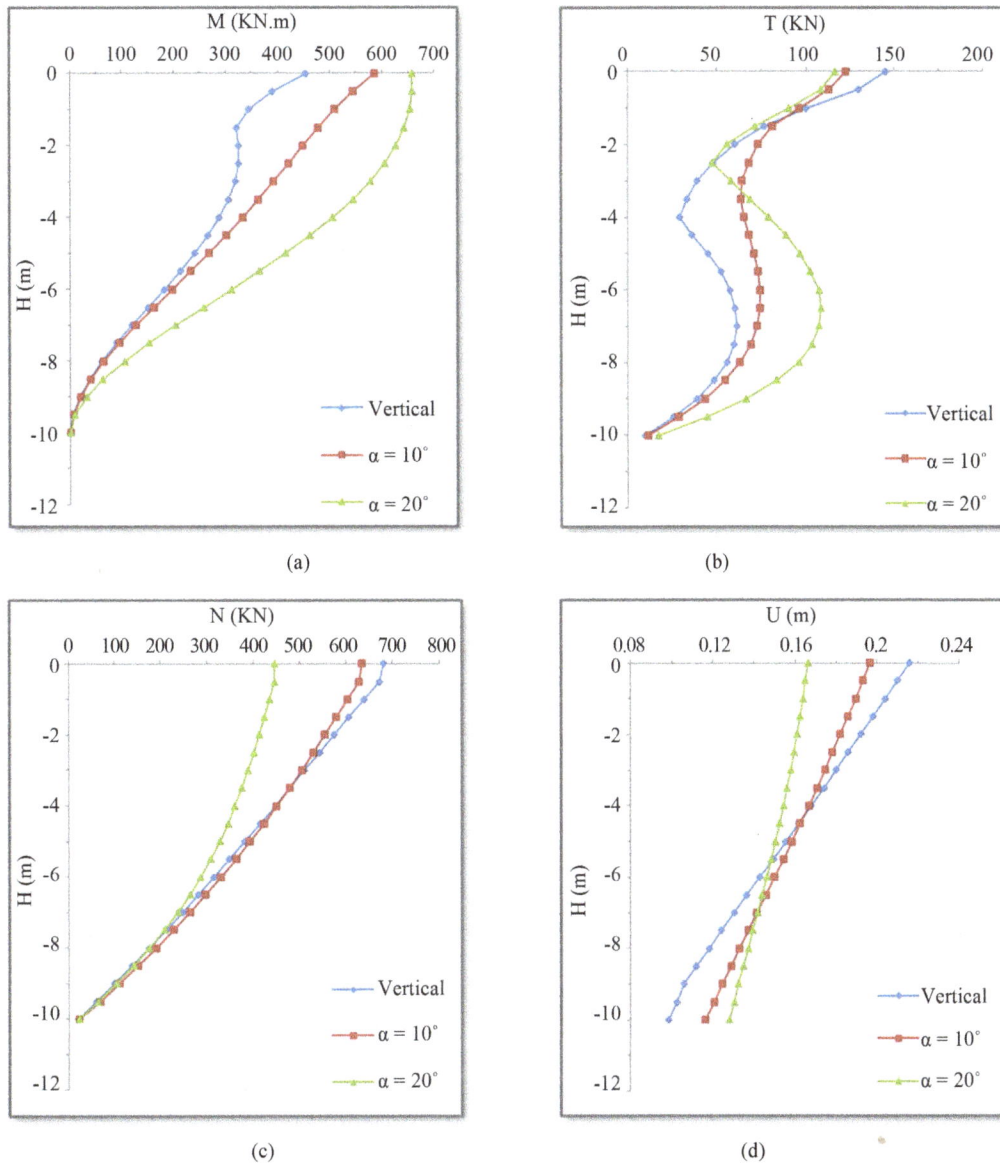

(a)

(b)

(c)

(d)

Figure 7. Influence of inclination of piles on the seismic response of pile groups (Registration of Turkey, Vg = 40 cm/s, f = 0.9 Hz, Ag = 0.247 g). (a) Bending Moment, (b) Shear force, (c) Normal force, (d) Displacement.

increase in the rigidity of the system. The inclination of piles can be beneficial for both the dynamic behavior and the behavior of the superstructure. It depends on the interaction of the frequency of the seismic load with the frequencies of the soil-pile-structure. The inclination increases the lateral stiffness of the foundation which, unfortunately, can cause a significant increase in the load transmitted to the foundation of the superstructure. Despite the improved performance of inclined piles, the bending forces at the top of piles are still very significant.

Acknowledgements

We thank the Lebanese National Council of Scientific Research for the funding of this work.

REFERENCES

[1] M. Sadek and I. Shahrour, "Three-Dimensional Finite Element Analysis of the Seismic Behaviour of Inclined Micropiles," *Soil Dynamics and Earthquake Engineering*, Vol. 24, 2004, pp. 473-485.

[2] M. Sadek and I. Shahrour, "Influence of the Head and Tip Connection on the Seismic Performance of Micropiles," *Soil Dynamics and Earthquake Engineering*, Vol. 26, No. 6, 2006, pp. 461-468.

[3] R. W. Boulanger, C. J. Curras, D. W. Wilson and A. A. Abghari, "Seismic Soil-Pile-Structure Interaction Experiments and Analyses," *Journal of Geotechnical and Geoenvironmental Engineering*, Vol. 125, No. 9, 1999, pp. 750-759.

[4] Y. Chung, "Etude Numérique de L'interaction Sol-Pieu-

Structure sous Chargement Sismique," Thèse de Doctorat, Université de Sciences et Technologie de Lille, 2000.

[5]   N. Gerolymos, A. Giannakou, I. Anastasopoulos and G. Gazetas, "Evidence of Beneficial Role of Inclined Piles: Observations and Summary of Numerical Analyses," Springer Science and Business Media B. V., 2008.

[6]   N. Gerolymos, S. Escoffier, G. Gazetas and J. Garnier, "Numerical Modeling of Centrifuge Cyclic Lateral Pile Load Experiments," *Earthquake Engineering and Engineering Vibration*, Vol. 8, No. 1, 2009, pp. 61-76.

[7]   Itasca Consulting Group, FLAC, "Fast Lagrangian Analysis of Continua," Vol. I. User's Manual, Vol. II. Verification Problems and Example Applications, 2nd Edition (FLAC3D Version 3.0), Minneapolis, Minnesota, 2005.

[8]   J. P. Bardet, I. M. Idriss, O'Rourke, N. Adachi, M. Hama-da and K. Ishihara, "North America-Japan Workshop on the Geotechnical Aspects of the Kobe," Loma 138 Prieta, and Northridge Earthquake. Report No. 98-36 to National Science Foundation, Air Force Office of Scientific Research, and Japanese Geotechnical Society, Osaka, 1996.

[9]   K. Tokimatsu, H. Arai and Y. Asak, "Deep Shear Structure and Earthquake Ground Motion. Characteristics in Sumiyoshi Area, Kobe City, Based on Microtremor Measurements," *Journal of Structural Engineering (ASCE)*, Vol. 491, 1997, pp. 37-45.

[10] G. Gazetas and G. Mylonakis, "Seismic Soil-Structure Interaction: New Evidence and Emerging Issues. Emerging Issues Paper, Geotechnical Special Publication No 75, ASCE, 2111," *Soil Dynamics and Earthquake Engineering*, Vol. 26, No. 6, 2006, pp. 461-468.

16

Variation of Altitude Observed on the Occasion of the Tohoku Earthquake (M = 9.0) Occurred on March 11, 2011

Pietro Milillo[1], Tommaso Maggipinto[2], Pier Francesco Biagi[2]
[1]School of Engineering, University of Basilicata, Potenza, Italy
[2]Department of Physics, University of Bari, Bari, Italy

ABSTRACT

Since October 1, 2010, a GPS receiver is put into operation at Tokai (Japan) in an experiment on Neutrino Physics (T2K). A significant variation of the altitude was detected from the beginning of March 2011, so that it has made worthwhile to investigate the possibility that such variations could be correlated to the Tohoku earthquake. In order to investigate in details this possibility, we analyzed the GPS data collected during 2011 by GEONet the GPS Earth Observation Network (GEONET). GEONET is the GPS network of Japan and consists of 1240 permanent stations. Preliminary results of the analysis seemed to show ten days before the earthquake, some possible anomalous behaviors of the stations. These anomalous behaviors were particularly relevant for stations of the network near the epicentral area. While co-seismic and post-seismic variations are widely expected, the anomalies recorded about ten days before the earthquake could be seriously considered among short-term precursors of the earthquake. In order to confirm this possibility, more detailed studies have been performed. In particular, GEONET currently makes available only daily solutions of the stations coordinates. On the contrary, it is very important to improve the time resolution just to understand the features of the anomalies till the last hours before the Earthquake. For this reason, we have performed an analysis to evaluate the coordinates and movement on hourly basis so improving the time resolution.

Keywords: Earthquake; GPS; Time Series

1. Introduction

A wide variety of natural phenomena are detrimental for the natural environment and for the anthropic structures and the human being. Earthquakes are among the most hazardous natural phenomena and, in order to mitigate their effects, the singling out of their precursors is a task of primary importance. Generally speaking, the earthquake precursors can be classified into three categories, depending on the time scale we are concerned with: "long-term" (of the order of a few hundred years), "medium-term" (of the order of hundreds to a few years) and "short-term" (of the order of a few months to a few days). We will focus on the short-term precursors.

A seismic precursor is an "anomalous" variation in some physical-chemical parameters that occurs before the seismic event and that is clearly linked to it. The most recent results show that seismic precursors can be divided into two different categories: ground precursors [1-4], and atmospheric precursors [5]. The first refer to anomalies in physical-chemical parameters of the ground such as resistivity, gas content and ionic content of deep waters, earth magnetic field, ground deformation etc.; the second refer to anomalies in physical-chemical parameters of the atmosphere, such as temperature, gas composition, density etc. Generally, the ground precursors are highlighted by means of on-ground measurements while atmospheric precursors are revealed by satellites [5,6].

In this paper, we present the possibility of investigating one of the main ground precursors, *i.e.* the crustal deformation. In particular, we study the height variation of the ground observed on the occasion of the Tohoku

Earthquake (M = 9.0) occurred on March 11, 2011. The heights are evaluated by means of GPS technique and in this sense such ground precursors are monitored by means of satellite techniques. The height data refer to different geographical areas of Japan and the data analysis as revealed possible preseismic, coseismic and postseismic effects in the crustal altitude.

2. Data Description

We used data of GPS Earth Observation Network (GEONET), a dense GPS receiver network composed by about one thousand GPS observation sites with 25 - 30 km average distance between stations. This network, operated by Geospatial Information Authority of Japan (GSI), has been used for crustal-deformation monitoring and geodetic control.

Each site is equipped with a GPS receiver, the communication device and a backup battery, which is stored in a stainless steel pillar. A pillar is five meters high, standing on a concrete cubic basement two meters large. The receivers are tuned to sample dual band carrier phase data and code data every 30 seconds. All the collected data are archived into a database in RINEX (Receiver Independent Exchange) format. GPS data network solution process starts soon after data downloading of the stations is complete (max latency of 15'). The processing generally provides Quick, Rapid and Final solutions according to different analysis strategies [7]. Quick analysis is routinely executed every three hours with 6-hour observation data and it used especially for monitoring crustal movement. Rapid analysis is performed every day with 24-hour data and it is used for screening in case of emergencies. Final analysis is executed two weeks after observation and it is the most accurate analysis in GEONET. Quick and Rapid analysis use the IGS Ultra Rapid Orbits; while the final analysis is executed with IGS precise orbits [8]. In this work we used precise solutions and RINEX observation files. As first step we have processed the data to achieve daily solutions of the coordinates for a period of four months centered around the epoch of occurrence of the Earthquake (11th March 2011) obtaining a daily preliminary analysis. This kind of products have been used for a daily preliminary analysis in order to emphasize general behavior in occasion of the 11 March 2011 M = 9.0 Tohoku Earthquake. RINEX observation files have been used applying the sliding window technique hourly analysis in order to bring out more accurate trends 15 days before and after the Earthquake. The daily network processing of the GEONET network is conducted by GSI, with Bernese GPS processing software, also RINEX observation files used for hourly analysis have been analyzed with BERNESE

software [9,10].

3. Data Analysis

After the preliminary step in which we have analyzed the daily solution provided by GEONET, we have, as second step, unwrapped the data to perform their reprocessing in order to achieve hourly solutions and, therefore, improve the time resolution in order to understand if abrupt events or discontinuities have been occurred during the investigated period .

Starting from January 1st, 2011, the daily data related to the altitude of all GPS station have been examined 2 months before and after 11th March 2011 identifying different behaviors in GPS Network heights.

An example of daily data plot, related to one GEONET GPS station, is shown in **Figure 1**.

Hourly solutions have been analyzed with Bernese Software in order to confirm trend and improve standard deviation error. The hourly analysis has been performed an a sub-network that consists of 12 fiducial stations (see **Table 1**) not very far from the Tsukuba station (TSKB IGS) which acts as a reference station; so we have a total number of 13 stations. Sliding window parameters are: 12 hours window with 3 hours sliding time.

All Processing steps have been performed using the Bernese Processing Engine (BPE) [10,11]. Both Precise Point Positioning (PPP) and Double-Difference Processing (RNX2SNX) blocks have been used for this analysis. Bernese Input Parameters are shown in **Table 2**.

PPP is a special case of zero-difference processing, in this particular case it can be considered a preparatory step to double-differencing processing. PPP relies on precise orbit and clock information for deriving precise site coordinates and a receiver clock correction independently for each analyzed station and is based on undifferenced code and/or phase observations.

PPP is a fast and efficient way to produce station coordinates but it should be underlined that it is not

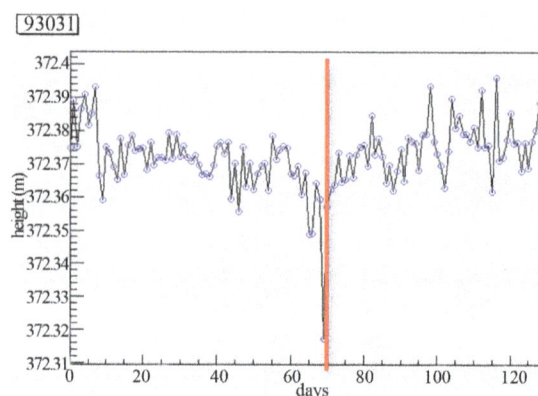

Figure 1. GPS-GSI daily data related to pre-seismic effect (Start day 1 January 2011).

Table 1. GPS network used for hourly analysis.

| 38 | 214 | 255 | 596 | 756 | 894 | 896 | 903 | 941 | 1100 | 1146 | 3009 | TSKB (IGS) |

Table 2. Bernese Input Parameters.

Bernese Input Parameters	
Precise Orbits	IGS
Tectonic Plate motion	**NNR-NUVEL-1A**
Solid Earth Tides, Pole Tides and Permanent Tides	**IERS Conventions 2003**
Ocean Tidal Loading Effects	**Yes**
Troposphere modeling estimation	**Niell dry part**

possible to reach the same network analysis precision.

The PPP approach in our case is useful to identify problems and reject eventually GPS stations from the selected network before the RNX2SNX.

The RNX2SNX is intended for a double-difference based analysis of RINEX GNSS observation data from a regional network. Station coordinates and troposphere parameters are estimated and stored in SINEX format to facilitate further processing and combination [10].

An important feature of this technique is that observation files with significant gaps or unexpectedly big residuals are automatically removed from the processing to ensure a robust processing and a reasonable network solution. The final network solution is a minimum constraint one, accomplished by three no-net translation conditions imposed on a set of ITRF2000 reference coordinates. The coordinates of all involved 12 fiducial stations are subsequently verified by means of a 7-parameter Helmert transformation in order to produce distortion-free transformations. In case of discrepancies, the network solution is recomputed based on a reduced set of fiducial stations. Height data as a function of time has been obtained after pre-processing step.

Non linear least square method with a function $x \cdot a + c$ (where a and c are computed parameters) has been computed on Daily and Hourly data in order to identify possible negative trend in occasion of the Seismic event.

4. Results and Discussion

4.1. Daily Results

After plotting all 1400 GPS Network stations daily height data starting from January 1st, 2011, four different behaviors have been identified (**Figure 2**):

- Co-seismic effect
- Post-seismic effect
- Pre-seismic effect
- A So-called "January" effect

Co-seismic effect is related to a large decrease of heights that occurs on the day of earthquake, while no other variation appears before or after the earthquake. This behavior has been revealed in many stations of the network generally located far enough from the epicentral area. These stations are reported in purple in the map (**Figure 3**).

Post-seismic effect concerns a permanent displacement that appears after the large decrease of altitude occurring the day of the earthquake. This behavior has been revealed in many stations of the network located mainly on the coastal direction near the epicentral area. Some of these stations are indicated in yellow in the map (**Figure 3**).

As for pre-seismic effect, about ten days prior the occurrence of the earthquake some suspect anomalous variation appears. This behavior was revealed in some stations of the network located mainly near the epicentral area.

Finally, during the first decades of January 2011 in many stations an evident variation of altitude (increase or decrease) appeared. In some case the variation was very large and this behavior was revealed mainly in stations located in the north of Japan. Red Squares in **Figure 3** indicate GPS stations characterized by this kind of effect.

4.2. Hourly Results

Figure 4 shows the complete hourly sequence from the first of March to the end of March. As expected the result is comparable with the GSI elaborated daily data sequence. The only difference which characterizes GSI and ASI data is a negative constant of about 10 cm after the seismic event, this could depend from different parameters used for BERNESE software elaboration, GSI does not give its set of parameters but it seems that the constant shift depends on the choice of the reference point. In hourly data TSK has been set as the reference point and all GPS station positions have been calculated respect to this station. If, after the main shock, there is a TSK station shift this would appear on all GPS station measurements as a systematic error. We looked at TSK IGS elaborated data in order to confirm this hypothesis and what we found was a shift of exactly 10 cm (**Figure 5**).

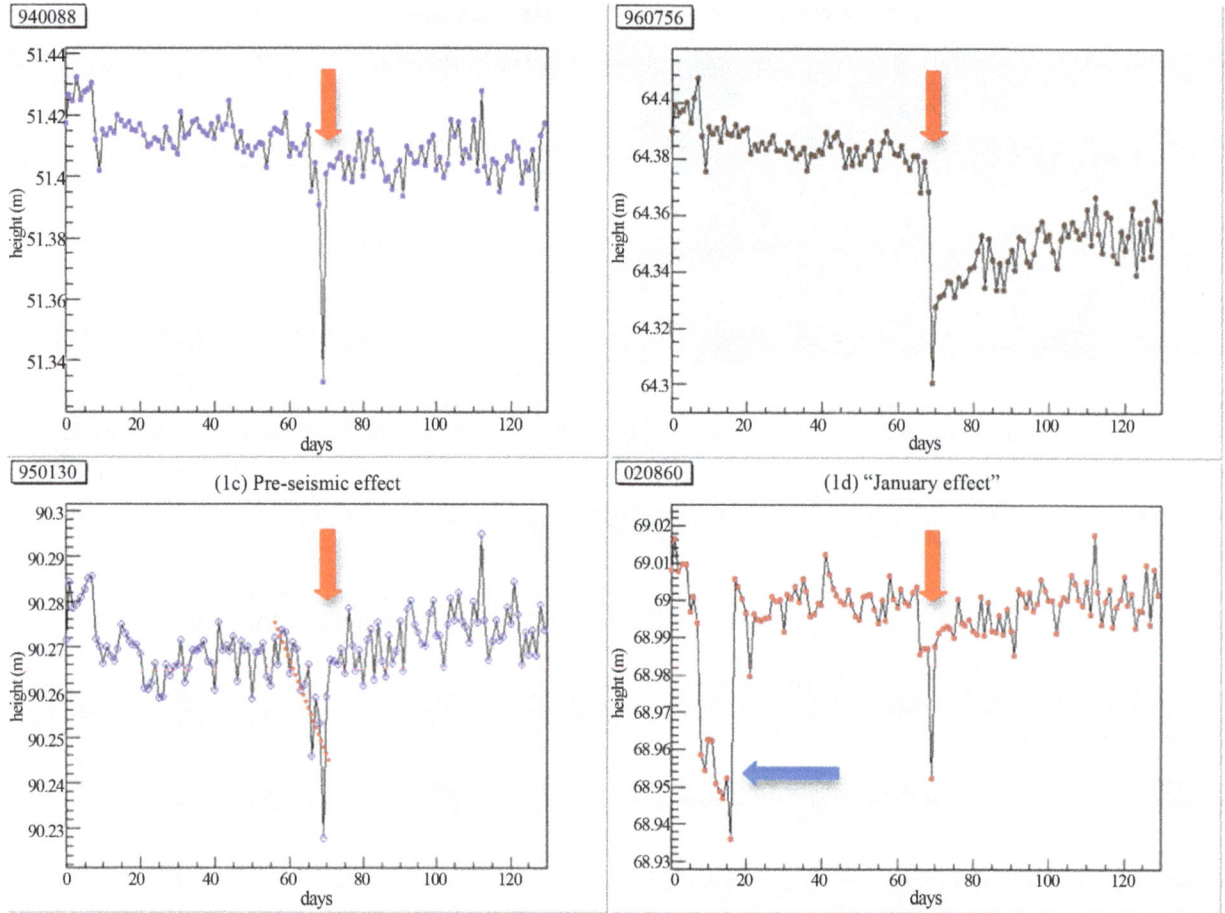

Figure 2. GPS-GSI daily data related to co-seismic effect, post-seismic effect, pre-seismic effect and January effect (Start day 1 January 2011).

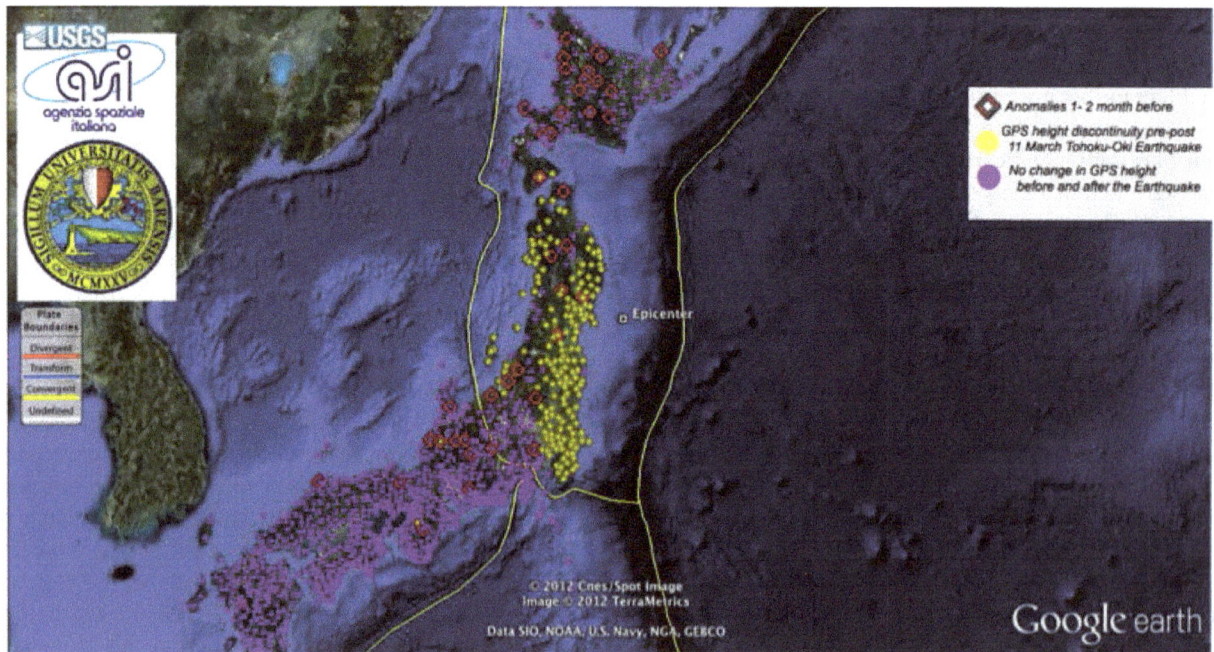

Figure 3. GPS-GSI daily data classified in terms of co-seismic effect (purple), post-seismic effect (yellow) and January anomalies (red-square).

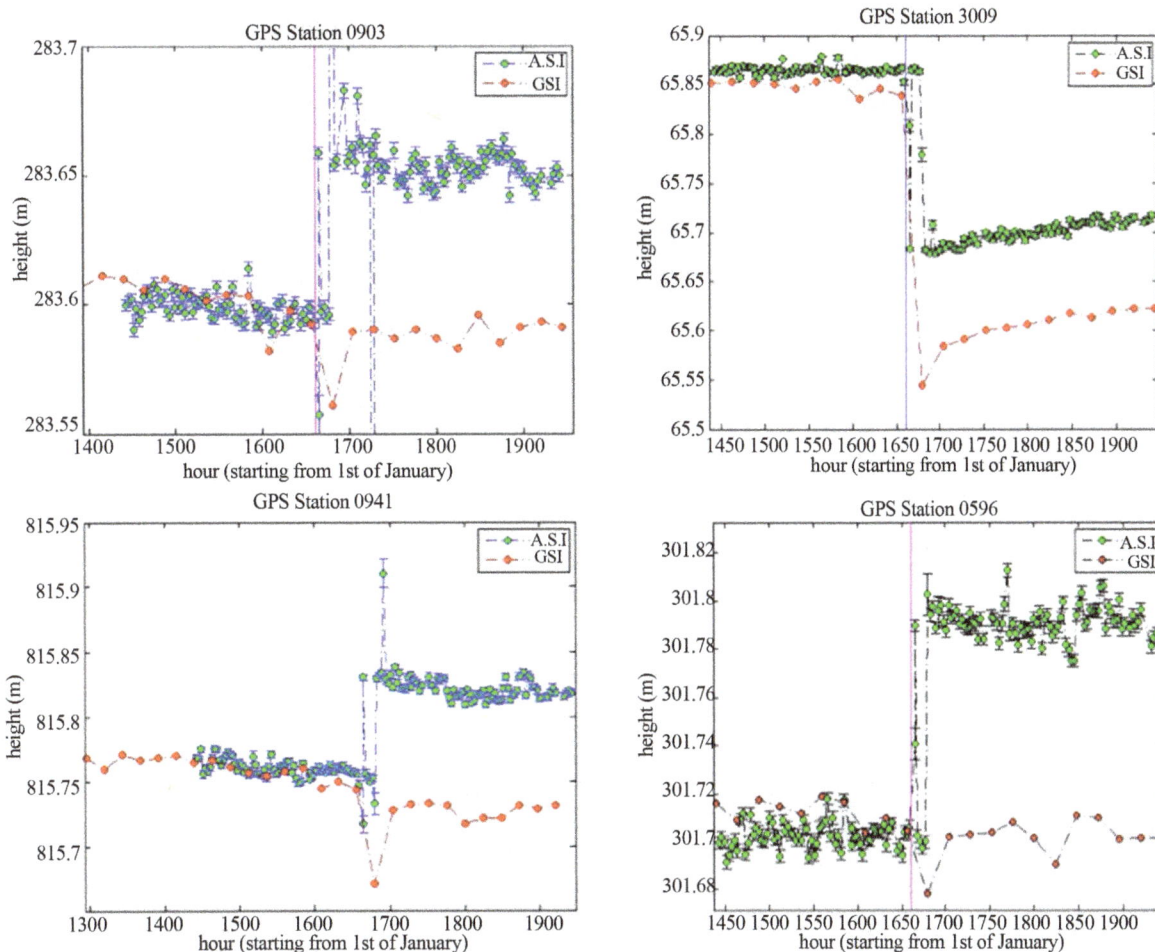

Figure 4. ASI and GSI Complete time series comparison.

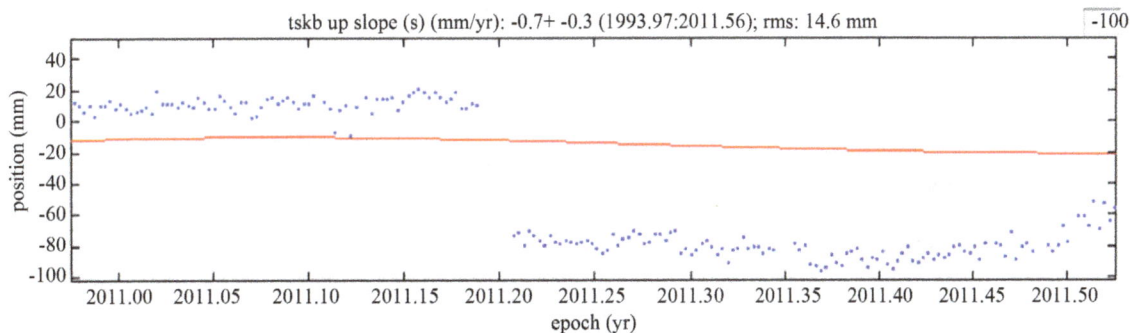

Figure 5. TSKB Altitude time series (JPL-SOPAC data [12]).

GSI and ASI GPS elaborated data substantially agree for what concerns data trends accordingly to the fact that GSI product is a data mediated over 24 hours while ASI product is a 12 hours window with a sliding window of 3 hours.

Only pre-seismic time series have been plotted in order to underline variation (**Figure 6**), trend fit has been added using a non linear least square method with a function

$$x \cdot a + c$$

a and c are computed parameters. Purple line indicates time when the earthquake occurred.

A variation cannot be seen clearly for all stations, a different plot has been used for a better visualization (**Figure 7**).

It can be noticed that the 2 cm data dispersion and data trend of ASI elaborated data, substantially agrees with

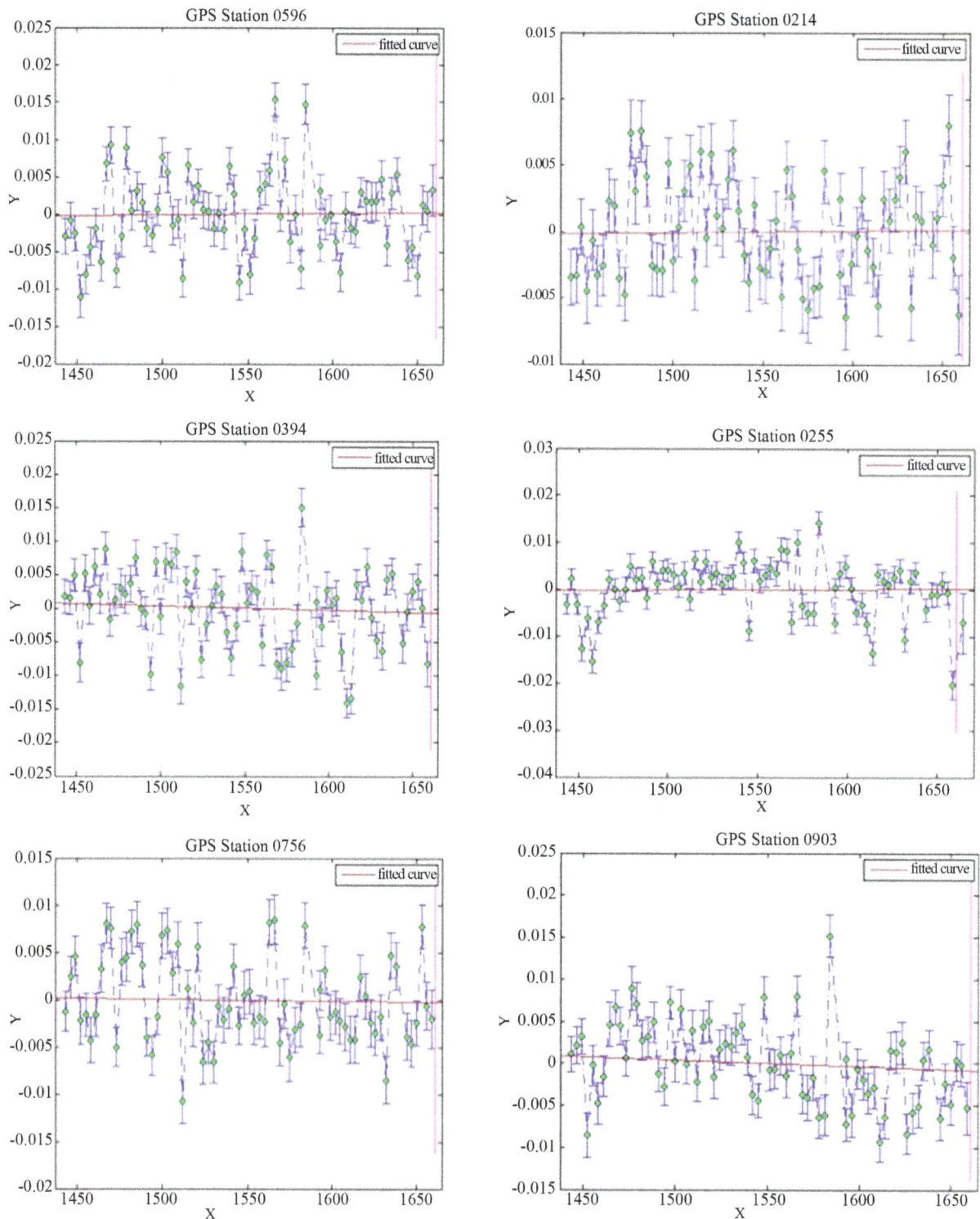

Figure 6. Pre-seismic time series of all GPS stations.

daily GSI data (**Figure 8**) in fact both data record a decline rate of 29.3 cm/year. According to the target, hourly analysis reduce RMS error from ±28.55 to ±8.63 cm/year giving statistical evidence to the pre-seismic decline rate.

The explanation of the not so evident pre-seismic decrease can be found in three equally important points:

• Only 2 stations (940 and 903 **Figure 7**) of the GPS network chosen for 3 hours sliding window analysis show possible pre-seismic anomalies in daily data,

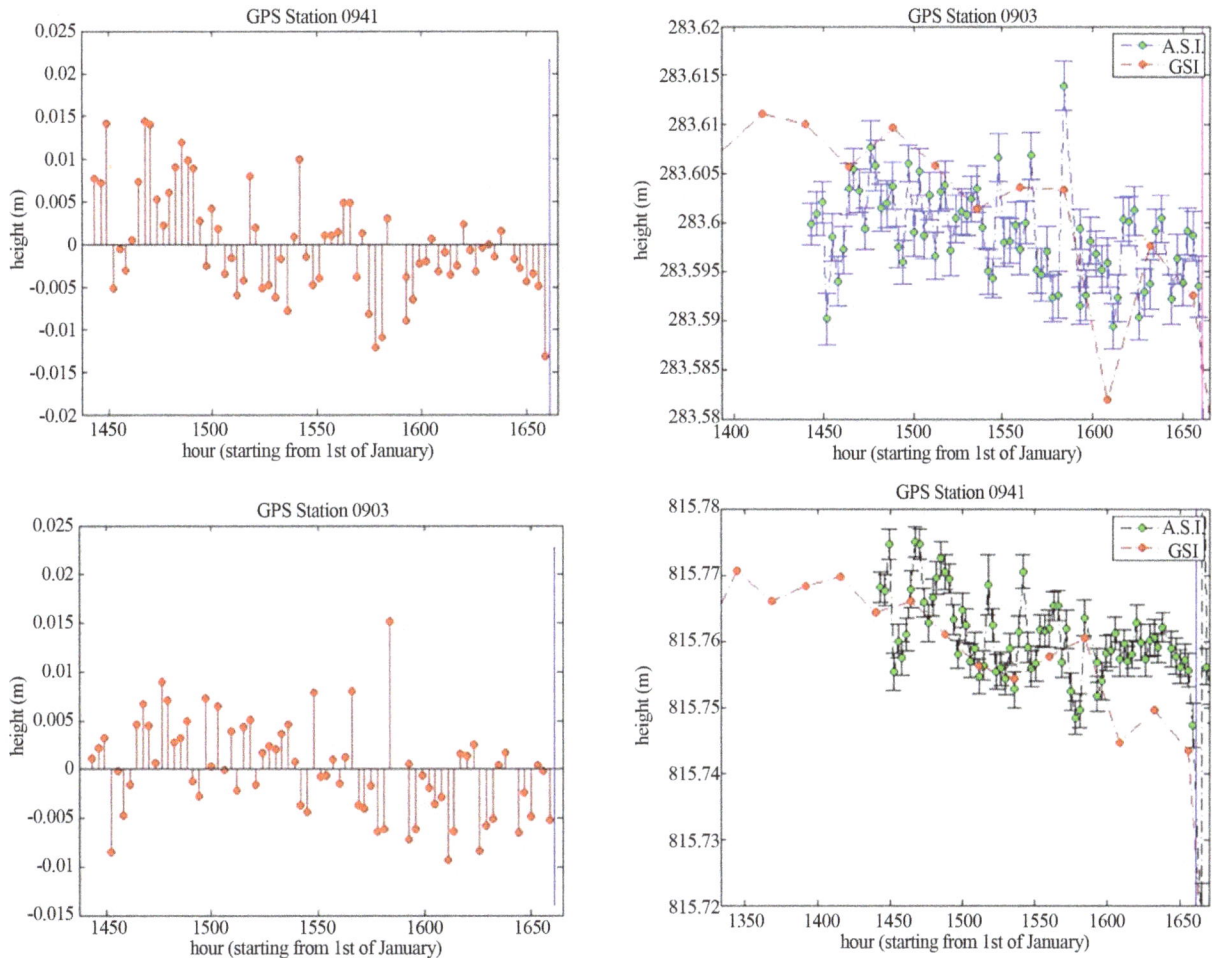

Figure 7. 903, 941 GPS stations pre seismic time series, GIS, A.S.I. Data comparison.

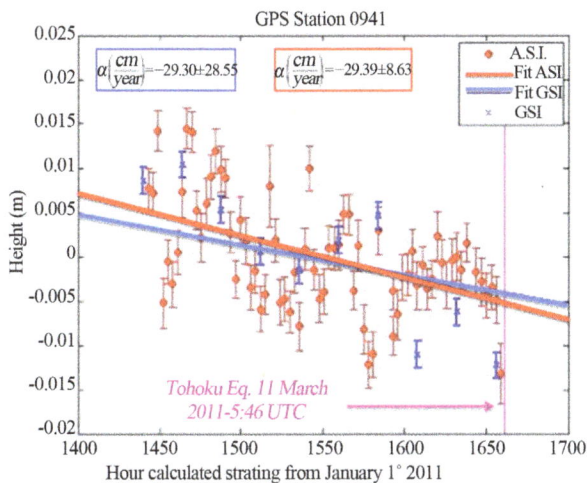

Figure 8. Dispersion and data trend of ASI elaborated data.

this negative trend is confirmed in hourly analysis with the same order of magnitude.

- Hourly analysis seems to be affected by a higher noise due to the 12 hour window used, GSI daily data

use a 24 hours window, this means that if GPS acquires a signal every 30 seconds, each value of altitude is averaged over 1440 more data than a single hourly altitude value. This inevitably introduces a certain amount of noise.

- Hourly analysis covers a time period of only 10 days before the earthquake, this could have hidden in part the decrease effect as in the case of stations 903 and 941 (**Figure 8**).

5. Conclusions

This study has confirmed that an earthquake occurrence is a complex phenomenon that, apart from obvious coseismic and postseimic effects, implies also preseismic effects that can be described as anomalies. In this paper, the anomaly investigated refers to the value of the crustal heights measured by means of GPS techniques. In accordance with other types of anomalies, such as radio [3,5, 13,14] or chemical anomalies, the presesimic effects are mainly located around the epicentral area.

We can conclude that GPS techniques can be of extreme importance in the task of earthquake prediction, and together with other type of anomalies related to seismic activity, we can have the basis for a multi-parametric approach in earthquake prediction.

REFERENCES

[1]   O. A. Molchanov, *et al.*, "Precursory Effects in the Subionospheric VLF Signals for the Kobe Earthquake, Physics of the Earth and Planetary Interiors," 1998.

[2]   Hayakawa, *et al.*, "On the Correlation between Ionospheric Perturbations as Detected by Subionospheric VLF/LF Signals and Earthquakes as Characterized by Seismic Intensity," *Journal of Atmospheric and Solar-Terrestrial Physics*, Vol. 72, No. 13, 2010, pp. 982-987.

[3]   http://www.izmiran.ru/projects/IK19/

[4]   Kasahara, *et al.*, "The Ionospheric Perturbations Associated with Asian Earthquakes as Seen from the Subionospheric Propagation from NWC to Japanese Stations," *Natural Hazards and Earth System Sciences*, Vol. 10, No. 8, 2010, pp. 581-588.
http://dx.doi.org/10.5194/nhess-10-581-2010

[5]   Hayakawa, *et al.*, "A Statistical Study on the Correlation between Lower Ionospheric Perturbations as Seen by Subionospheric VLF/LF Propagation and Earthquakes," *Journal of Geophysical Research*, Vol. 115, No. A9, Article ID: A09305.

[6]   K. Heki, "Ionospheric Electron Enhancement Preceding the 2011 Tohoku-Oki Earthquake," *Geophysical Research Letters*, Vol. 38, No. 8, Article ID: L17312.

[7]   http://www.gsi.go.jp/

[8]   http://acc.igs.org

[9]   http://www.denshi.e.kaiyodai.ac.jp/ja/

[10] Bernese, "GPS Software Reference Manual," University of Bern-Version 5.0, 2007.

[11] http://facility.unavco.org

[12] http://sopac.ucsd.edu/

[13] M. Hayakawa, *et al.*, "A Possible Precursor to the 2011 3.11 Japan Earthquake: Ionospheric Perturbations as Seen by Subionospheric VLF/LF Propagation. (in Phase of Pubblication)," 2011.

[14] Biagi, *et al.*, "Anomalies Observed in VLF and LF Radio Signals on the Occasion of the Western Turkey Earthquake (M = 5.7) at May 19, 2011 (in phase of pubblication)," 2011.

17

Seismic Precursory Phenomenology in Unusual Animal Behaviour in Val Pellice, Western Piedmont, in Comparison with Anomalies of Some Physical Parameters

Giovanna de Liso[1,2,3], Cristiano Fidani[1,4]

[1]"Seismic Precursors Study Center" (SPSC) Association, Via Servera 16, Torre Pellice, Italy
[2]Istituto di Alta Formazione Artistica e Musicale "G. F. Ghedini", Cuneo, Italy
[3]Voce Pinerolese, P. S. Donato 30, Pinerolo, Italy
[4]Central Italy Electromagnetic Network, CIEN, Fermo, Italy

ABSTRACT

The unusual animal behaviour, often observed before earthquakes in a moderate seismic area in Western Piedmont (NW Italy), can be connected with the anomalies of some physical parameters, recorded in a multiple parameters monitoring. Physical phenomena such as radioactive decay, gas emission, soil temperature increase, water pH variations in creeks and lakes, magnetic declination anomalies, air electricity and infra-sound, can generate damages to biological structures or, sometimes, death. A multiple parameters physical monitoring started recently in Western Piedmont, which is useful to propose the study of seismic precursors possibly linked to animal behaviour. Observations of unusual animal behaviour as vocal communications and movements were collected in the same area. A statistical analysis of strange behaviours in dogs and cats may indicate the probable early warning factor of the sense of smell.

Keywords: Unusual Animal Behaviour; Earthquake Early Warning; Radon Monitoring; Electromagnetic Monitoring; Sound Monitoring; Temperature Monitoring

1. Introduction

The unusual animal behaviours, often observed before earthquakes in a moderate seismic area in Western Piedmont (NW Italy), and the observation of anomalies of some physical parameters, induced the Authors to create the "Seismic Precursors Study Centre" (SPSC). SPSC is located in de Liso's house, in Torre Pellice (44°49'23"N, 7°12'04"E, Western Piedmont, NW Italy) at 699 m above sea level on Vandalino Mountain, not too far from an abandoned iron mine, near a particular "geological sanctuary" on "Castelluzzo" peak [1] and Bonnet, which are rich in augen-gneiss, of eruptive origin, biotite, ophiolite, zeolite, pechblenda and "Luserna Stone" (gneiss) [1].

SPSC is also close to a forest, at a distance of 70 m from the Biglione creek, 600 m from a graphite mine,

and 12 km far from the talc mines of Prali. The moderate seismic activity (the recent classification of the seismic risk for this area is 3S) in a region of intrusive rocks with geothermal activity, local emission of Radon[222] and sulphurous gases, "mistpoeffers", Earthquake Lights (EQL), can give Vandalino Mountain a good opportunity to be an excellent area to study pre-seismic phenomenology, with contemporaneous observations of unusual animal behaviours and monitoring of some physical parameters.

It must be remembered that Val Pellice was the theatre of historical earthquakes with great magnitudes, as the famous seismic event of April 2nd 1808, with a magnitude M recently estimated in 5.7 [2]; regarding this earthquake, we have an interesting relation of physicist G. Eandi and of Captain L. Garola about pre-seismic anomalous animal behaviours and weather anomalies [3], similar to those we observed now. Despite the fact that the earthquake destoryed most of the houses, only four were the victims. At the moment of the earthquake, most farmers were outside, working in the fields. Some people inside houses saved themselves, due to the agitation of their cows, a few seconds before the seismic shock. This information appears both in the oral tradition and in a few letters [4]. It is notable that two centuries later, as reported in the observational section of this work, observations regarding similar phenomena were made before earthquakes in the same region.

Generally, local seismic events are seldom announced by anomalous animal behaviours [5]. Unfortunately, we do not understand animal languages or their meaningful vocal modulations. Moreover, many animal alarms are not acknowledged. A number of animal and human observations in Northern Italy are reported, which can suggest physical observations potentially connected with seismic events. The study of seismic precursors is still in its infancy and the error margin on temporal and spatial forecast is still large and must be evaluated. Being so, a multiple parameter physical monitoring started recently in Western Piedmont, which is useful to propose the study of seismic precursors possibly linked to animal behaviours. Electric and magnetic detectors were operating together with alpha particle and acoustic detectors, while a collection of anomalous biological and meteorological observations was taken. In case of some local unusual animal behaviours, the occurrence of a local seismic event with epicentre distance ≤ 100 km can be suggested. Statistical information on warnings from local dogs and cats is resumed and discussed.

2. Observations

The observation of animal behaviours began in 1991, when it was noted that some unusual recurrences in animal behaviour occurred before earthquakes. The modality of observation of animal behaviours can be resumed in the following: two dogs and four cats live within SPSC. Being so, their behaviours and vocal communications can be directly observed by the first Author thanks to their great feeling and closeness. Since 2006, their behaviours were observed for five days a week, at least five hours a day for dogs and four hours a day for cats. Moreover, since 2006, observations in the wood near SPSC were undertaken every night (at 1:00 - 2:00 a.m. LT) to observe the wild animals that would eventually come to eat the food that was provided in a place, always the same. Vocal language of animals was heard and noted, especially of birds of prey. Since 2006, for five days a week, the dead arachnids in SPSC were counted at 6:30 a.m. LT. Since 1991, the first Author conducted irregular

observations of the behaviours of other wild animals that come near SPSC and hearings of farm animals (it is possible only to hear the vocal language of farm animals, as they live in farms or in pens not too far from SPSC). The observations of animals, which are resumed in **Table 1**, were recorded in a notebook and then compared "a posteriori" with a multiple physical monitoring and with seismic events. Observational results for animal behaviour and statistical features confirm the cases reported and discussed by Tsuneji Rikitake [6,7], even if not yet in a similar extended work.

These observations were supported by a multi-parameter monitoring gradually more complex since 1998 [8-10]. Starting in 2000, a daily monitoring for radioactivity and air temperature measurements have been undertaken at night in the forest near SPSC. The air temperature values were recorded from a little rock crack and water temperature measurements were intermittently recorded in the Biglione creek near SPSC. Since 2011, a continuous radon monitoring started in the basement of SPSC. Since 2012, irregular radon monitoring started by mean of dosimeters, located in some places inside and outside (garden) SPSC. Since May 2012, a station of the Central Italy Electromagnetic Network (CIEN), a ground thermometer and a meteorological station were installed in SPSC. Characteristic ELF signals are monitored in relation to seismic activity [11]. VLF and LF ranges allow monitoring several sub-ionospheric signals from different VLF and LF transmitters overlapping in the same channels, which is a necessary feature of a reliable system that is able to verify a single channel perturbation from at least two signals. Additionally, differently located sub-ionospheric channel monitoring stations of CIEN are able to realise overlapping [12]. Since July 2012, weekly measurements of water pH were started in the garden near SPSC and in the Biglione creek. The instruments in SPSC basement are resumed in **Table 2**.

3. Unusual Animal Behaviour before Earthquakes

Near SPSC there are several farms, so it is possible to observe farm and wild animals every day. Considering the behaviour, the animal language can be better understood if the farm and wild animals are observed in their habitual environment. When the epicentre distance to SPSC was inferior to 15 - 20 km, with a low magnitude (M ≤ 2), some unusual animal behaviours were always noted, a few days, a few hours and a few seconds before the seismic event, with different modalities for the three cases described as follows. But the same unusual animal behaviours were observed before earthquakes with epicentre distances to SPSC progressively increasing, if there was also a progressive increase in the magnitude.

Seismic Precursory Phenomenology in Unusual Animal Behaviour in Val Pellice, Western Piedmont, in Comparison with
Anomalies of Some Physical Parameters

151

Table 1. Study of animal behaviour.

Animal species observed	Observation place	Observation period	Number of animals	Monitoring modality	Typology of observation place
Dog	SPSC	1991-2006	1	5 hours a day, for 7 days a week, for 10 months a year; living together	Basement
Italian wolf	SPSC	2001-2006	1	5 hours a day, for 7 days a week, for 10 months a year; living together	Basement
Dog	SPSC	2007-2013	2	5 hours a day, for 5 days a week, for 12 months a year; living together	Basement
Cat	SPSC	1991-2006	6	4 hours a day, for 7 days a week, for 10 months a year; living together	Basement
Cat	SPSC	2007	6	4 hours a day, for 5 days a week, for 12 months a year; living together	Basement
Cat	SPSC	2008-2013	4	4 hours a day, for 5 days a week, for 12 months a year; living together	Basement
Insects, arachnids	SPSC and garden near SPSC	1991-2006	Numerous	1/2 hour a day, for 7 days a week, for 10 months a year; visual observation	Cellar, basement, garden, wall of de Liso's house
Insects, arachnids	SPSC	2007-2013	Numerous	1/2 hour a day, for 5 days a week, for 12 months a year; visual observation	Cellar, basement, garden, wall of de Liso's house
Dog	Farms near SPSC	1991-2006	Numerous, over 30	Irregular hearing observations for 7 days a week, for 10 months a year	Farms, distance to SPSC from 300 m. to 1500 m.
Dog	Farms near SPSC	2007-2013	Numerous, over 30	Irregular hearing observations for 5 days a week, for 12 months a year	Farms, distance to SPSC from 300 m. to 1500 m.
Owl, barn-owl	Wild area near SPSC	1991-2013	Numerous	Irregular visual and hearing observations	Forest near SPSC and near Biglione creek, distance to SPSC from 7 m. to 300 m.
Crow, rook	Wild area near SPSC	1991-2013	Numerous	Irregular visual and hearing observations	Forest near SPSC and near Biglione creek, distance to SPSC from 7 m. to 300 m.
Magpie	Wild area near SPSC	1998-2006	Numerous	2 time a day for 1/2 h every time, for 10 months a year; visual and hearing observations	Forest near SPSC and near Biglione creek, distance to SPSC from 7 m. to 300 m.
Magpie	Wild area near SPSC	2007-2013	Numerous	2 time a day for 1/2 h every time, for 12 months a year; visual and hearing observations	Forest near SPSC and near Biglione creek, distance to SPSC from 7 m. to 300 m.
Buzzard	Wild area near SPSC	1999-2006	2	2 time a day for 1/2 h every time, for 10 months a year, visual and hearing observations	Forest near SPSC and near Biglione creek, distance to SPSC from 7 m. to 300 m.
Buzzard	Wild area near SPSC	2007-2013	2	2 time a day for 1/2 h every time, for 12 months a year, visual and hearing observations	Forest near SPSC and near Biglione creek, distance to SPSC from 7 m. to 300 m.
Viper	Garden of SPSC	1998-2006	2	Irregular visual observations	SPSC, garden near SPSC
Salamanders, toads	Garden of SPSC	1991-2013	4	Irregular visual observations	SPSC, Garden near SPSC
Hedgehog	Garden of SPSC	1995-2013	2	Irregular hearing observations	Garden and forest near SPSC, distance to SPSC from 7 m. to 300 m.

Continued

Fox	Wild area near SPSC	1991-2013	2	Irregular visual and hearing observations	Garden and forest near SPSC, distance to SPSC from 7 m. to 70 m.
Wolf	Wild area near SPSC	1991	1	Irregular visual and hearing observations	Garden and forest near SPSC, distance to SPSC from 7 m. to 70 m.
Squirrel	Garden of SPSC	1991-2013	4	Irregular visual and hearing observations	Garden and forest near SPSC, distance to SPSC from 7 m. to 70 m.
Cow, sheep, goat, cock, chicken	Farms near SPSC	1991-2013	Numerous	Irregular hearing observations	Farms near SPSC, with farms distance to SPSC from 300 m. to 1500 m.
Donkey	Farms near SPSC	1999-2013	2	Irregular hearing observations	Farms near SPSC, first farm is 300 m. from SPSC, second farm is 500 m. from SPSC
Limacidae, earth-worm	Farms near SPSC	1991-2013	Numerous	Irregular observations	Garden and SPSC basement

Table 2. Instruments in SPSC and nearby area.

Physical parameter	Instrumentation	Measurement unit and sensitivity	Monitoring modality	Place of monitoring	Period of monitoring
Magnetic induction	1 TreField EM Meter	0.1 - 100 µT in logarithmic scale	2 time/day for 1/2 h every time	On gneiss in SPSC and nearby area	1999-2013
Magnetic declination δ	4 compasses Virginia 6036VA	sensitivity of ±0.5°	2 time/day for 1/2 h every time	2 on wood, 2 on iron surfaces in basement	1998-2013
β, γ particles	1 Geiger Ю нчмер SKM 05	Scale 0.1 - 99.9 µS/h; alarm at 0.5 µS/h; data every second	2 time/day for 1/2 h every time	SPSC/garden, on big rock crack	2003-2013
Radon^{222} α particles	1 Radon-meter detector (Geoex, model 1027)	Measurements in pC/l	Continuous: PC connection	basement, at 0.30 m from the floor, on gneiss	2011-2013
Temperature	1 analogical thermometer	Degrees Celsius (±0.1°C)	2 time/day at the same hours	In cellar and in Biglione creek	1999-2013
Temperature	1 thermometer TM-917 DICOM	from −100°C to + 132°C (±0.1°C), every 0.4 sec.	Continuous: PC connection	In cellar, at a depth of 2 meters in the soil;	2013
Temperature, pressure/humidity	PCE-FWS 20	Digital measurements: Celsius degrees, hPa, %	2 time/day at the same hours	On the roof of the house	2013
Water pH	Litmus papers	-------------	1 time/week	SPSC/garden/Biglione	2012-13
Infra-sounds	Infrasonic 200, Aetech,	5 samples/second	Continuous: PC connection	SPSC, at 1.20 m from the floor and 5 cm distance to a wall of the basement.	2013
EM signals: ELF, VLF and LF	CIEN electrodes	4 Hz - 50 kHz	Continuous: PC connection	SPSC garden	2012-2013

The anomalous animal behaviours were observed with the following modalities:

1) particular vocal language in a tripartite sequence (in 3.1, 3.2, 3.3);

2) non-vocal anomalous behaviour different from the usual pattern (in 3.4);

3) problems to health and safety (in 3.5).

3.1. Particular Vocal Language in a Tripartite Sequence: Animal Alarms with Shrill and High Sounds

The acoustic perception of vocal alarms can concern a large area, it gives concise information and it is easier to note. The animal vocal alarm is a particular vocal language, which seems aimed at its own species. It is an individual answer to a danger or a co-ordinate answer of the leading animal to the same danger. It can be supposed that the vocal animal alarm is a "thought answer", which expresses oneself as a dialogue with other animals or with humans. A few hours and sometimes days before seismic events, animal agitation is now well known in scientific literature [13,14]. It was supposed that this behaviour is possibly due to ultra-sounds emitted by rocks [14], electric and magnetic variation [14], or presence of

Seismic Precursory Phenomenology in Unusual Animal Behaviour in Val Pellice, Western Piedmont, in Comparison with
Anomalies of Some Physical Parameters

153

dangerous gases [5,8,14]. The emission of ultra-sounds before rock fractures was demonstrated with the experimental work on local characteristic "Luserna stone" [15, 16].

The first Author has individuated a particular tripartite sequence in the vocal alarms of domestic animals and birds: phase A, lasting up to 2 hours, with shrills and high sounds, from 30 minutes until 10 hours before the earthquake, then, when cries stop simultaneously, phase B follows, with a strange and worrisome silence; finally phase C, with animal cries normally 20 - 40 seconds before the earthquakes, a few times 5 - 10 seconds before, generally stopping few seconds before the shock. Phase C is corroborated by the observations of other researchers in case of other earthquakes [5,17-19]. The vocal animal alarms beginning up to 10 hours before local earthquakes, sometimes before distant earthquakes if the future magnitude is great, are contemporaneous to observations of drastic reduction of variations in intensity and declination of the magnetic field and of radioactivity values observed in SPSC [10]. **Figures 1(a)** and **(b)** resume an interesting relation between seismic epicentre distances to SPSC, magnitude and percentage of unusual animal behaviours, in relation to domestic animal cries and to bird songs, noted before the same seismic event. Unusual animal behaviours were taken into account only for those species whose normal behaviour is known.

During phase A, dogs, cattle, sheep, equines, bats, birds cry all together simultaneously, for a long time and sometime up to half an hour, with agitation, emitting shrills, howls or high sounds. Then, they stop their laments all together, with a stupefying synchrony [8,17]. This silence may last 3 - 5 hours before the local first seismic shock and it is strange and worrisome, like the quiet that precedes the storm. It is interesting to observe that dogs living in farms, as opposed to dogs that live in urban areas, at first ululate. These howls are similar to those they emit when they hear ambulances or church bells. Then, dog howls modulate into barking, at the same time, with short and repeated sounds. This barking is composed by two articulations with an ascendant order of frequencies for all dogs, but dogs of small size repeat the same vocalization more frequently in a minute, the two articulations forming a dissonant interval occur. In case of ambulances or church bells, only dogs cry, not other animals, and they stop their cries when ambulances or bells stop their sound: so dogs do not bark after (it is possible to know the journey of the ambulance by the howls of the different dogs along the way). In case of seismic precursors, the barks after the howls are very prolonged and contemporaneous.

Cocks also shrill, but with a vocalization composed by three sounds, on the same intonation, the last of them prolonged. This short scheme of three sounds is also re-

peated three times in sequences separated by short pauses, repeated many times with agitation. This agitation is similar to the same shown when there is a fox, a stone marten or a hawk. In these latter cases the tripartite phrase is interrupted by other phrases modulated differently with melodic variations on the last sound: in this case, cocks shrill on different moments according to their proximity to the "danger" and not all together and at the same time, as in case of a seismic event. Also magpies and crows chatter with agitation, but with cries similar to the bipartite sounds of dogs. This behaviour shows a tonal language for all these animal species and it is important for us to distinguish the musical sequence. **Figure 1(a)** shows the percentage of animal cries before seismic events on the y axis and the distance of hypocentre to SPSC on the x axis. The colours are related to the magnitude of the earthquake. Animal vocal alarms lasted from half an hour to one hour and half, if the magnitude of following seismic event was greater than 4.

3.2. Instrumental Observations Associated to Strange Animal Behaviours

During phase A, little variations of the magnetic intensity and drastic reductions in variations of the magnetic angle declination were recorded. Magnetic anomalies started with great variations, a few days or some weeks before the earthquake but the sudden commencement did not coincide with the beginning of phase A. After the simultaneous vocal stop, the worrisome silence follows, the phase B. During phase B, a stop in the magnetic and electric variations and a stop of radon emission were always noted. The continuous monitoring of infra-sounds operating in SPSC shows the background noise to be comprised in the values 0.1 - 3 Hz, with several factor having an influence on it, like the Biglione creek nearby, whose flow variations can increase it up to 7 Hz. During phase A, we were not able to record in SPSC any difference in the background noise that can be related to a seismic event. Then, 30 - 40 seconds before the seismic shock (phase C) infra-sounds of 3.5 - 5 Hz were recorded, lasting a few seconds, with a particular progressive "crescendo" and then a progressive "decrescendo" of intensity. The Data-logger elaboration of the data recorded by the Infrasonic 200 shows for this "crescendo and diminuendo" a "Moorish arch" shape for the graphic representation of frequencies.

Usually, territorial competitions, sexual calls, dangers by intruders are expressed by domestic animals and birds with vocalizations not at the same time; these shrills are more varied as succession of frequencies, with longer phrases between pauses. Ache and loneliness (for dogs, cats, cattle and equines) give long vocalizations on two or three descendent frequencies, repeated more and more with reducing intensity. These are the differences be-

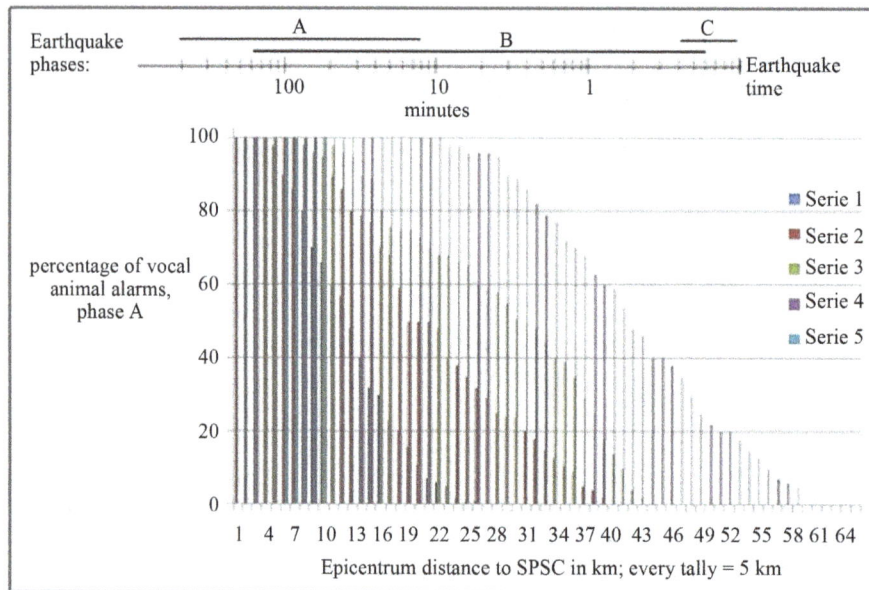

(a)

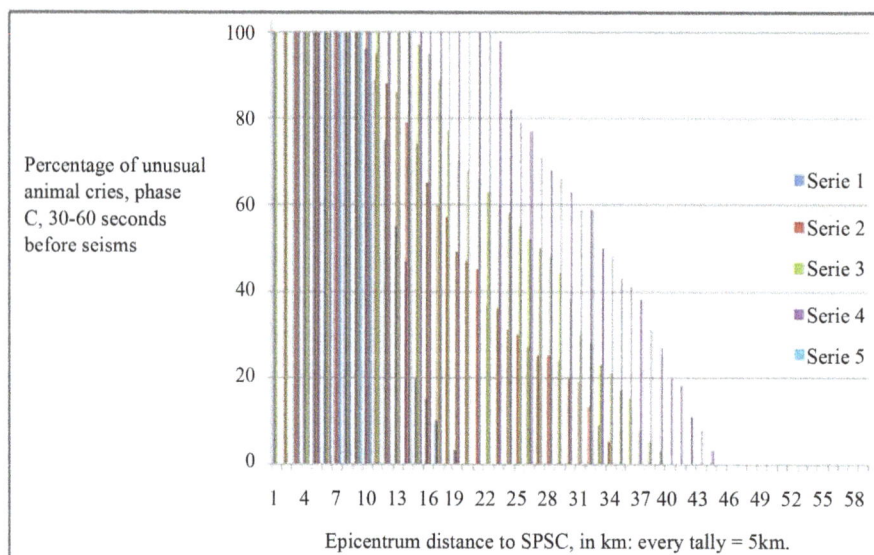

(b)

Figure 1. (a) percentage of unusual animal cries, phase A in relation to epicentre distance to SPSC and to magnitude; (b) percentage of unusual animal cries, phase C in relation to epicentre distance to SPSC and in relation to magnitude; series 1: $0 < M \leq 1$; series 2: $1 < M \leq 2$; series 3: $2 < M \leq 3$; series 4: $3 < M \leq 4$; series 5: $4 < M \leq 5$.

tween anomalous animal sounds during pre-seismic phases and normal animal sounds.

3.3. Statistical Behaviour of Dogs and Cats

A statistical study of unusual animal behaviours before earthquakes in Western Piedmont has been completed only for dogs and cats at this initial stage of study. Early warning signs as dog cries and cats hiding themselves were considered before 39 earthquakes around SPSC. The earthquakes were chosen according to their magnitudes and distances, approximately satisfying the Do-

brovolsky condition [20], where the Dobrovolsky radius is

$$RD = 10^{0.43M} \qquad (1)$$

considering seismic event distances $RE = 1.5\ RD$ [21] from SPSC. **Table 3** resumes the principal observations, consisting of a total of 55 early warning signs, with 39 examples of dog cries and 16 examples of cats hiding themselves. These warnings occurred during phase A, therefore considering a shorter time span compared to observations of all different animals in all the phases, as calculated in past works. Therefore, frequency distribu-

Seismic Precursory Phenomenology in Unusual Animal Behaviour in Val Pellice, Western Piedmont, in Comparison with Anomalies of Some Physical Parameters

155

Table 3. List of earthquakes near SPSC that can be linked with dog cries and cat hiding; Time (hh.mm.ss) is considered in LT; Depth is the depth of the earthquake hypocentre; Distance is the distance between the epicentre and the animal observation location; Dog and Cat columns represent observation times (in h) before the first shock.

	Date [mm/gg/aa]	Time [hh,mmss]	Depth [km]	Distance [km]	Magnitude [M_w]	Dogs [h]	Cats [h]
1	08/21/00	17.14'	7,4	91,7	4,6	7	7
2	07/18/01	22.47'	15	91,7	4,2	6,5	5,83
3	09/26/05	04.20'.38"	29	30,4	2,3	3	3,12
4	10/31/06	20.11'.08"	9,9	18,4	2,3	3,5	3,5
5	08/29/08	15.24'.23"	13	28,3	2,1	5,83	5,83
6	10/10/08	13.55'.39"	9,1	39,2	2,2	5	-
7	10/24/08	04.06'	10	97,5	4,1	7	7
8	03/19/09	12.39'.50"	43,3	51,7	4,1	8	-
9	03/23/09	20.01'.09"	10	36,4	2	6	-
10	04/19/09	13.39'	40	65	3,9	8	8
11	05/14/10	07.55'	9	35	2	5,33	-
12	09/29/10	03.46'.42"	10	62	3,3	7,67	-
13	11/11/10	18.24'.18"	13,5	14	2,1	8	-
14	11/11/10	21.57'.30"	13,2	24	2,2	2	-
15	11/14/10	20,31'58"	12,5	27,9	2,2	7,67	-
16	12/18/10	21,53'	18	18	2,3	10	8,67
17	12/22/10	11,02'	24	6	2,6	8	6,67
18	04/19/11	21,04'41"	18,5	45	2,7	5,5	5,5
19	07/23/11	16,56'53"	20	25,3	2,2	4,5	4,5
20	07/25/11	12,31'20"	25	17,2	4,3	8	6,67
21	01/26/12	3,23'52"	10,2	23,3	2	5,33	-
22	02/02/12	18,22'39"	10,2	32	3,9	8,78	-
23	02/05/12	7,26'14"	8,5	51,6	2,1	8	-
24	02/05/12	16,42'	8,4	31	2,1	7	-
25	02/26/12	22,37'55"	6,9	49,4	4,4	8,75	-
26	02/26/12	23,39'34"	6	48,9	3,3	9,78	-
27	02/27/12	16,31'20"	7,4	51,8	3,5	5,83	-
28	02/29/12	5,44'09"	11,8	16,5	2,7	8,03	-
29	09/29/12	21,06'32"	10	39,4	2	4,67	-
30	10/03/12	9,20'43"	10,02	43	3,9	8,08	8,08
31	10/04/12	17,27'47"	16,1	39	2,3	6,33	6,33
32	10/05/12	15,09'01"	11,7	38	2,4	6	-
33	12/31/12	12,48'06"	18,1	39,5	2,7	6	-
34	04/07/13	3,13'11"	10	54	3,3	8	6,67
35	04/13/13	17,30'38"	10,1	35	2,3	7	-
36	04/13/13	19,15'37"	10	39	2,1	7	-
37	04/17/13	13,37'39"	14,1	28,2	2,1	7,23	-
38	05/05/13	5,33'35"	9,7	40	2	6,33	-
39	05/06/13	9,5455"	9,4	30	2,1	8,72	8,72

tion, as shown in **Figure 2(a)**, was calculated relative to simple time t, as already shown in Rikitake [22]: its shape is well described by a Weibull distribution [23]. Following Rikitake [22], a function R(t) = 1 − F(t) was defined, where F(t) is the cumulative probability, **Figure 2(b)**, of an earthquake occurring during a period from 0 to t. A plot of Ln (Ln (1/R(t))) versus Ln (t) is shown in **Figure 2(c)**. The straight-line fitting in the figure, neglecting the lower values of Ln (t) for which a different distribution is probably valid [22], confirms that the recorded data can be roughly governed by a Weibull distribution with coefficients $k = 7.58 \ 10^{-6}$ and $m = 5.636$. Mean and standard deviation can be calculated through the Gamma function [23]

$$E = \left[k/(m+1) \right]^{-1/(m+1)} \Gamma \left[(m+2)/(m+1) \right] \quad (2)$$

and

$$\sigma = E \left\{ \Gamma \left[(m+3)/(m+1) \right] / \Gamma^2 \left[(m+2)/(m+1) \right] - 1 \right\}^{1/2} (3)$$

respectively with $E = 7.3$ hours and $\sigma = 1.3$ hours, anticipating of some hours the general animal precursors [22], but according to previous results for dogs and cats [24].

Data from the phase C were excluded as they were considered to reveal dog and cat agitation at the time of P-wave arrival. Taking into account earthquake data, short timed warnings from dogs and cats resulted independent from the earthquake magnitude, according to other studies [13], even if small earthquakes can alarm dogs and cats later than in the case of greater earthquakes. Such small earthquakes have generally greater depths than other small earthquakes which usually alarm dogs and cats according to average values. Finally, great earthquakes influence dog and cat behaviour for any depths with more or less the same mean warning time. The warning time distributions were also plotted in separate charts for dogs (**Figure 3**) and cats (**Figure 4**). The straight line shown in **Figure 3(c)** confirms that the recorded data for dogs (**Figure 3(a)**) can also be governed by a Weibull distribution with coefficients $k = 7.73 \ 10^{-6}$ and $m = 5.590$, which corresponds to $E(t) = 7.4$ hours and $\sigma = 1.3$ hours. And, the straight line shown in **Figure 4(c)** confirms that the recorded data for cats (**Figure 4(a)**) can also be governed by a Weibull distribution with coefficients $k = 5.60 \ 10^{-5}$ and $m = 4.723$, which corresponds to $E = 6.9$ hours and $\sigma = 1.4$ hours. **Figures 3(b)** and **4(b)** are the cumulative probabilities.

Even if a warning time difference of half an hour in dogs and cats is within the margin of error, it could reflect differences in some kind of perception between them: cats have the better sense of hearing, while dogs

have the better sense of smell. Cats can detect higher frequencies of sound than many other mammals including dogs. The upper range of hearing in cats is about 60 to 65 kHz, which allows them to hear both their kittens and the ultrasonic calls of rodents. Smell is the most developed sense in dogs and overcomes the sense of smell of cats, as the formers have more than 10 times (up to 300 millions) the number of odour sensitive cells of the latters. Being so, dogs are able to sense gases coming from the ground before cats and begin to cry, but they are not able to escape from SPSC because of their size. When cats smell the same gases, they are able to escape or hide. In order to be confirmed, such hypotheses need the mentioned gases to be identified and measured and to verify that such gases produce the observed behaviours in dog and cats. At the moment, only Radon measurements are active in SPSC, confirming an increase of Radon concentrations many hours before an earthquake and a sudden decreasing a few hours before the shock. It is well known that Radon comes out of the ground driven by other gases [25], sustaining the hypotheses above. A further confirmation should come from ultrasound measurements, now still absent in SPSC.

3.4. Non-Vocal Anomalous Behaviours Different from the Usual Pattern

The most evident non-vocal animal behaviour is the advanced awakening from hibernation, possibly due to an increase in temperature, to magnetic variations or to emissions of dangerous gas. During winter, the precursor increase in soil and water temperature is a better forewarning as it is easier to notice. A rise in temperature up to 5°C - 6°C above the mean seasonal values at SPSC was often recorded before seismic events. This could explain the premature awakening of animals in hibernation (bats, insects, amphibious and reptiles) and premature larvae development. This unusual animal behaviour is a short term precursor, usually 12 - 15 hours before. Damages to people and animal health were observed at SPSC, in connection with measured high radon values. A rapid evolution of pathologies, especially dermatological effects, was reported at SPSC when the radioactive emission was much higher than the local average values.

Unusual Flight Behaviour with Magnetic Sudden Variations: it is very important to individuate the right moment of the sudden magnetic commencement because it is the moment of unusual flight behaviour observations. SPSC observations confirm that when geomagnetic perturbation is not due to sun activity, but it is due to the variation of the magnetic permeability of rocks under stress, we can observe unusual flight of some birds and a few problems of balance for dogs, cats and also human subjects. The sudden beginning can be followed for up to